Advanced Characterization and On-Line Process Monitoring of Additively Manufactured Materials and Components

Advanced Characterization and On-Line Process Monitoring of Additively Manufactured Materials and Components

Editors

Giovanni Bruno
Christiane Maierhofer

MDPI • Basel • Beijing • Wuhan • Barcelona • Belgrade • Manchester • Tokyo • Cluj • Tianjin

Editors

Giovanni Bruno
Bundesanstalt für
Materialforschung und
–prüfung (BAM)
Germany

Christiane Maierhofer
Bundesanstalt für
Materialforschung und
–prüfung (BAM)
Germany

Editorial Office
MDPI
St. Alban-Anlage 66
4052 Basel, Switzerland

This is a reprint of articles from the Special Issue published online in the open access journal *Metals* (ISSN 2075-4701) (available at: https://www.mdpi.com/journal/metals/special_issues/characterization_monitoring).

For citation purposes, cite each article independently as indicated on the article page online and as indicated below:

LastName, A.A.; LastName, B.B.; LastName, C.C. Article Title. *Journal Name* **Year**, *Volume Number*, Page Range.

ISBN 978-3-0365-5813-4 (Hbk)
ISBN 978-3-0365-5814-1 (PDF)

Contents

About the Editors

Giovanni Bruno

Prof. Dr. Giovanni Bruno, Bundesanstalt für Materialforschung und -prüfung and University of Potsdam, Germany, studied nuclear engineering at the University of Bologna, graduating in 1989 and completing a second degree in physics in 1998; he obtained his doctorate (1997) in materials science at the University of Ancona, Italy. He was a research assistant at the Open University (Milton Keynes, UK), at the HMI, Berlin, and at the Institut Laue-Langevin in Grenoble, France. From 2005 to 2012, Giovanni worked for Corning Incorporated, first as group leader for material characterization in Avon, France, and then as project manager for R&D in Corning, NY, USA. At the end of 2012, he became head of division at BAM, Berlin, and Professor at the University of Potsdam, Germany. His scientific activities are focused on residual stress analysis by means of X-ray and neutron diffraction and on the investigation of the mechanical behavior and damage in materials using diffraction, X-ray computed tomography and refraction methods.

Christiane Maierhofer

Dr. Christiane Maierhofer, Bundesanstalt für Materialforschung und -prüfung, led Division 8.7 (thermographic methods) in the department of non-destructive testing at BAM from 2017 to 2022. Previously, she worked on acoustic NDT methods, always at BAM. Her scientific activities focused on the online monitoring of manufacturing processes, in particular additive manufacturing, and on the development of active and passive thermographic methods for materials characterization and non-destructive testing. She sadly passed away in May 2022.

Preface to "Advanced Characterization and On-Line Process Monitoring of Additively Manufactured Materials and Components"

Additive manufactured (AM) metallic materials are becoming more common in research and development. This is not completely true for AM components in the industry since they need to be screened for quality control and meet stringent requirements. The quality of AM components depends on the microstructure and internal stress resulting from the manufacturing process. Consequently, important aspects of quality control are not only defect analysis, microstructural investigations and determination of residual stress, but also online monitoring. This Special Issue is concerned with these aspects and the non-destructive testing techniques associated with the assessment of the AM part quality

Giovanni Bruno and Christiane Maierhofer
Editors

metals

MDPI

Editorial

Advanced Characterization and On-Line Process Monitoring of Additively Manufactured Materials and Components

Giovanni Bruno * and Christiane Maierhofer

BAM, Bundesanstalt für Materialforschung und -Prüfung, Unter den Eichen 87, 122025 Berlin, Germany
* Correspondence: giovanni.bruno@bam.de

1. Introduction

Additive manufacturing (AM) techniques have risen to prominence in many industrial sectors. This rapid success of AM is due to the freeform design, which offers enormous possibilities to the engineer, and to the reduction of waste material, which has both environmental and economic advantages. Even safety-critical parts are now being produced using AM. This enthusiastic penetration of AM in our daily life is not yet paralleled by a thorough characterization and understanding of the microstructure of materials and of the internal stresses of parts. The same holds for the understanding of the formation of defects during manufacturing. While simulation efforts are sprouting and some experimental techniques for on-line monitoring are available, still little is known about the propagation of defects throughout the life of a component (from powder to operando/service conditions). This Issue was aimed at collecting contributions about the advanced characterization of AM materials and components (especially at large-scale experimental facilities such as Synchrotron and Neutron sources), as well as efforts to liaise on-line process monitoring to the final product, and even to the component during operation. The goal was to give an overview of advances in the understanding of the impacts of microstructure and defects on component performance and life at several length scales of both defects and parts.

Citation: Bruno, G.; Maierhofer, C. Advanced Characterization and On-Line Process Monitoring of Additively Manufactured Materials and Components. *Metals* **2022**, *12*, 1498. https://doi.org/10.3390/met12091498

Received: 22 August 2022
Accepted: 6 September 2022
Published: 9 September 2022

2. Characterization and Process Monitoring

This Issue was born with a further precise scope: BAM funded in 2018 two large internal projects on characterization of materials and on-line process monitoring in additive manufacturing (AM) of metals (therefore including PBF, LMD and WAAM techniques). Therefore, we aimed to spark the debate on those two important aspects, starting from the output of such projects. In particular, we fostered a) the discussion about the influence of the microstructure and residual stress in AM of metals on the performance of materials and components and b) the investigation of possible ways to predict the appearance of defects in printed parts by on-line monitoring during manufacture. One particular aspect of point a) above was the use of advanced characterization techniques, especially based on large-scale facilities (synchrotron radiation and neutrons).

Indeed, many aspects of the generation, determination, and effects of residual stress (RS) in metallic AM materials and components are discussed in this Special Issue [1–3], whereby such stresses are determined by neutron or synchrotron X-ray diffraction. A review paper on the subject is also published in this Special Issue [4]. Moreover, advanced imaging techniques, in particular laboratory and synchrotron X-ray computed tomography, are used to disclose the defects generated by AM processes and some strategies for their mitigation [5–7].

Another axis of investigation in AM is the use of on-line monitoring techniques and their coupling with post-mortem microstructural analysis. This Special Issue contains a number of important contributions to the solution of the problems of how to extract defect distributions from temperature profiles in the manufactured parts during printing [8,9]. Not only are X-ray computed tomography data compared with infrared thermographic investigations, but also aspects of the calibration and registration of such techniques are thoroughly discussed [10,11].

Interestingly enough, authors contributed to demonstrating how more 'classic' non-destructive testing techniques can also well give invaluable insights into the problem of defect characterization [12], thereby complementing the high-end (but somehow expensive) characterization techniques.

Finally, the discussion is extended to component level, whereby defects [13] and residual stress [14] are determined in relevant industrial cases.

3. Conclusions

The Special Issue opens a few important points for discussion in the scientific community, such as the correlation between on-line measurements and defects in the final AM printed part, and the proper determination of residual stress in complex materials and components, such as additively manufactured metallic parts. It demonstrates that advanced and classic characterization techniques are both needed to solve the problems of defect and microstructure determination in the above-mentioned materials, together with on-line monitoring techniques and data fusion.

Author Contributions: G.B. conception and writing, C.M. conception. G.B. and C.M. funding and administration. All authors have read and agreed to the published version of the manuscript.

Funding: Part of this research received BAM internal funding in the frame of the BAM internal Projects ProMoAM and AGIL (see https://www.bam.de/Content/EN/Standard-Articles/Topics/Materials/article-agil.html and https://www.bam.de/Content/EN/Press-Releases/2019/Materials/2019-03-27-detecting-defects-with-thermography.html, accessed on 5 September 2022).

Acknowledgments: Christiane Maierhofer passed away in June 2022. We all acknowledge her inspiring, leading and supportive role in designing the Issue, supervising her students, supporting her peers, and carrying out the role of editor. We wish she would still be with us to further help and inspire us.

Conflicts of Interest: The authors declare no conflict of interest. Moreover, the funders had no role in the design of the study; in the collection, analyses, or interpretation of data; in the writing of the manuscript; or in the decision to publish the results.

References

1. Ulbricht, A.; Altenburg, S.J.; Sprengel, M.; Sommer, K.; Mohr, G.; Fritsch, T.; Mishurova, T.; Serrano-Munoz, I.; Evans, A.; Hofmann, M.; et al. Separation of the Formation Mechanisms of Residual Stresses in LPBF 316L. *Metals* **2020**, *10*, 1234. [CrossRef]
2. Muiruri, A.; Maringa, M.; du Preez, W.; Masu, L. Effect of Stress-Relieving Heat Treatment on the High Strain Rate Dynamic Compressive Properties of Additively Manufactured Ti6Al4V (ELI). *Metals* **2020**, *10*, 653. [CrossRef]
3. Mishurova, T.; Artzt, K.; Haubrich, J.; Requena, G.; Bruno, G. Exploring the Correlation between Subsurface Residual Stresses and Manufacturing Parameters in Laser Powder Bed Fused Ti-6Al-4V. *Metals* **2019**, *9*, 261. [CrossRef]
4. Schröder, J.; Evans, A.; Mishurova, T.; Ulbricht, A.; Sprengel, M.; Serrano-Munoz, I.; Fritsch, T.; Kromm, A.; Kannengießer, T.; Bruno, G. Diffraction-Based Residual Stress Characterization in Laser Additive Manufacturing of Metals. *Metals* **2021**, *11*, 1830 [CrossRef]
5. Ulbricht, A.; Mohr, G.; Altenburg, S.J.; Oster, S.; Maierhofer, C.; Bruno, G. Can Potential Defects in LPBF Be Healed from the Laser Exposure of Subsequent Layers? A Quantitative Study. *Metals* **2021**, *11*, 1012. [CrossRef]
6. Jaenisch, G.-R.; Ewert, U.; Waske, A.; Funk, A. Radiographic Visibility Limit of Pores in Metal Powder for Additive Manufacturing *Metals* **2020**, *10*, 1634. [CrossRef]
7. Chioibasu, D.; Mihai, S.; Mahmood, M.A.; Lungu, M.; Porosnicu, I.; Sima, A.; Dobrea, C.; Tiseanu, I.; Popescu, A.C. Use of X-ray Computed Tomography for Assessing Defects in Ti Grade 5 Parts Produced by Laser Melting Deposition. *Metals* **2020**, *10*, 1408 [CrossRef]
8. Mohr, G.; Altenburg, S.J.; Ulbricht, A.; Heinrich, P.; Baum, D.; Maierhofer, C.; Hilgenberg, K. In-Situ Defect Detection in Laser Powder Bed Fusion by Using Thermography and Optical Tomography—Comparison to Computed Tomography. *Metals* **2020**, *10*, 103. [CrossRef]
9. Mohr, G.; Sommer, K.; Knobloch, T.; Altenburg, S.J.; Recknagel, S.; Bettge, D.; Hilgenberg, K. Process Induced Preheating in Laser Powder Bed Fusion Monitored by Thermography and Its Influence on the Microstructure of 316L Stainless Steel Parts. *Metals* **2021**, *11*, 1063. [CrossRef]

10. Mohr, G.; Nowakowski, S.; Altenburg, S.J.; Maierhofer, C.; Hilgenberg, K. Experimental Determination of the Emissivity of Powder Layers and Bulk Material in Laser Powder Bed Fusion Using Infrared Thermography and Thermocouples. *Metals* **2020**, *10*, 1546. [CrossRef]
11. Oster, S.; Fritsch, T.; Ulbricht, A.; Mohr, G.; Bruno, G.; Maierhofer, C.; Altenburg, S.J. On the Registration of Thermographic In Situ Monitoring Data and Computed Tomography Reference Data in the Scope of Defect Prediction in Laser Powder Bed Fusion. *Metals* **2022**, *12*, 947. [CrossRef]
12. Xue, Z.; Xu, W.; Peng, Y.; Wang, M.; Pelenovich, V.; Yang, B.; Zhang, J. Measuring the Depth of Subsurface Defects in Additive Manufacturing Components by Laser-Generated Ultrasound. *Metals* **2022**, *12*, 437. [CrossRef]
13. Spurek, M.A.; Luong, V.H.; Spierings, A.B.; Lany, M.; Santi, G.; Revaz, B.; Wegener, K. Relative Density Measurement of PBF-Manufactured 316L and AlSi10Mg Samples via Eddy Current Testing. *Metals* **2021**, *11*, 1376. [CrossRef]
14. Mishurova, T.; Sydow, B.; Thiede, T.; Sizova, I.; Ulbricht, A.; Bambach, M.; Bruno, G. Residual Stress and Microstructure of a Ti-6Al-4V Wire Arc Additive Manufacturing Hybrid Demonstrator. *Metals* **2020**, *10*, 701. [CrossRef]

metals

MDPI

Article

On the Registration of Thermographic In Situ Monitoring Data and Computed Tomography Reference Data in the Scope of Defect Prediction in Laser Powder Bed Fusion

Simon Oster [1,*], Tobias Fritsch [1], Alexander Ulbricht [1], Gunther Mohr [1], Giovanni Bruno [1,2], Christiane Maierhofer [1] and Simon J. Altenburg [1]

[1] Bundesanstalt für Materialforschung und-Prüfung (BAM, Federal Institute for Materials Research and Testing), Unter den Eichen 87, 12205 Berlin, Germany; tobias.fritsch@bam.de (T.F.); alexander.ulbricht@bam.de (A.U.); gunther.mohr@bam.de (G.M.); giovanni.bruno@bam.de (G.B.); christiane.maierhofer@bam.de (C.M.); simon.altenburg@bam.de (S.J.A.)

[2] Institute of Physics and Astronomy, University of Potsdam, Karl-Liebknecht-Straße 24/25, 14476 Potsdam, Germany

* Correspondence: simon.oster@bam.de

Abstract: The detection of internal irregularities is crucial for quality assessment in metal-based additive manufacturing (AM) technologies such as laser powder bed fusion (L-PBF). The utilization of in-process thermography as an in situ monitoring tool in combination with post-process X-ray micro computed tomography (XCT) as a reference technique has shown great potential for this aim. Due to the small irregularity dimensions, a precise registration of the datasets is necessary as a requirement for correlation. In this study, the registration of thermography and XCT reference datasets of a cylindric specimen containing keyhole pores is carried out for the development of a porosity prediction model. The considered datasets show variations in shape, data type and dimensionality, especially due to shrinkage and material elevation effects present in the manufactured part. Since the resulting deformations are challenging for registration, a novel preprocessing methodology is introduced that involves an adaptive volume adjustment algorithm which is based on the porosity distribution in the specimen. Thus, the implementation of a simple three-dimensional image-to-image registration is enabled. The results demonstrate the influence of the part deformation on the resulting porosity location and the importance of registration in terms of irregularity prediction.

Keywords: selective laser melting (SLM); laser powder bed fusion (L-PBF); additive manufacturing (AM); process monitoring; infrared thermography; X-ray micro computed tomography (XCT); defect detection; image registration

Citation: Oster, S.; Fritsch, T.; Ulbricht, A.; Mohr, G.; Bruno, G.; Maierhofer, C.; Altenburg, S.J. On the Registration of Thermographic In Situ Monitoring Data and Computed Tomography Reference Data in the Scope of Defect Prediction in Laser Powder Bed Fusion. *Metals* **2022**, *12*, 947. https://doi.org/10.3390/met12060947

Academic Editor: Yongho Sohn

Received: 22 April 2022
Accepted: 25 May 2022
Published: 31 May 2022

1. Introduction

The industrial use of metal-based additive manufacturing (AM) processes has rapidly increased in recent years [1]. In comparison to traditional manufacturing, AM technologies such as laser powder bed fusion (L-PBF) offer the benefit of producing parts of highly complex geometry directly from the 3D CAD model while reducing the material waste [2]. L-PBF counts as one of the most established AM techniques and stands out due to its ability to produce features in high spatial resolution of tens of microns [3,4]. However, the occurrence of irregularities such as internal porosity, cracks, or surface roughness during manufacturing poses a risk to the final component quality [4,5]. Poor process parametrization (i.e., by scan velocity and laser power) was found to be an influential factor for the formation of irregularities [5]. Thermography as a radiometric nondestructive testing method has been utilized to monitor the part's local thermal history. This is performed by extracting thermal features from the spatial and temporal temperature distribution of the melt pool and part surface. From the obtained feature distribution, local areas of thermal deviation can be identified in which porosity is likely to form [6]. As a reference technique for the determination of the spatial distribution of porosity, X-ray micro computed tomography (XCT) is widely applied [7]. The correlation of thermographic feature data and XCT reference data facilitates the prediction of porosity likelihood [6]. The in situ detection of porosity has received increasing attention in the scientific community in recent years which is evident from the rising number of publications [8].

An important aspect for the prediction of irregularities such as porosity is the registration of the in situ monitoring and the reference XCT data. Image registration can be understood as the spatial alignment of two or more images. This mainly includes the goal of finding a transformation that aligns the features of interest visible in the image data [9]. A registration function can be obtained by applying a spatial transformation on a moving image that is registered with a fixed image. Here, a similarity measure or a cost function between the two images is optimized typically in several iteration steps [10,11]. A common example of a registration function is the affine transformation model, which allows translation, rotation, scaling, and skew of the moving image. This high number of degrees of freedom with respect to image transformation is not always necessary. For many applications, it might be sufficient to utilize only a rigid model that allows translation and rotation [11]. The evaluation of the registration accuracy is challenging and often limited to a qualitative validation by the user [12].

In terms of predicting internal porosity from sensor data, methods of artificial intelligence such as machine learning (ML) algorithms can be applied [13]. A requirement for a successful prediction is the accurate spatial allocation between sensor signal and resulting porosity information. Otherwise, the model is trained on spatially mismatched data and basically learns irrelevant data patterns. In L-PBF, the occurring irregularities have small dimensions. For example, in a study by Sinclair et al. [14], keyhole pores with diameters in the range of 10–60 µm were quantified. From that, it can be concluded that for a prediction of single keyhole pores, even small allocation errors resulting from the registration may significantly reduce the performance of the prediction model.

Furthermore, the differences in data format and dimensionality resulting from the different measurement methods are challenging for registration. The layer-wise acquired thermograms from in situ thermography contain information about the thermal radiation from the different object surfaces visible in the field of view of the camera (i.e., melt pool, solidified material, unmolten powder, and machine surroundings) [15]. Due to the projection of the 3-dimensional (3D) scene to the 2-dimensional (2D) focal plane array of the camera sensor, the height information of the specimen surface is lost. Thermographic in situ monitoring in L-PBF will result in 4-dimensional (4D) data consisting of a time series of 2D thermograms for each manufactured layer. Here, the measurable signal is limited by the camera dynamics as well as the chosen spatial and temporal resolution. In contrast, XCT provides a spatially high resolved 3D object representation of the manufactured specimen. The 3D object reconstruction is created from 2D projection images captured by a flat panel X-ray detector. The projection images are reconstructed using algorithms such as the Feldkamp algorithm for cone beam geometry [16]. The reconstruction includes an interpolation process on the gray value voxel grid. Artifacts, such as scatter [17], cone beam [18], and beam hardening [19], may decrease the spatial resolution and the registration procedure. Furthermore, the XCT data contains all shape deformations of the part caused by the manufacturing process which remain in the part after removal from the dummy cylinder, such as shrinkage and warping [20].

In the literature [6,8,21–27], a range of methods is utilized to align thermographic in situ monitoring data and XCT reference data as a preparational step for irregularity prediction in the scope of L-PBF. Here, the insertion of artificial voids offers the advantage of predefined void location and shape. This benefits the identification of the defects in the determined in situ signal [8]. Mireles et al. [21] integrated artificial voids in the design of a metal part manufactured by electron beam melting. The correlation between reference XCT data and data obtained by an off-axis infrared (IR) camera was performed utilizing the known defect position within the part. Lough et al. [22] qualitatively compared single lateral slices in a cylindrical test specimen on the basis of voids that were produced by decreasing the laser power. In a study by Coeck et al. [23], the registration of IR data and XCT data was performed utilizing the large size of the present lack of fusion (LoF) voids in the observed specimen as reference object. Due to the low number of voids (45 voids in a 10 mm^3 volume) and their large dimensions, a straightforward spatial assignment between voids and melt pool monitoring data was feasible. The data time series derived from the off-axis photodiodes of the melt pool monitoring system were mapped to the 3D position of the laser scanner. Single data points from the obtained point cloud were allocated to a void event if the distance between the data point centroid and the void centroid was below 500 μm. Forien et al. [24] overlapped X-ray radiography scans of single tracks with coaxial pyrometry in situ monitoring data using a manual translation. Like [23], the time series data obtained by pyrometry was mapped to the 3D laser scanner position. Voids were segmented manually and correlated with the pyrometry signal in a radius of 65 μm around the void centroids.

Apart from manual mapping of signal and reference data, image registration algorithms can be utilized to automatically align multiple images for further analysis [9]. A 3D image registration was performed by Mohr et al. [25] to overlap optical tomography and XCT datasets. Here, the open-source software elastix 4.9 (University Medical Center Utrecht, Utrecht and contributors, The Netherlands) was utilized to apply an affine transformation. In a recently published study, Lough et al. [6] performed a voxel-based quantitative analysis using a layer-by-layer registration along the z axis of the specimen which included down-sampling of the XCT dataset. Here, the datasets were aligned manually along the z axis. Afterwards, an automated translation algorithm was applied. Taherkani et al. [26] registered XCT data according to the specimens CAD file using the 3D image analysis software Dragonfly Pro v4.0 (Object Research Systems Inc., Montreal, QC, Canada). As geometric reference, horizontal and vertical grooves were integrated into their specimen design. The alignment of the CT data with the melt pool monitoring data was carried out on the basis of artificial voids integrated in the specimen. Gobert et al. [27] utilized an affine mapping function to register CT data and powder bed image data acquired by a high resolution digital single-lens camera. The registration was based on minimization of the root mean square error between geometrical reference points in the datasets.

The literature [4,5] emphasizes that irregularity prediction in L-PBF is an important task to predict the service life of the produced part. The development of accurate prediction models requires a precise image registration of the acquired monitoring data and the reference data, especially due to the small dimensions of the occurring irregularities. To the authors' knowledge, few systematic investigations concerning registration methods in the scope of irregularity prediction in L-PBF have been performed. Some authors [21,22] rely on large artificial voids integrated in the specimen design to simplify their detection in the sensor data and in the ground truth data. If registration algorithms are applied, their accuracy is usually not further specified, even though it is essential information to evaluate the measurement uncertainty. Furthermore, the influence of occurring shape distortions such as warping or shrinkage in the datasets are usually not included in the registration approach.

This study focuses on the registration of feature datasets extracted from in situ thermography and an XCT dataset of a cylindric specimen that contained keyhole pores [28]. The registration is performed as a preliminary step to facilitate a highly accurate prediction of the present irregularities using ML methods (to be reported in a sister paper). The introduced registration methodology focuses on dataset preprocessing to enable the application of a simple 3D image-by-image registration. A systematic description of the singular data processing steps and the challenges arising from the different formats and dimensionalities of the datasets is given. In the context of the XCT dataset, especially the influence of the present shrinkage and material elevation on the registration accuracy is examined. Here, a novel method to adaptively adjust the part deformation is developed based on the pore distribution in the specimen. The obtained registration accuracy is evaluated and future challenges in the context of irregularity prediction in L-PBF are derived.

2. Materials and Experimental Procedures

A cylindrical specimen was produced on a commercial L-PBF machine (SLM 280 HL, SLM Solutions Group AG, Lübeck, Germany) using AISI 316L stainless steel powder. The powder was specified as follows: apparent density of 4.58 g/cm^3, D_{mean} = 34.69 µm, D_{10} = 18.22 µm, D_{50} = 30.5 µm, and D_{90} = 55.87 µm. The specimen design included a surrounding staircase structure as registration landmark (see Figure 1) inspired by a study by Gobert et al. [27]. The entire specimen was built upon the milled surface of a dummy cylinder to prevent cutting losses. The inner cylinder (diameter: 7 mm, height: 12 mm) consisted of six sections. The sections were manufactured with varying processing parameters to introduce keyhole porosity into the material. The parameter variation was performed by decreasing the scanning velocity. This resulted in increased volumetric energy densities (VED) [28], see Table 1. A hatch distance of 120 µm and a layer thickness of 50 µm were utilized. Furthermore, a cross and a letter landmark were added to the specimen's top surface as further geometric landmarks.

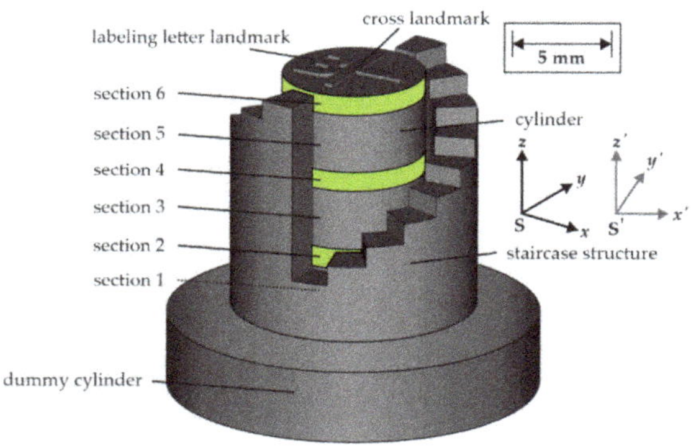

(**a**)

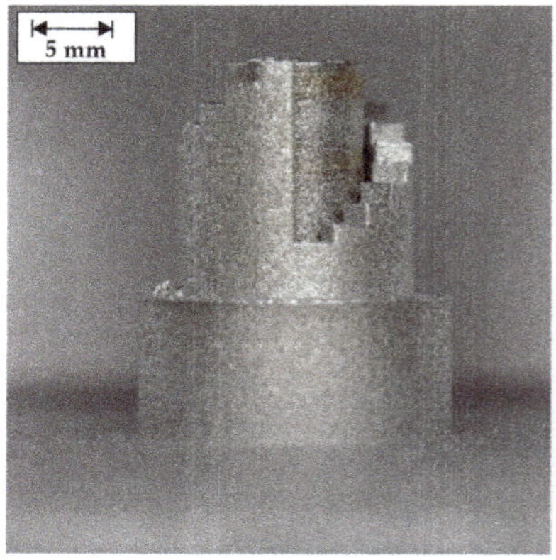

(**b**)

Figure 1. (**a**) Specimen design including staircase structure, top surface cross landmark, labeling letter landmark, dummy cylinder, and the introduced coordinate systems **S** and **S**′. The green areas mark sections that were manufactured using increased VED. In this view, section 1 is hidden behind the staircase which is indicated by the dashed line. (**b**) Manufactured specimen on dummy cylinder Adapted from Ref. [28].

Table 1. Overview of the manufacturing conditions of the different cylinder sections (corresponding to specimen "B" in [28]). In the last column, the relative increase of the VED in comparison to sections 1, 3, and 5 is given by the percentage value in brackets.

Section	Layer Count	Laser Power P in W	Scan Velocity v in mm/s	VED in J/mm^3
1	1–60	275	700	65.45
2	61–80	275	560	81.84 (+25%)
3	81–140	275	700	65.45
4	141–160	275	467	98.21 (+50%)
5	161–220	275	700	65.45
6	221–240	275	400	114.45 (+75%)

The in situ monitoring setup consisted of three infrared cameras that were mounted off-axis outside of the machine, utilizing a custom-made optical entrance. The thermal radiation of the process was guided to the cameras by a system of gold-coated mirrors and beam splitters that were optically adapted to the spectral sensitivity of the respective camera system. For further details concerning the powder specifications, scanning strategy, and the machine setup, refer to a previously conducted study [28]. In this study, the in situ monitoring data of the deployed short-wave infrared camera (Goldeye CL-033 TEC1 from Allied Vision Technologies GmbH, Stadtroda, Germany) was utilized for the investigation.

From the short-wave infrared camera, a 4D dataset consisting of thermograms with the size of 90 pix^2 (pixel scale of approximately 100 µm/pix) was obtained from 240 manufacturing layers during the process monitoring. The dataset size was $90 \times 90 \times n_{im,l} \times 240$, where $n_{im,l}$ depicts the number of the image that was acquired during the exposure of a single layer l. $n_{im,l}$ could vary slightly for the different layers due to variations in the starting time of the recording and durations of the layer illumination. It was of the order of approximately 8000 images. All thermograms were temperature calibrated using a single point calibration method [29]. Ten different features were identified from the spatial and temporal temperature information present in the thermograms (Table 2). Detailed insights concerning the feature extraction can be found in [28]. A coordinate system **S** (Figure 1) was introduced to describe the respective 3D voxel position of each feature value in the specimen data. The features were distinguished into melt pool-based and time-dependent temperature features. Both feature classes differed in spatial information density. Melt pool-based features were extracted for each image and spatially assigned to the *x*-*y* position of the pixel with the highest temperature visible in the image. This pixel represented the position where the laser spot was located on the specimen surface. Due to the temporal and spatial resolution of the camera, the melt pool feature data were distributed sparsely in each respective layer. The sparsity was dependent on the scan velocity and resulted in data point distances along a single scan track ranging from approximately 110 µm (in section 6) to 200 µm (in sections 1, 3, and 5) with a hatch distance of 120 µm. The feature extraction resulted in 3D datasets $\mathbf{F}^i$ of the size $3 \times n_{mp,l} \times 240$. Here, the first dimension represented the individual feature value and its associated *x*-*y* position in the observed layer (in total 240 layers). The index i in $\mathbf{F}^i$ corresponds to the observed melt pool feature and $n_{mp,l}$ corresponds to the number of melt pool images in the image series of the respective layer. The time-dependent features were calculated from the temporal temperature information of each image pixel from a single layer manufacturing. Hence, the spatial information density was limited only by the spatial resolution of the camera. As a result, 3D datasets $\mathbf{F}^j$ of the size $90 \times 90 \times 240$ were generated for the respective time-dependent feature j. According to the spatial resolution of the camera and the nominal layer height, a voxel size of $100 \times 100 \times 50$ µm^3 was present in all datasets.

Table 2. Extracted features from in-situ monitoring thermograms.

Feature Class	Feature
Melt pool-based features [1]	Area Length Width Eccentricity Perimeter Mean temperature Maximum temperature
Time-dependent temperature features	Time over threshold of 1200 K Time over threshold of 1680 K Time over threshold of 2400 K

[1] Corresponding to the geometry and temperature distribution of the apparent melt pool blob visible in the thermogram data.

Subsequent to manufacturing, XCT was performed on the specimen and the surrounding staircase after separation from the base plate using the commercial CT-scanner GE v | tome | x 180/300 (GE Sensing and Inspection Technologies GmbH, Wunstorf, Germany) [28]. A voltage of 222 kV and a current of 45 µA were used to acquire 3000 projections at an acquisition time of 2 s. To improve the signal-to-noise ratio of the projections, three images were taken at each of the 3000 angular positions, and their average was used for the 3D reconstruction. The reconstruction of the XCT projection data was performed by using the filtered back-projection algorithm [16], resulting in a raw dataset $\mathbf{V}^{raw}$ with the spatial dimension of 2024 × 2024 × 2024 and a voxel size of 10 μm^3. The achieved voxel size enabled the quantitative analysis of features of a size above 20 μm^3. A further coordinate system $\mathbf{S}'$ (Figure 1) given by the axis of the original XCT data was utilized to describe the XCT voxel positions. Subsequent to the reconstruction, a beam hardening correction [30] was performed. Furthermore, the cylinder axis was aligned parallel to the z' axis. This was carried out using ImageJ Fiji [31] and MATLAB (MathWorks Inc., Natick, MA, USA). Here, contiguous x'-y' slices were extracted from $\mathbf{V}^{raw}$ and a circle fit [32] was applied to the circular shape visible in the slice. Afterwards, the circle fit centroids of all slices were calculated. From the calculated course of the centroids along the z' axis, the angle of cylinder axis inclination was determined. The alignment was performed by manual rotation of the cylinder around the x' and y' axis. The resulting angle deviation after the axis alignment was calculated to approximately 0.005°. This resulted in a maximum height deviation at the specimen surface of approximately 1 μm (regarding height and diameter of the CAD of the specimen). The image data were further processed by adjusting the brightness and the contrast. A 3D dataset $\mathbf{V}^{proc}$ with the size of 711 × 711 × 1260 voxels was obtained containing the density information of inner cylinder and the landmark structures on the specimen top. In the following, a local thresholding algorithm introduced by Phansalker et al. [33] was applied to the data to distinguish between material and voids. The binarized dataset is denoted as $\mathbf{V}^{proc,bin}$. An overview of the datasets that are used in this study is given in Table 3.

Table 3. Overview of the obtained datasets from thermography and XCT. $n_{mp,l}$ corresponds to the number of images acquired during the manufacturing of a single layer and is in the order of approximately 8000 images.

Dataset	Source	Content	Dimensions	Voxel Size in μm^3
$\mathbf{F}^i$	SWIR camera	Values of i-th melt pool-based feature	3 × $n_{mp,l}$ × 240	100 × 100 × 50
$\mathbf{F}^j$	SWIR camera	Values of j-th time-dependent temperature features	90 × 90 × 240	100 × 100 × 50
$\mathbf{A}^k$	SWIR camera	Preprocessed and interpolated values of k-th feature	935 × 980 × 1200	10 × 10 × 10
$\mathbf{V}^{raw}$	XCT	Density information (raw)	2024 × 2024 × 2024	10 × 10 × 10
$\mathbf{V}^{proc}$	XCT	Density information (cropped, increased contrast)	711 × 711 × 1260	10 × 10 × 10
$\mathbf{V}^{proc,bin}$	XCT	Porosity information (cropped)	711 × 711 × 1260	10 × 10 × 10
$\mathbf{V}^{final}$	XCT	Porosity information (cropped, adjusted to CAD)	711 × 711 × 1200	10 × 10 × 10

3. Registration Methodology and Results

The aim of this study was the registration of the obtained thermogram feature datasets and the corresponding XCT dataset to produce an accurately aligned data basis for future irregularity prediction modeling. Preprocessing methods were used to adjust each dataset to match the original specimen geometry given by the CAD. Based on a sophisticated preprocessing workflow, the registration was simplified to a simple 3D image-to-image algorithm. In the following, the performed processing steps concerning thermogram feature data, the reference XCT data, and the image registration are described, and the obtained results are presented. A schematic overview of all performed steps is given in Figure 2.

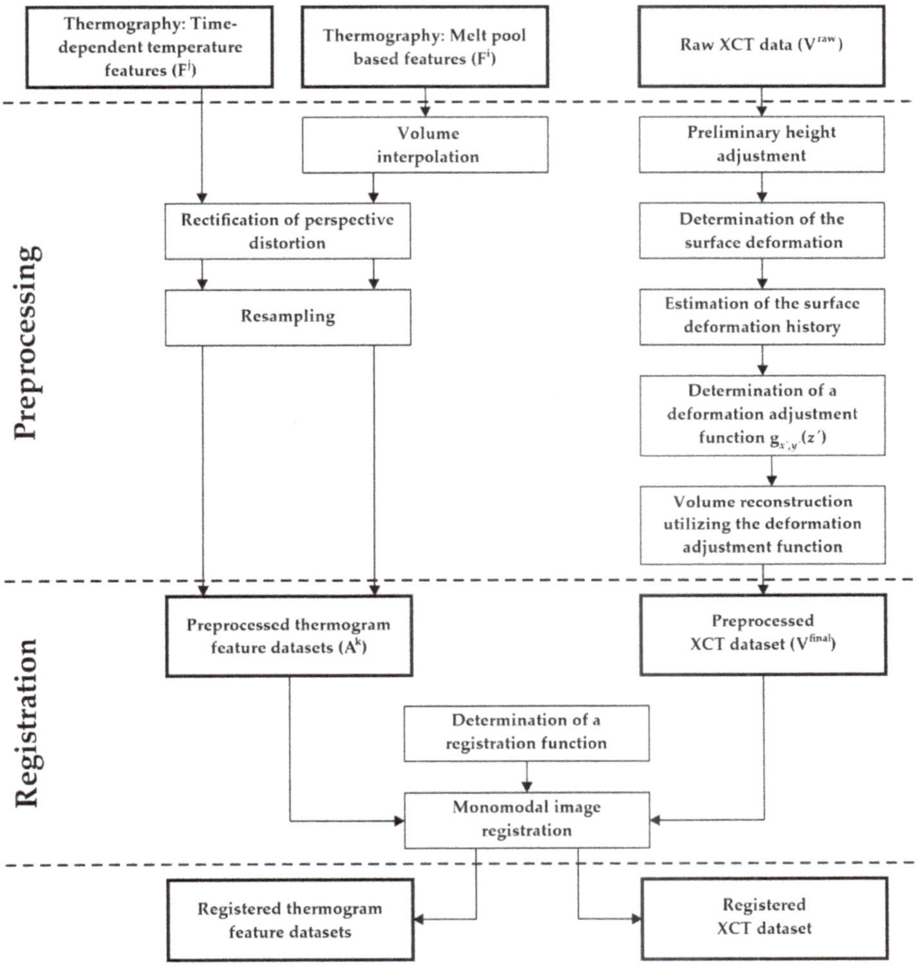

Figure 2. Schematic overview the processing steps for the registration of thermographic and XCT datasets. The dashed lines indicate which steps correspond to the data preprocessing and which correspond to the registration. Datasets are indicated by bold frames.

3.1. Preprocessing of Thermogram Feature Dataset

A first evaluation of the obtained melt pool-based feature dataset $\mathbf{F}^i$ and the time-dependent feature dataset $\mathbf{F}^j$ revealed three pre-registration challenges: First of all, due to the sparse nature of $\mathbf{F}^i$, a volume interpolation was necessary to perform an image-to-image registration [34]. Secondly, perspective distortion was found in both datasets resulting from the optical setup [28] used for the process monitoring. Such distortion led to inconsistent voxel scales in the x and y axis. Thirdly, the datasets were resampled to the voxel scale 10 μm^3 of the XCT reference data to enable the precise spatial overlap of all datasets. One 3D linear interpolation algorithm was implemented to solve all three pre-registration challenges. Thus, missing pixel values in the sparse melt pool-based features were interpolated (see Figure 3a), the present imaging error was rectified, and the voxel size was adjusted. The distorted image scales s_x = 103.8 μm/pix and s_y = 108.9 μm/pix were calculated from pre-manufacturing recordings of a calibration target (grid pattern), while the scale in the z direction s_z was given by the layer thickness of 50 μm. The size of the rectified target volume **A** (width w_A, depth d_A, and height h_A) was derived from the former thermogram size 90 pix^2 and the overall layer count (240 layers):

$$w_A = 90 \cdot \frac{s_{x'}}{s_x} \tag{1}$$

$$d_A = 90 \cdot \frac{s_{y'}}{s_y} \tag{2}$$

$$h_A = 240 \cdot \frac{s_{z'}}{s_z} \tag{3}$$

Here, $s_{x'}$ = 10 µm, $s_{y'}$ = 10 µm and $s_{z'}$ = 10 µm denote the voxel size of the XCT data.

From the initial datasets $\mathbf{F}^i$ and $\mathbf{F}^j$, k = i + j = 10 datasets $\mathbf{A}^k$ with a size of 935 × 980 × 1200 voxels were interpolated. The results of the interpolation of melt pool-based feature data from a single layer are depicted in Figure 3b.

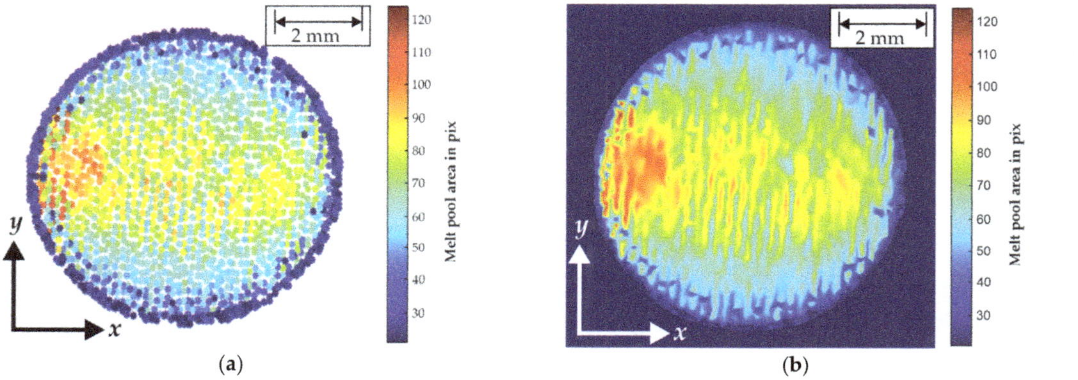

Figure 3. Preprocessing of thermogram feature data (here: melt pool blob size) of layer 59. (**a**) Incremental data points of melt pool blob size containing image distortion. (**b**) Rectified and resampled thermogram with an adjusted scale of 10 µm/pix.

3.2. Preprocessing of XCT Dataset

The XCT dataset contained the porosity information present in the specimen. The dimensions and the shape of the cylindric specimen in the XCT dataset deviated significantly from the shape of the original CAD. A first comparison of the specimen to the CAD model height showed a vertical shrinkage. Furthermore, the observation of the top surface revealed a severe shape deformation. The surface rim was elevated in comparison to its center. This resulted in significant height deviations (see Figure 4). The maximum height differences of approximately 400 µm between surface rim and center (which corresponds to 8 manufacturing layers) would produce major errors if a simple lateral slicing along the x'-y' plane was performed for correlation with the monitoring data. Alongside this, the registration landmarks (letter and cross) that were later utilized to obtain a registration function were inspected since they can be clearly separated from the main cylindrical specimen. The bordering areas of the landmarks exhibited a local height increase. This represented a distortion of the original surface (see Figure 4). The surface deformation posed a major difficulty for the registration because it was unclear how the deformation was formed over the course of the manufacturing process.

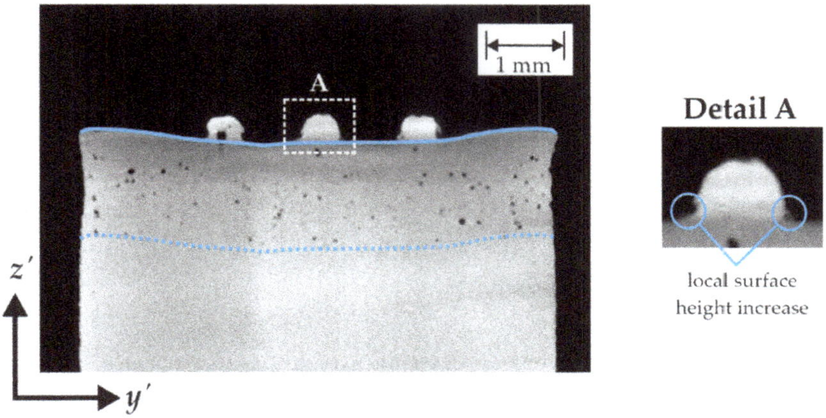

Figure 4. Cross-sectional slice of XCT reference volume at a depth of $x' = 3$ mm. Gray value variations in the bulk material are image artifacts originated by the cone-beam reconstruction of strong absorbing material. Height elevations on the surface rim and lateral shrinkage in the section of increased VED are visible. The approximated top surface shape (continuous blue line) is repeated below (dashed blue line) for a better comparison with the pore distribution. In Detail A, the areas of local surface height increase close to the landmarks are marked by blue circles.

Ulbricht et al. [7] found a similar surface deformation in an equally designed specimen that was manufactured using the same material and machine. Furthermore, they found indications of a comparable surface deformation at multiple stages of the manufacturing process. These previous deformations were estimated from the observation of LoF voids at the transition from a void-free section and followed by a section with artificial voids (notches) within the specimen. The authors suggested that the deformation could have been present during the entire manufacturing process. In the current study, the observation of the pore distribution in the transition from section 5 to 6 indicated a similar effect (see Figure 4). Therefore, a method was developed to adjust the surface in the dataset based on the deformation information obtained from the pore distribution. The method assumed that all previous surface deformations that had formed during manufacturing had a qualitatively similar shape to the visible deformation of the specimen surface. Furthermore, it was assumed that the previous surface deformation can be reconstructed from the depth distribution of the lowest keyhole pores at the onset of a region with increased VED, i.e., section 6. Finally, the preprocessing method for the XCT dataset consisted of five steps (compare Figure 2):

1. Preliminary height adjustment;
2. Determination of the surface deformation;
3. Estimation of the surface deformation history;
4. Determination of a deformation adjustment function;
5. Volume reconstruction utilizing the deformation adjustment function.

3.2.1. Preliminary Height Adjustment

Due to the height difference between the manufactured specimen represented in the XCT dataset and the original CAD, a preliminary height adjustment was performed. The shrinkage was estimated from the height of the staircase treads that were built as registration landmarks. The staircase was built upon the milled surface of the dummy structure and scanned by XCT along with the cylindric specimen. The stair treads were located at specific layer heights, starting at a height of 3 mm, and repeating every 1 mm. Even though the staircase was built without VED variation (utilization of standard VED of 65.45 J/mm^3), it was found that the highest staircase tread had an approximately equal height to the rim of the top surface. Furthermore, it was found that section 1 exhibited a height decrease of approximately 230 µm (7.66%) in comparison to the CAD model. This was probably caused by solidification shrinkage [25]. In the remaining sections 2 to 6, the height decrease was noticeably lower and amounted in total to approximately 60 µm (0.67%). A 3D linear interpolation to the original CAD height was utilized to perform the height adjustment. From the large difference in height decrease, it was decided to split the whole dataset into two subsets (the first subset containing section 1, and the second subset containing sections 2 to 6) and to perform separate height adjustments. Afterwards, the two subsets were vertically fused again.

The voxel size was effectively locally distorted by the performed height adjustment. In the case of a quantitative void volume analysis, this must be considered. For this study, this effect was of no further significance.

3.2.2. Determination of the Surface Deformation

For the quantification of the surface deformation, the voxels representing the specimen surface were extracted in the first place. For further refinement of the surface information of $\mathbf{V}^{\text{proc,bin}}$, the extraction was performed on the adjusted intensities of $\mathbf{V}^{\text{proc}}$. Here, a customized thresholding algorithm was utilized. It is based on the ISO50% value $T_{\text{ISO50\%}}$ which was calculated as a global threshold from the intensity values of $\mathbf{V}^{\text{proc}}$. The ISO50% value represents the average between the highest peak of background voxels and the highest peak of material voxels in a histogram of all voxel gray values [35]. The surface height was calculated for every single x'-y' position. To suppress errors arising from possible reconstruction artifacts located above the surface, the algorithm was extended by further thresholding conditions. A subset $\mathbf{V}^{\text{proc,ROI}}$ was extracted from a region of interest (ROI) in $\mathbf{V}^{\text{proc}}$ which included the entire surface information. $I_{x',y',z'}$ corresponds to the respective gray value of a voxel in $\mathbf{V}^{\text{proc,ROI}}$ where the indices x', y', and z' correspond to the voxel position in the respective axis. A surface edge was determined if the following criteria were fulfilled:

$$I_{x',y',z'} \geq T_{\text{ISO50\%}} \quad (4)$$

$$I_{x',y',z'+1} < T_{\text{ISO50\%}} \quad (5)$$

$$\sum_{i=1}^{\text{nvox}} I_{x',y',z'-i} \geq n_{\text{vox}} \cdot T_{\text{ISO50\%}} \quad (6)$$

n_{vox} is the number of voxels below the observed edge voxel which were considered for the surface determination. The parameter was manually tuned to 10, which produced good results concerning the suppression of reconstruction artifacts. The 2D array of calculated surface data points (Figure 5a) were denoted as $\mathbf{H}^{\text{surf}}$. Subsequently, the cross landmark, the labeling letter, and their bordering areas were removed since the found local surface increase was obstructive for determination of the surface information. Here, arithmetic image multiplication with polygon masks was used to perform the removal of the cross landmark and the labeling letter. An interpolation algorithm [36] was used to reconstruct the missing surface parts. Additional smoothing was applied to remove local height deviations, i.e., spatter elements that were connected to the surface. The determination accuracy was quantified by a mean absolute error (MAE) of approximately 4 µm. The resulting surface reconstruction $\mathbf{H}^{\text{surf,rec}}$ is depicted in Figure 5b.

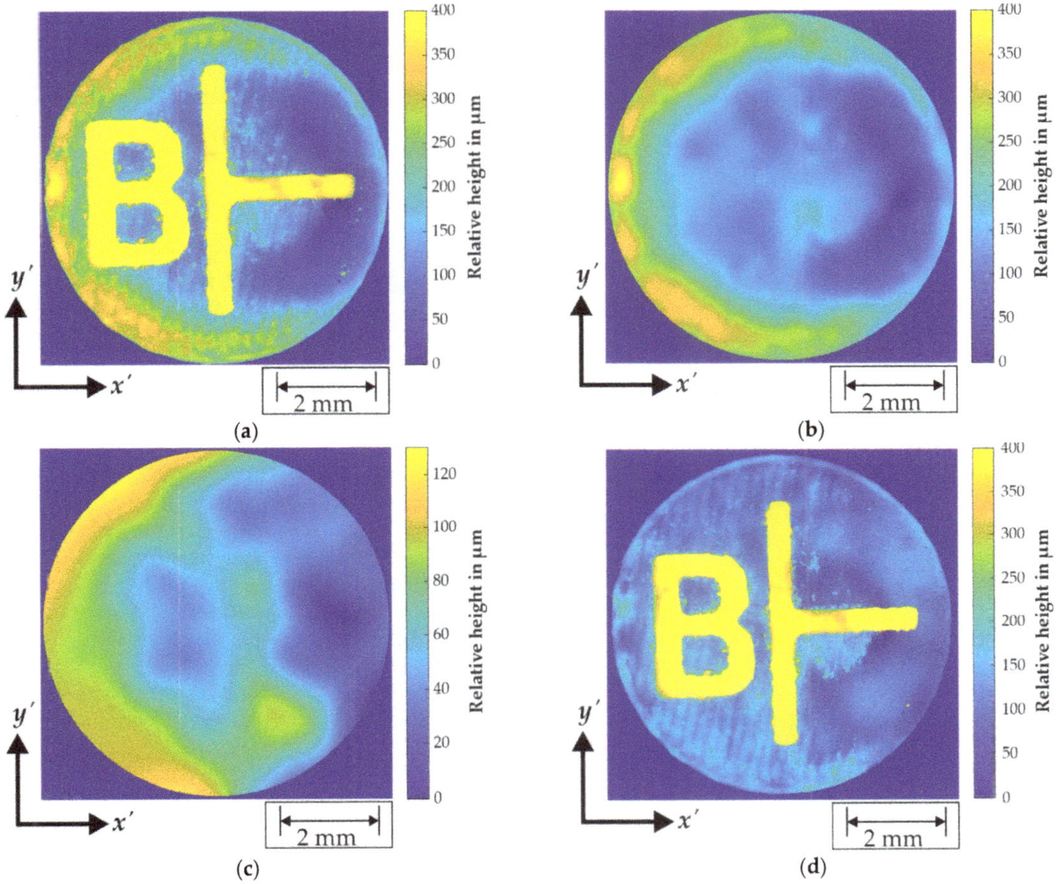

Figure 5. (**a**) Part surface extracted by thresholding algorithm. (**b**) Reconstructed part surface after removal of registration landmarks. (**c**) Reconstructed surface from lowest pores in the transition zone (between sections 5 and 6) at an approximately average specimen height of 10.7 mm. (**d**) Part surface after application of deformation adjustment function extracted by thresholding algorithm.

3.2.3. Estimation of the Surface Deformation History

The distribution of the pores in the transition zones between sections of standard VED followed by sections of increased VED was the basis for the estimation of the former surface deformation history. Such pores in the transition zones are called boundary pores hereafter. Due to the increased VED utilized in sections 2, 4, and 6, predominantly keyhole pores were present. Keyhole pores are likely to form at the bottom of the melt pool and be entrapped in its lower part [37]. Inspired by the findings of Ulbricht et al. [7], it is assumed that the topography of the former surface can be reconstructed from the statistically distributed pores at the boundary of the transition zone. The first step to estimate the previous surface deformation was the extraction of the boundary pores. The transition zones from section 1 to 2 and from section 3 to 4 were not regarded due to the low information density resulting from the low number of pores in these sections. The decreased number of pores contributed to the lower increase of VED [38] utilized in these sections (+25% in section 2 and +50% in section 4). Only in the transition from section 5 to 6 was a sufficient number of pores present. A data subset from height $z' = 1050$ to 1090 vox (equal to a specimen height of 10.5 µm to 10.9 µm) that contained the lowest boundary pores was extracted from

$\mathbf{V}^{\text{proc,bin}}$. The 3D position of each respective pore centroid in the subset was calculated (MATLAB function "regionprops3") and stored in a 2D array. The pores located at the rim of the obtained disc-shaped point cloud were extracted (MATLAB function "boundary"). For the surface reconstruction, only the pores of the bottom part of the rim were utilized since they were located in the transition zone. These centroids were denoted as $\mathbf{H}^{\text{pores}}$ and were located at an average centroid height of approximately $z' = 1070$ voxel. Here, the same interpolation algorithm as used for the top surface [36] was applied to reconstruct the surface shape. This shape was denoted as $\mathbf{H}^{\text{pores,rec}}$. Additional smoothing was added to remove local inhomogeneities. $\mathbf{H}^{\text{pores,rec}}$ is depicted in Figure 5c. Good agreement was found between the reconstructions $\mathbf{H}^{\text{pores,rec}}$ and $\mathbf{H}^{\text{surf,rec}}$. Furthermore, the standard deviation (STD) σ^{pores} of the original centroid heights in $\mathbf{H}^{\text{pores,rec}}$ was calculated to 33 µm, which is significantly lower than the STD σ^{surf} of $\mathbf{H}^{\text{surf,rec}}$ (95 µm).

3.2.4. Determination of a Deformation Adjustment Function

Based on $\mathbf{H}^{\text{pores,rec}}$, a deformation adjustment function $g_{x',y'}(z')$ was determined for the compensation of the surface deformation history. As stated above, the method assumed that all surface deformations during the manufacturing had an approximately similar shape like the surface deformation $\mathbf{H}^{\text{surf,rec}}$ determined in Section 3.2.2. However, the comparison of σ^{pore} and σ^{surf} implied that the average height amplitude of the deformed surfaces increased with growing z'. This was taken into account by the introduction of a compression factor c. The idea behind c was the adaption of the average height amplitude in dependency to z'. Due to the given pore distribution, only three reference surfaces were available for the calculation of c: (i) The non-deformed milled surface of the dummy base volume $\mathbf{H}^{\text{plate}}$ at $z' = 0$ vox with $\sigma^{\text{plate}} = 0$, (ii) the surface reconstruction $\mathbf{H}^{\text{pores,rec}}$ from the boundary pores at an average height of $z' \approx 1070$ voxel with $\sigma^{\text{pores}} = 33$ µm, and (iii) the surface reconstruction $\mathbf{H}^{\text{surf,rec}}$ with $\sigma^{\text{surf}} = 95$ µm of the specimen top. Due to this lack of information, only a sectional linear interpolation approach was feasible for the calculation of c (see Figure 6a).

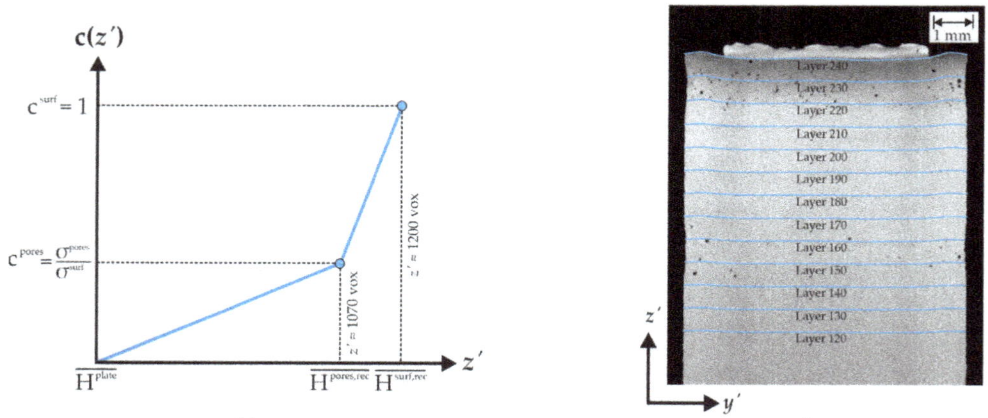

(a) (b)

Figure 6. (**a**) Qualitative illustration of the linear sectional fit for the determination of the compression factor c. (**b**) Cross-sectional slice of the XCT data at a depth of 3.5 mm. The height deviations corresponding to the manufacturing layers calculated from $g_{x',y'}(z')$ are depicted as blue lines.

The compression factor c is given by:

$$c(z') = \begin{cases} \frac{m_1 z' + b_1}{\sigma^{\text{surf}}} : \overline{\mathbf{H}^{\text{plate}}} < z' < \overline{\mathbf{H}^{\text{pores,rec}}} \\ \frac{m_2 z' + b_2}{\sigma^{\text{surf}}} : \overline{\mathbf{H}^{\text{pores,rec}}} \leq z' < \overline{\mathbf{H}^{\text{surf,rec}}} \end{cases} \tag{7}$$

$m_1 = 3.902 \times 10^{-3}$, $m_2 = 4.479 \times 10^{-3}$ correspond to the slope and $b_1 = 0$, $b_2 = -4.33$ correspond to the intercept of $g_{x',y'}(z')$ in the respective interpolation section (see Figure 6a). The sections were chosen corresponding to the average height of the substrate plate $\overline{\mathbf{H}^{plate}} = 0$, the average height of the extracted pore centroids $\overline{\mathbf{H}^{pores,rec}}$ and the average surface height $\overline{\mathbf{H}^{surf,rec}}$. To obtain values between 0 and 1, a normalization with σ^{surf} was performed.

The local height deviation at an arbitrary voxel position resulting from the surface deformation was calculated by the deformation adjustment function $g_{x',y'}(z')$:

$$g_{x',y'}(z') = \Delta\mathbf{H}_{x',y'}^{surf,rec} \cdot c(z') \tag{8}$$

$$\Delta\mathbf{H}_{x',y'}^{surf} = \mathbf{H}_{x',y'}^{surf,rec} - \overline{\mathbf{H}^{surf,rec}} \tag{9}$$

Here, the local height deviation at voxel position x' and y' is given by $\Delta H_{x',y'}^{surf,rec}$. Exemplary values of $g_{x',y'}(z')$ for chosen layer heights are depicted in Figure 6b.

3.2.5. Volume Reconstruction Utilizing the Deformation Adjustment Function

Equation (8) gives an incremental floating number which corresponds to the height deviation of the considered voxel with the height z'. The volume reconstruction was performed by the generation of a new volume $\mathbf{V}^{final}$ from the distorted volume $\mathbf{V}^{proc,bin}$. Here, for each new voxel in $\mathbf{V}^{final}$, the position of a corresponding voxel in $\mathbf{V}^{proc,bin}$ was calculated and its associated binary value was assigned to the new voxel:

$$\mathbf{V}^{rec}(x',y',z') = \mathbf{V}^{proc,bin}(x',y',z^*) \tag{10}$$

$$\text{with } z* = \left\lceil z' + g_{x',y'}(z') \right\rfloor \tag{11}$$

z* was rounded to avoid non-integer values. Since the specimen height was locally adjusted to the CAD height, the number of voxels containing density information was larger in $\mathbf{V}^{final}$ than in $\mathbf{V}^{proc,bin}$. Therefore, single voxel values from $\mathbf{V}^{proc,bin}$ were duplicated during the assignment to $\mathbf{V}^{final}$. The decision to duplicate a voxel was determined by the rounding operation of z*. The maximum percentage of duplicated voxel at a single x'-y' position was approximated to ~3% from the maximum height difference between the real specimen and the CAD (~400 µm). The final size of $\mathbf{V}^{final}$ was 711 × 711 × 1200 voxels. For evaluation, the adjusted top surface of $\mathbf{V}^{final}$ was extracted by the thresholding algorithm described Section 3.2.2 (Figure 5d). After the removal of the cross and the label landmark (see Section 3.2.2), the STD of the surface height of $\mathbf{V}^{final}$ was calculated to 30 µm. This is a significant decrease compared to the surface of $\mathbf{V}^{proc}$ (STD of 95 µm). Furthermore, a comparison of the pore distribution in the transition zone pre- and post-adjustment is depicted in Figure 7. In the hypothetical case of a specimen free of surface deformations, the pores were expected to appear at an approximately similar height z' in the specimen under the assumption of a comparatively small statistical depth variation of keyhole pores. The former dataset is distinguished by pores being distributed heterogeneously, appearing at first especially in the right half of the specimen in Figure 7a with a transition to the left half with growing z'. The height difference between the first appearance of keyhole pores in the cylinder right half to a pore distribution over the entire specimen cross-section amounts to approximately 10 to 11 voxels. In the case of the adjusted dataset, this height difference is decreased to approximately 5 to 6 voxels, especially when disregarding the pores present at the specimen rim. These were presumably LoF voids resulting from the interface between bulk and contour scans [7,39–41]. Furthermore, the described transition in the former dataset from right specimen half to left specimen half is no longer present here.

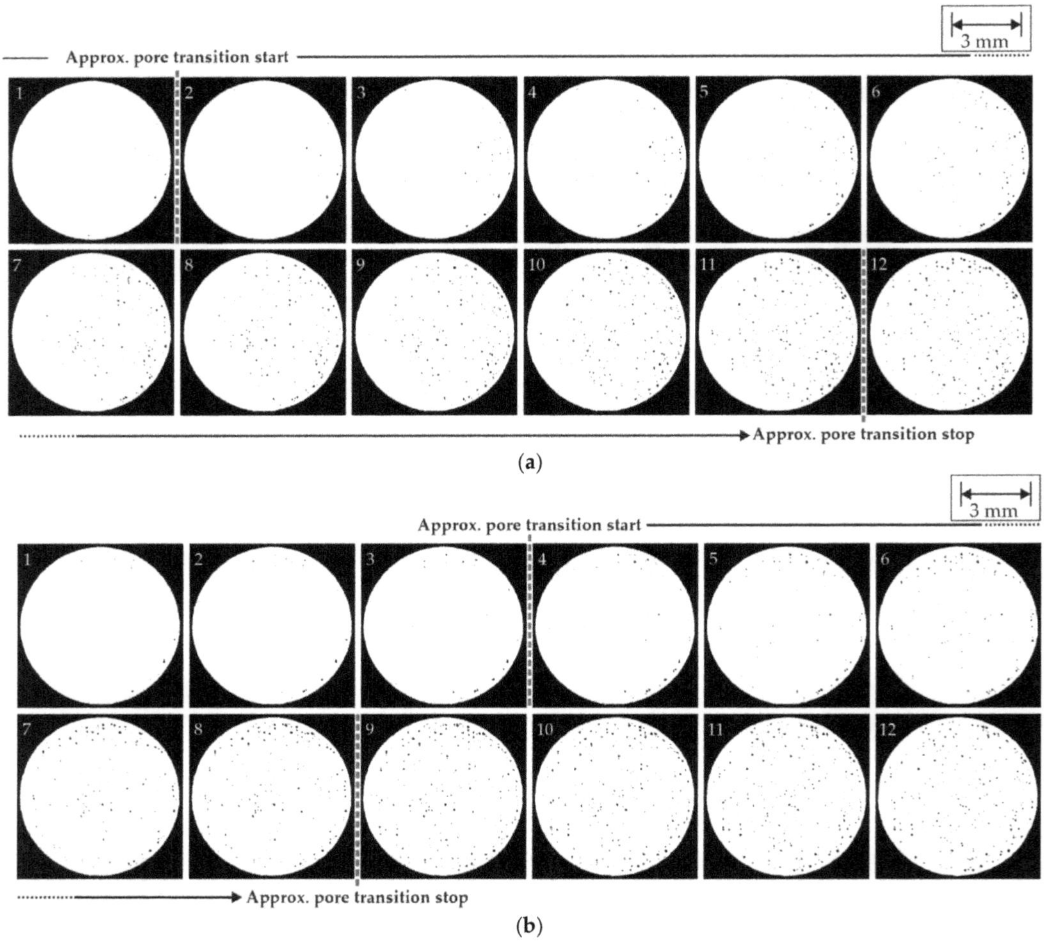

Figure 7. Pore distribution between section 5 and 6 starting at a specimen height of approximately 1050 voxels (10.5 mm) increasing by $z' = 1$ voxel (10 μm) per image slice (**a**) before deformation adjustment ($\mathbf{V}^{proc,bin}$) and (**b**) after deformation adjustment ($\mathbf{V}^{final}$). While the pore appearance in (**a**) is mainly starting on the right specimen half and transitions over the course of 10 image slices to the left half, the pore appearance in (**b**) begins mainly in the center and transitions faster (approximately in 5 image slices) from singular pores to a widespread distribution over the entire cross-section. The dashed red lines indicate the approximate starting and stopping points of the pore transition and are added to increase the readability of the figure. The starting points were manually approximated based on the appearance of a sufficient number of clearly identifiable pores in the cylinder bulk. The stopping points were manually approximated as the images in which pores appeared in major parts of the bulk.

Due to the performed surface adjustment, the utilization of a simple 3D image-by-image registration algorithm was enabled.

3.3. Image Registration

The spatial image registration was performed using the MATLAB Registration Estimator Toolbox. Here, several different feature-based and intensity-based algorithms for image registration are available. For this study, a monomodal intensity-based algorithm [10] was chosen because the datasets were acquired by different imaging techniques and therefore, had different spatial resolutions and different geometric features. The thermograms were labeled as moving images which were spatially transformed to be registered on the XCT fixed images [11]. A registration function T was generated using images of the cross landmark on the specimen top surface. The cross landmark was visible in both thermography and XCT image data (see Figure 8). In the case of the XCT data, the cross was extracted from the cross-section slice of $\mathbf{V}^{\mathrm{final}}$. Regarding the thermogram data, the extraction of the cross landmark was performed from a sum of intensity image from the thermograms of the landmark manufacturing. Here, the pixel temperatures of all 2D thermograms from the manufacturing series of specimen layer were summed up and normalized afterwards. Layer 242 was chosen for the sum of intensity image since the landmark structures were the most clearly identifiable here. Both images were further processed to improve the clearness of the landmark geometry. At first, noisy background elements were removed from the image using arithmetical image multiplication and a polygonal mask. The low spatial resolution of the thermographic camera demanded the use of morphological and median filters to clarify the cross edges. Furthermore, the introduced preprocessing in Section 3.1 was applied on the thermography cross landmark image. Finally, the image was binarized using a global threshold.

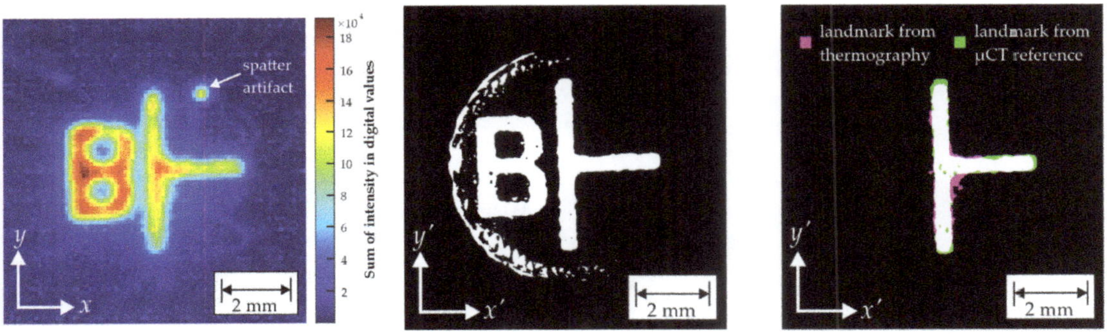

(a) (b) (c)

Figure 8. (**a**) Sum of temperature image from the raw 4D thermogram dataset. (**b**) x'-y' slice of $\mathbf{V}^{\mathrm{bin}}$ at a height of $z' = 1205$ vox. (**c**) Extracted and processed cross labels from (**a**,**b**) after the spatial registration was performed. Shown here is a monomodal translation registration with a quality of 0.958 calculated by the MATLAB Registration Estimator Toolbox. White areas correspond to overlapping area, while pink and green correspond to deviating areas.

Four different transformation models ("similarity", "affine", "rigid", and "translation") were available in the MATLAB toolbox for the monomodal registration to produce a registration function. These models exhibit different degrees of freedom concerning the available transformations. The highest degree of freedom is given by the similarity model that allows translation, rotation, shearing, and scaling of the moving image. The affine model does not allow shearing, while only translation and rotation are available when using the rigid model. In the case of the translation model, only image translation is applied on the moving image. Due to the performed resampling and rectification from the preprocessing of the cross landmark, the moving image already had the same image scale as the XCT image and present distortions were adjusted. Furthermore, the size of the sum of intensity image of the cross landmark in the thermogram dataset may be artificially enlarged by thermal expansion and the choice of the binarization threshold. Therefore, shearing and scaling are unnecessary degrees of freedom that might introduce registration inaccuracy. Two registration functions based on the rigid and the translation model were determined. In the following, the obtained registration functions were applied on the slices of the preprocessed thermogram datasets $\mathbf{A}^{\mathrm{k}}$ and the preprocessed XCT dataset $\mathbf{V}^{\mathrm{final}}$ [34]:

$$T\left(A_z^{k,2D}\right) \leftrightarrow V_{z'}^{\text{final},2D} \text{ with } z = z' \quad (12)$$

Here, $A_z^{k,2D}$ correspond to a 2D x-y slice of $\mathbf{A}^k$ and $V_{z'}^{\text{final},2D}$ to the 2D x'-y' slice of $\mathbf{V}^{\text{final}}$. As stated in Section 1, the evaluation of the registration accuracy is often challenging. In this study, a simple method was utilized to evaluate the registration results. A full dataset of layer maps from thermography (chosen feature: melt pool area) was registered with the corresponding XCT dataset. Afterwards, the contours of the registered x-y slices were approximated by a circle fit [32]. In the case of the thermogram dataset, a binarization using a manually chosen threshold was necessary for the detection of the boundary edges. From the circle fit, the average distance ΔD between the circle centroids of the registered datasets was calculated for all slices:

$$\Delta D = \frac{1}{n_{\text{slice}}} \sum_{i=1}^{n_{\text{slice}}} \sqrt{(x_z^{\text{cen}} - x'^{\text{cen}}_{z'})^2 + (y_z^{\text{cen}} - y'^{\text{cen}}_{z'})^2} \text{ with } z = z' \quad (13)$$

Here, the centroid position of the thermogram feature image is given by x_z^{cen} and y_z^{cen}, while the centroid position of the XCT image is given by $x'^{\text{cen}}_{z'}$ and $y'^{\text{cen}}_{z'}$. ΔD was an indicator for the translation error that was present after the registration. Furthermore, the MAE of average difference of the circle radius ΔR was calculated as a measure of the scaling error. The results are shown in Table 4.

Table 4. Geometric errors resulting from different transformation modes of registration in the corresponding sections. Here, ΔD corresponds to the MAE of the distance between the circle fit centroids and ΔR to the MAE of the difference between the circle fit radii. The STD of both sizes is given by $\sigma_{\Delta D}$ and $\sigma_{\Delta R}$, respectively.

Transformation Model (Including Degrees of Freedom)	Section	ΔD in µm	$\sigma_{\Delta D}$ in µm	ΔR in µm	$\sigma_{\Delta R}$ in µm
Rigid (translation and rotation)	1	23	±12	21	±8
	2	42	±20	38	±6
	3	23	±11	25	±7
	4	65	±29	59	±11
	5	25	±12	30	±11
	6	80	±31	57	±23
Translation (translation)	1	27	±13	19	±8
	2	43	±19	37	±5
	3	29	±12	23	±7
	4	53	±22	55	±9
	5	27	±12	28	±10
	6	64	±27	55	±23

4. Discussion

It is important to place this investigation in the overall context of irregularity prediction in L-PBF parts. The precise prediction of irregularities based on the obtained thermographic in situ monitoring data was the objective. The registration is a necessary step to align monitoring and reference datasets. Therefore, the achieved registration precision is a crucial information because it limits the effective volume size in which irregularities can be accurately predicted.

The preprocessing of the thermogram data included, in the case of melt pool-based features, an interpolation from a sparse point cloud of voxels to a resampled 3D volume dataset. This was performed to adjust the thermogram voxel sizes to the XCT voxel size. Thus, the datasets could be registered more precisely with a theoretical accuracy lower than the pixel resolution of the IR camera, 100 µm^2. Single voxel values of the interpolated datasets should be treated with care because it is uncertain if they necessarily reflect the exact thermal history of the temporally high dynamic L-PBF process. It can be concluded that the temporal and spatial resolution of the raw thermograms is important information that should be considered for the prediction of irregularities. Especially if the original thermogram datasets have a lower spatial information density than the registered datasets, a further resampling of the registered datasets to a larger voxel size might be necessary.

In the context of XCT data preprocessing, it was found that an adjustment of the detected surface deformation is of vital importance for the overall registration accuracy. The obtained results from the introduced adjustment method are promising since they show a clear decrease in shape deformation (compare Figures 5d and 7). Nonetheless, it is necessary to critically analyze the single XCT preprocessing steps that were performed.

The basic method of reconstructing former surface deformations from the internal void distribution in the part was strongly motivated by the findings of Ulbricht et al. [7] and their observations related to the surface deformation. Here, multiple indications of the former surface deformations were visible in the XCT data due to the implementation of artificial notches in their design. In the specimen investigated in the current study, no such notches were present since their implementation might have interfered with the presence of keyhole pores whose formation was forced by the chosen design. Therefore, the available information for surface reconstruction was limited. Only pores in one section could be used for the reconstruction of the surface deformation history. From this lack of information arose the decision to utilize a linear sectional approach for the calculation of the compression factor c. This choice might be an oversimplification in terms of describing the surface deformation history. The potential need of more complex fitting approaches such as higher order polynomials or exponential functions to describe the surface deformation history will be investigated in future research. In future experimental designs, the insertion of artificial voids as references instead of boundary pores could be used for an easier determination of the deformation history.

The chosen simple fitting approach for the compression factor c well compensated the error induced by the surface deformation (Figures 5d and 7). The results show that the specimen deformation was determined accurately by the chosen customized thresholding and surface reconstruction algorithm. Here, the choice of other registration landmarks which are not positioned on the specimen top could improve this result even further, since the step of surface reconstruction of missing areas would be omitted.

In order to describe the surface deformation history, the assumption was made that all surface deformations during manufacturing had an approximately similar shape to the deformation of the specimen top. The results indicate that this assumption is promising. The deformation reconstructed from the pore centroids at an approximate specimen height of $z' = 1070$ voxel is in good agreement with the determined top surface deformation. This is remarkable since the reconstruction is based on the positional information of the present keyhole pores.

The formation of keyhole pores is connected to unstable conditions in the melt pool keyhole [37]. As a result, keyhole pores can occur spatially irregularly during the manufacturing of a scan track (compare Figure 7). Additionally, the melt pool depth varies to a certain extent. As a result, the z' position of pores which result from a single layer manufacturing can fluctuate. The melt pool depth was shown to depend on the laser processing parameters. Mohr et al. [25] found melt pool depths of 213 ± 19 μm for a VED of 65.5 J/mm^3 and 471 ± 54 μm for a VED of 152.7 J/mm^3 using the same machine and material that was utilized in the present study. It can be assumed that the average melt pool depths from the VED utilized in section 6 (114.45 J/mm^3) lay in between these results. However, regarding the pores that were utilized for the former surface deformation reconstruction, the positional fluctuations are limited by the manufacturing of overlying layers. Here, the keyhole of the new layer can interact with the lower pores which may lead to the escape of the entrapped gas or a recombination with new pores. Therefore, the positional fluctuation is effectively limited to the layer thickness (50 μm).

The heterogeneous spatial pore distribution and the fluctuation in the vertical pore location were identified as interfering factors for the reconstruction of the former surface deformation. Nonetheless, the surface deformation calculated by pore distribution (Figure 5c) showed similar shape tendencies as the top surface deformation (Figure 5a). From that, it can be concluded that the history of surface deformation is effectively linked to the resulting pore location. Precise knowledge of the surface deformation history is, therefore, essential for an allocation of in situ sensor data to the porosity information obtained from XCT. To the best of the authors' knowledge, these findings are new and underline the importance of shape adjustment in the registration procedure for irregularity prediction.

It needs to be remarked that the density information of the manufactured specimen is effectively locally distorted by the shape adjustment to match with the "ideal" specimen geometry given by the CAD. Therefore, the quantitative void sizes in the adjusted dataset should be analyzed with care since they might differ from the results in the raw dataset. However, a quantitative void analysis can be enabled again if the applied deformation adjustment is reverted.

In terms of the image registration, the allowed transformation options of the investigated registration models were rotation and translation. Due to the performed preprocessing of both datasets, it was decided that shearing and scaling as additional transformations were unnecessary degrees of freedom. The results (Table 4) show that the lowest errors were achieved when applying the rigid model in section 3. In the section of increased VED, lower accuracy was achieved. This was presumably caused by the lateral shrinkage present in these sections (visible in Figures 4 and 6b). The shrinkage seemed to prevent the algorithm to produce even better results. The comparison between the registration functions showed that the translation model performed better in the section of increased VED. The additionally allowed rotation of the rigid model appears to be counterproductive here and the results indicate that both XCT and thermography datasets were already aligned well from preprocessing. The lower performance from the rigid transformation might result from the shape deviations between the cross landmark images that might have induced an unnecessary rotation.

The results show that already a simple transformation function with a low number of degrees of freedom is sufficient for the registration of the preprocessed datasets. For data that are not preprocessed, it might be a reasonable choice to choose a model that contains shearing and scaling. Here, the quality of the geometric landmark is a crucial factor. Furthermore, the results show the potential of the developed registration method if no lateral shrinkage is present. An extension of the algorithm to improve the registration accuracy if lateral shrinkage is present will be the objective of future studies.

5. Conclusions

In this study, a 3D image-to-image registration was performed on datasets of thermal features extracted from in situ thermography and a corresponding XCT dataset. The registration was performed as a prerequisite for irregularity prediction. Extensive data preprocessing was conducted to obtain similar data dimensionalities to enable the utilization of the chosen, simple registration method. The preprocessing of the thermal feature dataset included the compensation of image distortion, the interpolation of missing datapoints and a resampling to the voxel scale of the XCT data. In the case of the XCT dataset, vertical shrinkage was preliminary corrected from the height information of a staircase landmark structure. Furthermore, a novel shape adjustment method was introduced to eliminate the surface deformation history that was found in the entire part. An image registration function was derived from the utilization of geometric landmarks located on the specimen top surface. The registration accuracy was assessed among the obtained geometrical errors in the registered datasets. From the results of the performed image registration, the following conclusions are drawn:

- Thermally induced warping and solidification shrinkage, especially in the form of surface deformations, are a major challenge for the image registration because it prevents the application of simple registration methods. In this study, it was demonstrated that the distribution of boundary keyhole pores within the observed specimen can be utilized to reconstruct the surface deformation for a specific point in time within the manufacturing. Furthermore, it was shown that an adjustment function based on the approximated surface deformation history enables the adjustment of the part deformation and the application of a simple 3D image-to-image registration.

- The geometrical references that were included in the specimen design proved to be very beneficial for the data registration. The surrounding staircase structure provided information for the preliminary adjustment of the vertical part shrinkage. The cross label added on the specimen top could be utilized to generate a registration function. A drawback of this structure was its disruptive influence during the reconstruction of the top surface information.
- The performed registration resulted in a translation error of 23 μm ± 12 μm and a scaling error of 21 μm ± 8 μm (rigid model in section 3). These errors are significantly lower than the spatial resolution achieved by the IR camera. From the utilization of the adjustment function, it was found that high improvement was achieved in comparison to the unadjusted dataset. Based on these findings, it is possible for the first time to consider the registration accuracy in irregularity prediction modeling.

In future studies, the registered datasets will be used to generate a ML-based model for the prediction of irregularities within the produced part. Here, the obtained registration error information will be incorporated to determine the spatial resolution in which the porosity can be reasonably predicted. Apart from that, we aim to improve the algorithm to make it more robust if lateral shrinkage is present in the part. The insertion of artificial voids within the specimen is a promising option to reconstruct the surface information at different specimen heights. A further option in this regard is the in situ measurement of the part surface topography by laser profilometry that can be integrated into the process. By that, a more accurate deformation adjustment can be achieved which will further increase the registration accuracy and ultimately the irregularity prediction accuracy.

Author Contributions: Conceptualization, S.O., T.F. and S.J.A.; methodology, S.O., T.F., A.U., G.M. and S.J.A.; software, S.O.; investigation, S.O., G.M. and T.F.; data curation, S.O., T.F. and A.U.; writing—original draft preparation, S.O. and T.F.; writing—review and editing, S.O., T.F., A.U., G.M., G.B., C.M. and S.J.A.; visualization, S.O.; supervision, G.B., C.M. and S.J.A.; project administration, G.B., C.M. and S.J.A. All authors have read and agreed to the published version of the manuscript.

Funding: This work was funded by the BAM Focus Area Materials project ProMoAM "Process monitoring of Additive Manufacturing".

Institutional Review Board Statement: Not applicable.

Informed Consent Statement: Not applicable.

Data Availability Statement: Not applicable.

Acknowledgments: We are thankful for the financial support and the fruitful cooperation with all partners.

Conflicts of Interest: The authors declare no conflict of interest.

References

1. Al Rashid, A.; Khan, S.A.; Al-Ghamdi, S.G.; Koç, M. Additive Manufacturing: Technology, Applications, Markets, and Opportunities for the Built Environment. *Autom. Constr.* **2020**, *118*, 103268. [CrossRef]
2. Tapia, G.; Elwany, A. A Review on Process Monitoring and Control in Metal-Based Additive Manufacturing. *J. Manuf. Sci. Eng.* **2014**, *136*, 060801. [CrossRef]
3. Frazier, W.E. Metal Additive Manufacturing: A Review. *J. Mater. Eng. Perform.* **2014**, *23*, 1917–1928. [CrossRef]
4. Fritsch, T.; Farahbod-Sternahl, L.; Serrano-Muñoz, I.; Léonard, F.; Haberland, C.; Bruno, G. 3d Computed Tomography Quantifies the Dependence of Bulk Porosity, Surface Roughness, and Re-Entrant Features on Build Angle in Additively Manufactured In625 Lattice Struts. *Adv. Eng. Mater.* **2021**, *2021*, 2100689. [CrossRef]
5. McCann, R.; Obeidi, M.A.; Hughes, C.; McCarthy, É.; Egan, D.S.; Vijayaraghavan, R.K.; Joshi, A.M.; Acinas Garzon, V.; Dowling, D.P.; McNally, P.J.; et al. In-Situ Sensing, Process Monitoring; Machine Control in Laser Powder Bed Fusion: A Review. *Addit. Manuf.* **2021**, *45*, 102058. [CrossRef]

6. Lough, C.S.; Liu, T.; Wang, X.; Brown, B.; Landers, R.G.; Bristow, D.A.; Drallmeier, J.A.; Kinzel, E.C. Local Prediction of Laser Powder Bed Fusion Porosity by Short-Wave Infrared Imaging Thermal Feature Porosity Probability Maps. *J. Mater. Processing Technol.* **2022**, *302*, 117473. [CrossRef]
7. Ulbricht, A.; Mohr, G.; Altenburg, S.J.; Oster, S.; Maierhofer, C.; Bruno, G. Can Potential Defects in Lpbf Be Healed from the Laser Exposure of Subsequent Layers? A Quantitative Study. *Metals* **2021**, *11*, 1012. [CrossRef]
8. Grasso, M.; Remani, A.; Dickins, A.; Colosimo, B.M.; Leach, R.K. In-Situ Measurement and Monitoring Methods for Metal Powder Bed Fusion: An Updated Review. *Meas. Sci. Technol.* **2021**, *32*, 112001. [CrossRef]
9. Oliveira, F.P.; Tavares, J.M. Medical Image Registration: A Review. *Comput. Methods Biomech. Biomed. Eng.* **2014**, *17*, 73–93. [CrossRef]
10. Gao, Z.; Gu, B.; Lin, J. Monomodal Image Registration Using Mutual Information Based Methods. *Image Vis. Comput.* **2008**, *26*, 164–173. [CrossRef]
11. Klein, S.; Staring, M.; Murphy, K.; Viergever, M.A.; Pluim, J.P. Elastix: A Toolbox for Intensity-Based Medical Image Registration. *IEEE Trans. Med. Imaging* **2010**, *29*, 196–205. [CrossRef] [PubMed]
12. Alpert, N.M.; Berdichevsky, D.; Levin, Z.; Morris, E.D.; Fischman, A.J. Improved Methods for Image Registration. *Neuroimage* **1996**, *3*, 10–18. [CrossRef] [PubMed]
13. Guo, W.G.; Tian, Q.; Guo, S.; Guo, Y. A Physics-Driven Deep Learning Model for Process-Porosity Causal Relationship and Porosity Prediction with Interpretability in Laser Metal Deposition. *CIRP Ann.* **2020**, *69*, 205–208. [CrossRef]
14. Sinclair, L.; Leung, C.L.A.; Marussi, S.; Clark, S.J.; Chen, Y.; Olbinado, M.P.; Rack, A.; Gardy, J.; Baxter, G.J.; Lee, P.D. In Situ Radiographic and Ex Situ Tomographic Analysis of Pore Interactions During Multilayer Builds in Laser Powder Bed Fusion. *Addit. Manuf.* **2020**, *36*, 101512. [CrossRef]
15. Mohr, G.; Nowakowski, S.; Altenburg, S.J.; Maierhofer, C.; Hilgenberg, K. Experimental Determination of the Emissivity of Powder Layers and Bulk Material in Laser Powder Bed Fusion Using Infrared Thermography and Thermocouples. *Metals* **2020**, *10*, 1546. [CrossRef]
16. Feldkamp, L.A.; Davis, L.C.; Kress, J.W. Practical Cone-Beam Algorithm. *J. Opt. Soc. Am. A* **1984**, *1*, 612–619. [CrossRef]
17. Schörner, K.; Goldammer, M.; Stephan, J. Scatter Correction by Modulation of Primary Radiation in Industrial X-Ray Ct: Beam-Hardening Effects and Their Correction. In Proceedings of the International Symposium on Digital Industrial Radiology and Computed Tomography—Mo.3.2, Berlin, Germany, 20–22 June 2011.
18. Ametova, E.; Ferrucci, M.; Dewulf, W. A Tool for Reducing Cone-Beam Artifacts in Computed Tomography Data. In Proceedings of the 7th Conference on Industrial Computed Tomography (iCT 2017), Leuven, Belgium, 7–9 February 2017.
19. Roche, R.C.; Abel, R.A.; Johnson, K.G.; Perry, C.T. Quantification of Porosity in Acropora Pulchra (Brook 1891) Using X-ray Micro-Computed Tomography Techniques. *J. Exp. Mar. Biol. Ecol.* **2010**, *396*, 1–9. [CrossRef]
20. Shah, P.; Racasan, R.; Bills, P. Comparison of Different Additive Manufacturing Methods Using Computed Tomography. *Case Stud. Nondestruct. Test. Eval.* **2016**, *6*, 69–78. [CrossRef]
21. Mireles, J.; Ridwan, S.; Morton, P.A.; Hinojos, A.; Wicker, R.B. Analysis and Correction of Defects within Parts Fabricated Using Powder Bed Fusion Technology. *Surf. Topogr. Metrol. Prop.* **2015**, *3*, 034002. [CrossRef]
22. Lough, C.S.; Wang, X.; Smith, C.C.; Landers, R.G.; Bristow, D.A.; Drallmeier, J.A.; Brown, B.; Kinzel, E.C. Correlation of Swir Imaging with Lpbf 304l Stainless Steel Part Properties. *Addit. Manuf.* **2020**, *35*, 101359. [CrossRef]
23. Coeck, S.; Bisht, M.; Plas, J.; Verbist, F. Prediction of Lack of Fusion Porosity in Selective Laser Melting Based on Melt Pool Monitoring Data. *Addit. Manuf.* **2019**, *25*, 347–356. [CrossRef]
24. Forien, J.-B.; Calta, N.P.; DePond, P.J.; Guss, G.M.; Roehling, T.T.; Matthews, M.J. Detecting Keyhole Pore Defects and Monitoring Process Signatures During Laser Powder Bed Fusion: A Correlation between in Situ Pyrometry and Ex Situ X-ray Radiography. *Addit. Manuf.* **2020**, *35*, 101336. [CrossRef]
25. Mohr, G.; Altenburg, S.J.; Ulbricht, A.; Heinrich, P.; Baum, D.; Maierhofer, C.; Hilgenberg, K. In-Situ Defect Detection in Laser Powder Bed Fusion by Using Thermography and Optical Tomography—Comparison to Computed Tomography. *Metals* **2020**, *10*, 103. [CrossRef]
26. Taherkhani, K.; Sheydaeian, E.; Eischer, C.; Otto, M.; Toyserkani, E. Development of a Defect-Detection Platform Using Photodiode Signals Collected from the Melt Pool of Laser Powder-Bed Fusion. *Addit. Manuf.* **2021**, *46*, 102152. [CrossRef]
27. Gobert, C.; Reutzel, E.W.; Petrich, J.; Nassar, A.R.; Phoha, S. Application of Supervised Machine Learning for Defect Detection During Metallic Powder Bed Fusion Additive Manufacturing Using High Resolution Imaging. *Addit. Manuf.* **2018**, *21*, 517–528 [CrossRef]
28. Oster, S.; Maierhofer, C.; Mohr, G.; Hilgenberg, K.; Ulbricht, A.; Altenburg, S.J. Investigation of the Thermal History of L-Pbf Metal Parts by Feature Extraction from in-Situ Swir Thermography. In Proceedings of the Thermosense: Thermal Infrared Applications XLIII, online, 12–16 April 2021; p. 117430C. [CrossRef]
29. Scheuschner, N.; Strasse, A.; Altenburg, S.J.; Gumenyuk, A.; Maierhofer, C. In-Situ Thermographic Monitoring of the Laser Metal Deposition Process. In Proceedings of the II International Conference on Simulation for Additive Manufacturing, Pavia, Italy, 11–13 September 2019.
30. Herman, G.T. Correction for Beam Hardening in Computed Tomography. *Phys. Med. Biol.* **1997**, *24*, 81–106. [CrossRef]
31. Schindelin, J.; Arganda-Carreras, I.; Frise, E.; Kaynig, V.; Longair, M.; Pietzsch, T.; Preibisch, S.; Rueden, C.; Saalfeld, S.; Schmid B.; et al. Fiji: An Open-Source Platform for Biological-Image Analysis. *Nat. Methods* **2012**, *9*, 676–682. [CrossRef]

32. Chernov, N. Circle Fit (Pratt Method). Available online: https://www.mathworks.com/matlabcentral/fileexchange/22643-circle-fit-pratt-method (accessed on 1 November 2021).
33. Phansalkar, N.; More, S.; Sabale, A.; Madhuri, J. Adaptive Local Thresholding for Detection of Nuclei in Diversity Stained Cytology Images. In Proceedings of the 2011 International Conference on Communications and Signal Processing, Kerala, India, 10–12 February 2011. [CrossRef]
34. Markelj, P.; Tomazevic, D.; Likar, B.; Pernus, F. A Review of 3d/2d Registration Methods for Image-Guided Interventions. *Med. Image Anal.* **2012**, *16*, 642–661. [CrossRef]
35. Kiekens, K.; Welkenhuyzen, F.; Tan, Y.; Bleys, P.; Voet, A.; Kruth, J.P.; Dewulf, W. A Test Object with Parallel Grooves for Calibration and Accuracy Assessment of Industrial Computed Tomography (Ct) Metrology. *Meas. Sci. Technol.* **2011**, *22*, 115502. [CrossRef]
36. D'Errico, J. Surface Fitting Using Gridfit. Available online: https://www.mathworks.com/matlabcentral/fileexchange/8998-surface-fitting-using-gridfit (accessed on 5 July 2021).
37. Hojjatzadeh, S.M.H.; Parab, N.D.; Guo, Q.; Qu, M.; Xiong, L.; Zhao, C.; Escano, L.I.; Fezzaa, K.; Everhart, W.; Sun, T.; et al. Direct Observation of Pore Formation Mechanisms During Lpbf Additive Manufacturing Process and High Energy Density Laser Welding. *Int. J. Mach. Tools Manuf.* **2020**, *153*, 103555. [CrossRef]
38. Wang, L.; Zhang, Y.; Chia, H.Y.; Yan, W. Mechanism of Keyhole Pore Formation in Metal Additive Manufacturing. *Npj Comput. Mater.* **2022**, *8*, 22. [CrossRef]
39. Ertay, D.S.; Ma, H.; Vlasea, M. Correlative Beam Path; Pore Defect Space Analysis for Modulated Powder Bed Laser Fusion Process. In Proceedings of the 2018 International Solid Freeform Fabrication Symposium, Austin, TX, USA, 13–15 August 2018. [CrossRef]
40. Du Plessis, A. Effects of Process Parameters on Porosity in Laser Powder Bed Fusion Revealed by X-ray Tomography. *Addit. Manuf.* **2019**, *30*, 100871. [CrossRef]
41. Jost, E.W.; Miers, J.C.; Robbins, A.; Moore, D.G.; Saldana, C. Effects of Spatial Energy Distribution-Induced Porosity on Mechanical Properties of Laser Powder Bed Fusion 316l Stainless Steel. *Addit. Manuf.* **2021**, *39*, 101875. [CrossRef]

metals

Article

Measuring the Depth of Subsurface Defects in Additive Manufacturing Components by Laser-Generated Ultrasound

Zhixiang Xue [1], Wanli Xu [1], Yunchao Peng [2], Mengmeng Wang [2], Vasiliy Pelenovich [1], Bing Yang [1] and Jun Zhang [1,*]

[1] School of Power and Mechanical Engineering, Wuhan University, Wuhan 430072, China; zhixiangxue1119@whu.edu.cn (Z.X.); wanlixu@whu.edu.cn (W.X.); pelenovich@mail.ru (V.P.); toyangbing@whu.edu.cn (B.Y.)

[2] Pipe China Network Corporation Eastern Oil Storage and Transportation Co., Ltd., Xuzhou 221000, China; pengyc02@pipechina.com.cn (Y.P.); wangmm02@pipechina.com.cn (M.W.)

* Correspondence: zhangjun2010@whu.edu.cn; Tel.: +86-193-7105-7915

Abstract: A new method to measure the depth of subsurface defects in additive manufacturing components is proposed based on the velocity dispersion analysis of Lamb waves by the wavelet-transform of laser ultrasound. Firstly, the mode-conversion from laser-generated surface waves to Lamb waves caused by subsurface defects at different depths is studied systematically. Secondly, an additive manufactured 316L stainless steel sample with six subsurface defects has been fabricated to validate the efficiency of the proposed method. The measured result of the defect depth is very close to the real designed value, with a fitting coefficient of 0.98. The defect depth range for high accuracy measurement is suggested to be lower than 0.8 mm, which is enough to meet the inspection of layer thickness during additive manufacturing. The result indicates that the proposed method based on laser-generated ultrasound (LGU) velocity dispersion analysis is robust and reliable for defect depth measurement and meaningful to improve the processing quality and processing efficiency of additive/subtractive hybrid manufacturing.

Keywords: additive manufacturing; subsurface defects; laser ultrasound; stainless steel

Citation: Xue, Z.; Xu, W.; Peng, Y.; Wang, M.; Pelenovich, V.; Yang, B.; Zhang, J. Measuring the Depth of Subsurface Defects in Additive Manufacturing Components by Laser-Generated Ultrasound. *Metals* **2022**, *12*, 437. https://doi.org/10.3390/met12030437

Academic Editors: Giovanni Bruno and Christiane Maierhofer

Received: 7 February 2022
Accepted: 25 February 2022
Published: 1 March 2022

1. Introduction

Metal additive manufacturing (AM) has disruptive applications in many industries, including the aerospace, biomedical, and automotive industries [1]. Compared with traditional manufacturing methods, this layer-by-layer manufacturing technology has many advantages in the customization of products with complex geometric structures [2]. However, mainstream AM methods have interlayer defects such as inclusions and lack-of-fusion buried in the subsurface of the printing layer [3]. To remove the random defects, additive/subtractive hybrid manufacturing is proposed with performing additive and subtractive manufacturing (SM) alternatively until the whole part is fabricated [4]. The online detection and location of defects are indispensable for the SM processing. The more accurate the measurement of the defects' position, the faster SM can repair the defective part. Therefore, the online monitoring method is meaningful to significantly improve processing quality and processing efficiency.

The laser-generated ultrasound (LGU) has been widely used in various manufacturing fields due to its advantages of being non-contact, broadband, and high-resolution [5]. LGU is also considered to be a potential method for the online detection of metal additive manufacturing samples [6]. Current research mainly focuses on the detection of surface and subsurface defects by LGU Rayleigh waves [7]. Zeng Y. produced three kinds of artificial defects including crack, flat bottom hole, and through hole defects and carried out an LGU inspection and finite element analysis on these three kinds of artificial defects [8]. In the defect evaluation, Wang C. used the LGU Rayleigh wave to measure the thickness of the subsurface defects with rectangular sides. The two ends of the defects were detected separately to quantify the width of the sub-surface groove defects [9]. Chen D. used the phase evolution of LGU Rayleigh waves to detect subsurface defects [10]. Although LGU Rayleigh wave has many advantages in detecting surface and subsurface defects, there are still few applications in measuring defect depth. In a previous finite element analysis, it had been found that ultrasonic surface waves are modulated by near-surface defects, resulting in a waveform conversion from surface waves to Lamb waves. There was a finite element simulation study of surface defects with laser phased array Rayleigh waves [11]. It used the phased array principle to enhance the diffraction wave signal of the LGU detection of cracks and defects [12]. Zhou Z. performed finite element analysis on large-scale surface gaps in LGU inspection and studied the interaction between the Rayleigh wave generated by the laser and surface cracks [13]. Therefore, if we can use appropriate signal processing methods to systematically study the Lamb conversion law, it is possible to propose a defect depth method.

The Lamb waves have velocity dispersion characteristics, which means that the propagation speed of the Lamb wave is changed with frequency, sample thickness, and elastic properties. Based on the principle of velocity dispersion, many researchers focused on the estimation of a material's properties from velocity dispersion analysis using computer-aided signal processing [14]. Previous studies had shown that the attenuation, velocity, frequency, and dispersion characteristics of the Lamb wave generated by the laser are closely related to the anisotropy and viscoelastic properties of the material [15]. Fourier transform and wavelet-transform are two methods to analyze the velocity and dispersion characteristics of Lamb waves. In the application of the Fourier transform, Farouk B. studied the influence of symmetry and discontinuity on the Lamb wave modes [16,17]. This is because of the multi-modal characteristics caused by the velocity dispersion characteristics of the Lamb waves, and it can be quantitatively displayed using the Fourier transform [18]. Although Fourier transform can achieve better results, wavelet-transform has better performance in the field of time-frequency analysis [19]. Amir M. combined wavelet-transform, fast Fourier transform, and modal positioning theory with variable frequency wave speed and considered specific frequency ranges through fast Fourier transform and wavelet packet analysis [20]. In particular, the wavelet-transform enables the transient signal to identify required information and irrelevant information, even overlapping each other in frequency [21–23].

This paper presents a systematic study of the mode-conversion from the LGU surface wave to the Lamb waves caused by subsurface defects at different depths. A new method to measure the depth of subsurface defects is proposed based on the Lamb waves velocity dispersion analysis by wavelet-transform. A 316L stainless steel sample with six subsurface defects is fabricated to validate the efficiency of the proposed method.

2. Experimental Setup

The experimental setup is shown in Figure 1. A Nd: YAG pulsed laser (WEDGE 1064HB DB, Pavia, Italy) with a wavelength of 1064 nm and a pulse duration of 12 ns is used to generate ultrasonic waves (Table 1). The Laser receiver (QUARTET-1500 Bossanova, Los Angeles, CA, USA), with an operating wavelength of 532 nm and a bandwidth of 102 MHz, is applied to receive the ultrasonic waves. The stainless steel (316L) plates were fabricated by the selective laser melting method (SLM AmPro SP-500, Victoria, Australia) with 30 mm in length, 5 mm in thickness, and 30 mm in width (Figure 2). A series of notch defects with a fixed area of 3.0 mm × 0.5 mm and varying depths of 0.1, 0.2, 0.3, 0.5, 0.7, and 1 mm were fabricated as subsurface defects in the specimen (Figure 2). In this paper the pulse laser energy density E can be calculated by $E = (4 \times A \times e)/\left(\pi d^2\right)$, where A is the laser coefficient of the sample (here, 0.1 is adopted for A), e is the pulse laser energy with a value of 2 mJ, and d is the Spot diameter with a value of 150 µm. The calculated result of E is 11.3 mJ/cm^2, which is significantly lower than the stainless-steel ablation threshold of 450 mJ/cm^2 (Table 2) [24]. Therefore, the LGU is controlled by a thermoelastic mechanism. The sample surface rapidly expands and contracts in the laser heating zone, forming internal stress and strain, which propagates in the form of the elastic wave.

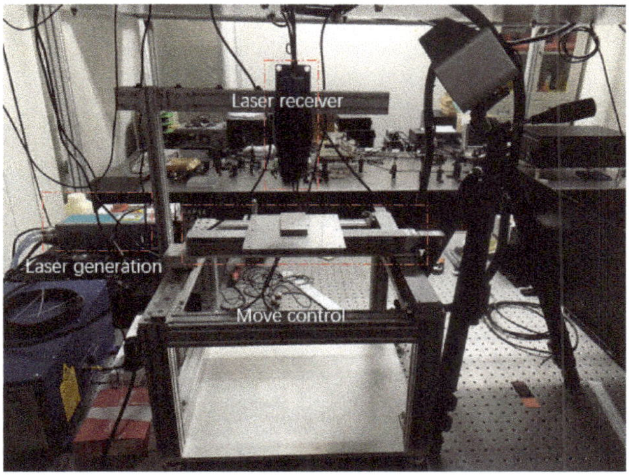

Figure 1. LGU testing system.

Table 1. WEDGE 1064 HB DB parameters.

Parameter	Wavelength/(nm)	Laser Coefficient of Sample (A)	Spot Diameter (d)/(μm)	Pulse Energy (e)/(mJ)
Value	1064	0.1	150	2

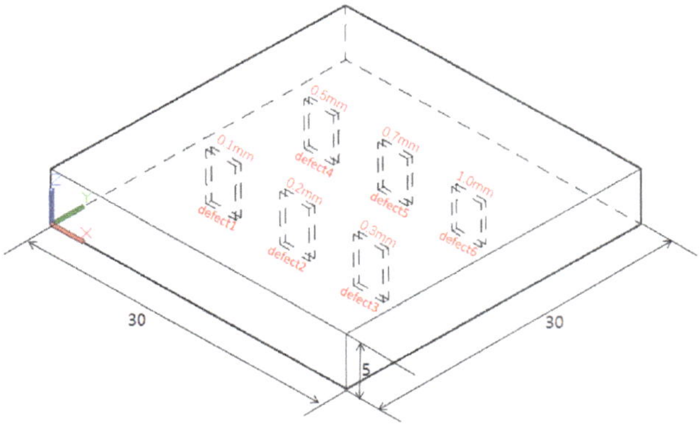

Figure 2. Sample schematic with six embedded notches.

Table 2. The stainless-steel (316L) parameters.

Parameter	Longitudinal Sound Velocity (c_l)/(m/s)	Shear Wave Velocity (c_s)/(m/s)	Rayleigh Wave Velocity (c_R)/(m/s)	Ablation Threshold/ (mJ/cm^2)
Value	5880	3230	2990	450

The schematic of the mechatronic system for generating and detecting the LGU is shown in Figure 3. The laser spots of the excitation and the reception maintains a distance of D, D = 2 mm. The scanning steps (d_x, d_y) are set to 0.1 mm. The acquired A-scans are arranged and stored into a three-dimensional matrix. The B-scan and C-scan images are plotted by extracting a sub-matrix from the acquisition data, which is helpful to find the horizontal position of the defects. Then, the depth of the defects is measured by the proposed method, explained in the next section.

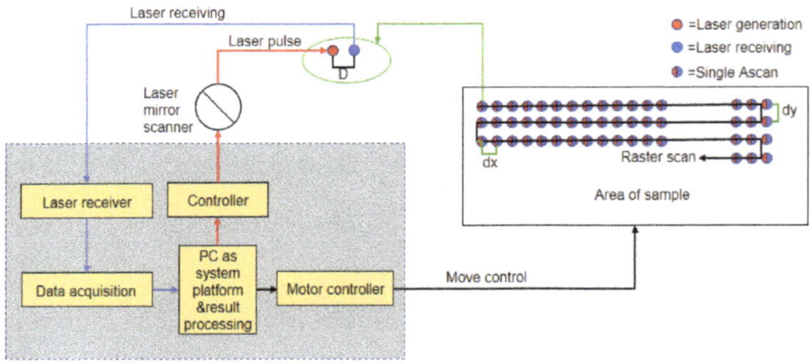

Figure 3. LGU system and scanning imaging strategy.

3. Method

The proposed method for defect depth measurement, based on the LGU signals, consists of three steps. Firstly, the defect depth is characterized by the phase velocity and central frequency of the Lamb waves based on the velocity dispersion principle. Secondly, time-frequency analysis of the LGU signal is used to obtain its frequency, in which the wavelet-transform is employed. Finally, the velocity of the LGU Lamb waves is calculated with the time of flight from the excitation spots to the reception spots.

3.1. Velocity Dispersion of the Lamb Waves

The quantitative relation between the defect depth and the characteristics of the Lamb wave is the key point for the depth measurement. According to the velocity dispersion characteristics of the Lamb wave, the dispersion curves describing the influence of the frequency and velocity on the defect depth could be used for measurement and calibration. The dispersion curves can be calculated by the Rayleigh-Lamb equation of the Lamb wave [25] as follows.

S mode:

$$\frac{tan k_s b}{tan k_l b} = -\frac{4{k_0}^2 k_l k_s}{({k_0}^2 - {k_s}^2)^2} \tag{1}$$

A mode:

$$\frac{tan k_s b}{tan k_l b} = -\frac{({k_0}^2 - {k_s}^2)^2}{4{k_0}^2 k_l k_s} \tag{2}$$

$${k_l}^2 = \left(\frac{\omega}{c_l}\right)^2 - {k_0}^2 \tag{3}$$

$${k_s}^2 = \left(\frac{\omega}{c_s}\right)^2 - {k_0}^2 \tag{4}$$

$$\omega = 2\pi f \tag{5}$$

Here,

k_0—wave number along the horizontal direction of the sample
b—1/2 sample thickness
ω—angular frequency
c_l—longitudinal wave velocity
c_s—shear wave velocity

According to the above equations and the stainless-steel (316L) parameters about c_l and c_s (Table 2), the dispersion curves of A and S mode Lamb waves are shown in Figure 4a. The dispersion curve of the A_0 mode is extracted and used for the depth measurement, as shown in Figure 4b. If the velocity and the frequency changes of the Lamb waves induced by the existing defect are measured according to the LGU experiment, the depth of the defect can be calculated by the dispersion curve of the A_0 mode.

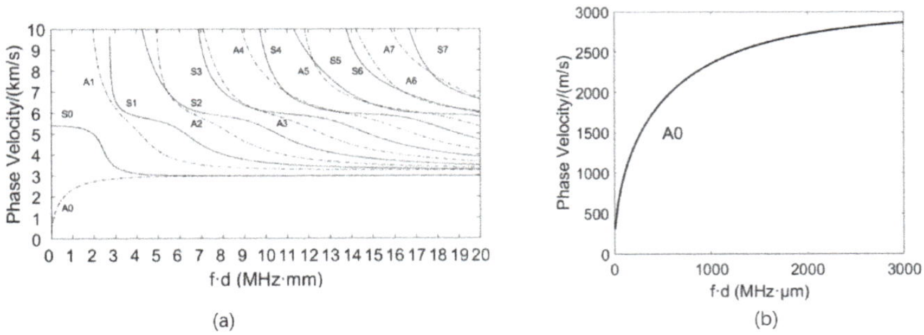

Figure 4. (**a**) Velocity dispersion curves (**b**) A_0 mode velocity dispersion curve.

3.2. Time and Frequency Measurement by Wavelet-Transform

To get the frequency of the defect signal, a time-frequency analysis based on the wavelet-transform is used. The wavelet-transform of an A-scan signal $f(t)$ is defined as [19]:

$$(\omega_{\psi}\mathrm{f})(a,b) = \frac{1}{\sqrt{a}} \int_{-\infty}^{+\infty} f(t)\overline{\psi}\left(\frac{t-b}{a}\right)dt \tag{6}$$

Here, $\psi(t)$ designates the basic wavelet, and $\overline{\psi}(t)$ means a complex conjugate. a and b are scale and shift parameters, respectively. $\psi(t)$ must satisfy the admissibility condition by this equation:

$$C_{\psi} = \int_{-\infty}^{+\infty} \frac{|\psi(\omega)|^2}{|\omega|} d\omega < \infty \tag{7}$$

At the same time, the basic wavelet must satisfy the following two constraints.

$$\int_{-\infty}^{+\infty} |\psi(t)| dt < \infty \tag{8}$$

$$\int_{-\infty}^{+\infty} \psi(t) dt = 0 \tag{9}$$

To obtain a continuous wavelet-transform (CWT), Morlet wavelet is used in this paper; then, the mother wavelet function $\psi(t)$ is expanded and translated by Equations (8) and (9)

$$\psi_{a,b}(t) = \frac{1}{\sqrt{|a|}} \psi\left(\frac{t-b}{a}\right) \quad a,\ b\,\epsilon\mathrm{R}.a \neq 0. \tag{10}$$

The generation of $\psi(t)$ depends on the parameters a and b. $\psi_{a,b}(t)$ is the wavelet basis function. As the core part of this method, we can perform better time-frequency analysis on the signal through wavelet-transform and extract the time-frequency pairs required by our method to obtain better depth measurement results.

3.3. Velocity Calculation

The accurate measurement of time-of-flight is the key point for velocity calculation if the propagation distance is fixed. However, the actual flight time cannot be read directly on the time axis, due to the different system delays in each LGU system. The Rayleigh wave sound velocity (C_R) of 316L stainless steel is used to calculate the actual flight time according to:

$$T_t = d_0/C_R \tag{11}$$

After reading the flight time of the A-scan signal without defects, the system delay can be calculated by subtraction:

$$T_0 = T_t - T \tag{12}$$

T_t—Ultrasonic actual flight time
T—Ultrasonic flight time read from A-scan
T_0—ystem delay r
d_0—Fixed distance between generation and receiving end

Then, the phase velocity of the Lamb wave A-scan signal with defects can be calculated:

$$C_p = d_0/(T - T_0) \tag{13}$$

4. Results and Discussion

The B-scan image of the defects by LGU inspection is shown in Figure 5a. The indications marked in the red frames are the defects of 0.1, 0.2, and 0.3 mm in depth. However, the B-scan image of LGU cannot provide accurate depth information due to the wide time-domain signal modulated by the defect and sample surface. The C-scan image of the LGU detection of the pre-made six defects is shown in Figure 5b. All defects are detected with depths of 0.1, 0.2, 0.3, 0.5, 0.7, and 1 mm. The horizontal position of the defects could be located by measuring the edge of these indications. But it's still difficult to accurately measure the depth of defects, as there is only qualitative evidence that the signal strength becomes weaker as the defect depth increases. However, the C-scan image is helpful to find and extract the defect's A-scan signal for further analysis.

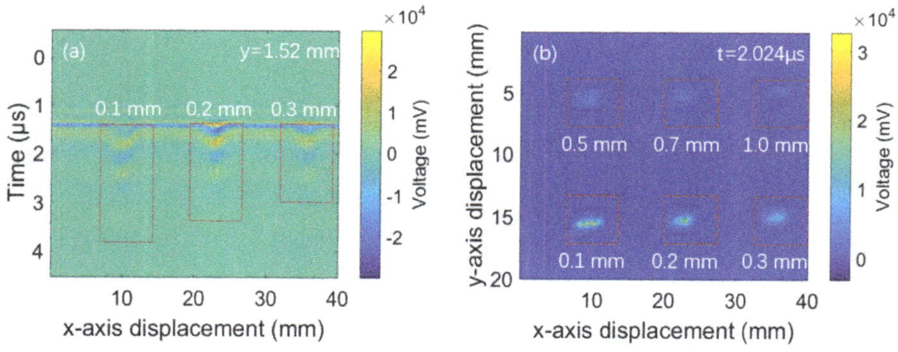

Figure 5. (**a**) B-scan and (**b**) C-scan plots of LGU inspection of defects.

Figure 6a–f are the A-scan signals of defects with six different depths extracted from the horizontal position provided by the C-scan image in Figure 5b. The black line in the figure represents the time domain A-scan signal with amplitude and time, and the blue line represents the spectrum domain signal with frequency and voltage obtained by the fast Fourier transform (FFT). Comparison of the A-scan signals reveals that the width of the signals decreases as the defect depth increases. The embedded notch of 0.1 mm and the sample surface forms a foil-like structure and induces typical Lamb waves signals, as shown in Figure 6a. When the depth of the notch becomes larger, the LGU signals are only modulated by the sample surface and induce typical surface waves signals, as shown in Figure 6f. It means that the mode conversion happens when the defects exist and the depth changes. Comparison of the spectrum domain signals shows a frequency shift, which occurs when the defect depth changes. Double frequency peaks appear when the Lamb wave and the surface wave coexist, as shown in Figure 6d, e. Further time-frequency analysis by means of the Wavelet-transform is introduced to distinguish the Lamb wave mode and quantify the correlated central frequency and time.

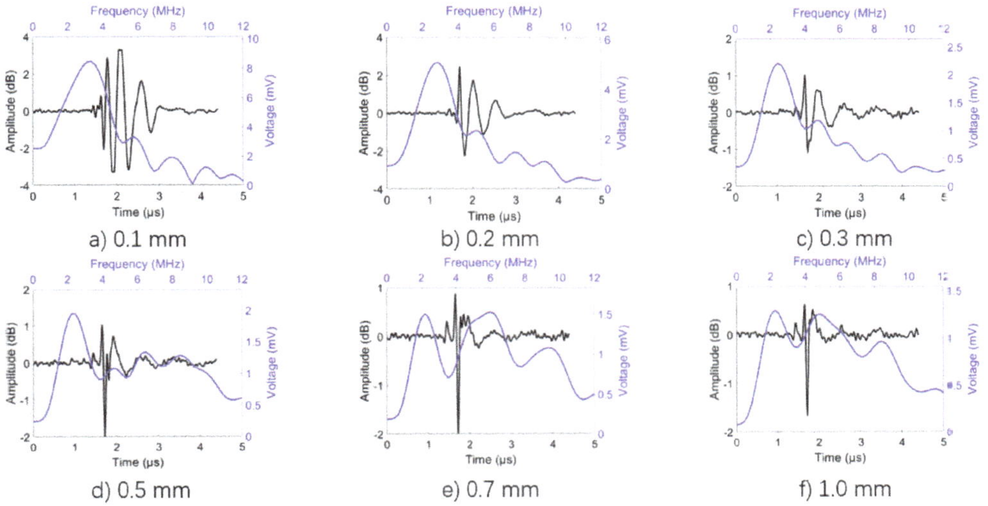

Figure 6. (**a**) 0.1 mm, (**b**) 0.2 mm, (**c**) 0.3 mm, (**d**) 0.5 mm, (**e**) 0.7 mm, and (**f**) 1.0 mm A-scan signal of six different depth defects and the spectrum signals obtained after Fourier transform.

Figure 7a–f are the time-frequency images of the six A-scan signals obtained by the wavelet-transformation. The yellow area represents the location of a wavelet energy concentration. It can be clearly seen from the figures that, as the depth increases, the wavelet energy packet gradually shifts from low to high frequency. This is because, when the depth is small, the Lamb wave is the main signal form. When the depth is 1 mm, the surface wave is the main signal form. However, the low-frequency wavelet energy concentration of the Lamb wave can still be seen. The time-frequency images also prove the frequency shift and mode conversion of LGU waves. The propagation time of the Lamb wave can be accurately recorded if the wave mode and frequency are fixed. Then, the velocity of the Lamb wave can be calculated using the Formulas (6)–(10).

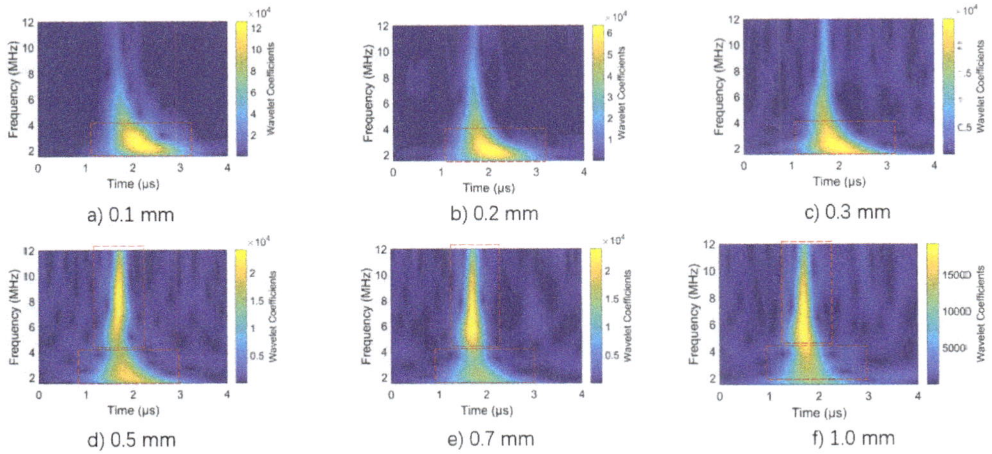

Figure 7. Time—frequency images of six defects with different depths of (**a**) 0.1 mm, (**b**) 0.2 mm, (**c**) 0.3 mm, (**d**) 0.5 mm, (**e**) 0.7 mm, and (**f**) 1.0 mm obtained by the A-scan wavelet-transform.

The theoretical and experimental results of the relationship between propagation velocity and defect depth are shown in Figure 8. In order to find the best matching curve for calibration, five frequencies of the low-frequency wavelet energy packets have been extracted for measurement accuracy evaluation. In Figure 8, the trends of defect depth and sound velocity curves measured from the experimental results are close to the theoretical value. It can be seen that, as the depth of the defect increases, the sound velocity increases, and the phase velocity is very close to the Rayleigh wave velocity when the depth exceeds 1 mm. The difference between the theoretical and experimental results is the smallest when the frequency is 2.2 MHz and the fitting coefficient reaches 0.98. This is because 2.2 MHz is the center frequency of the wavelet energy packet.

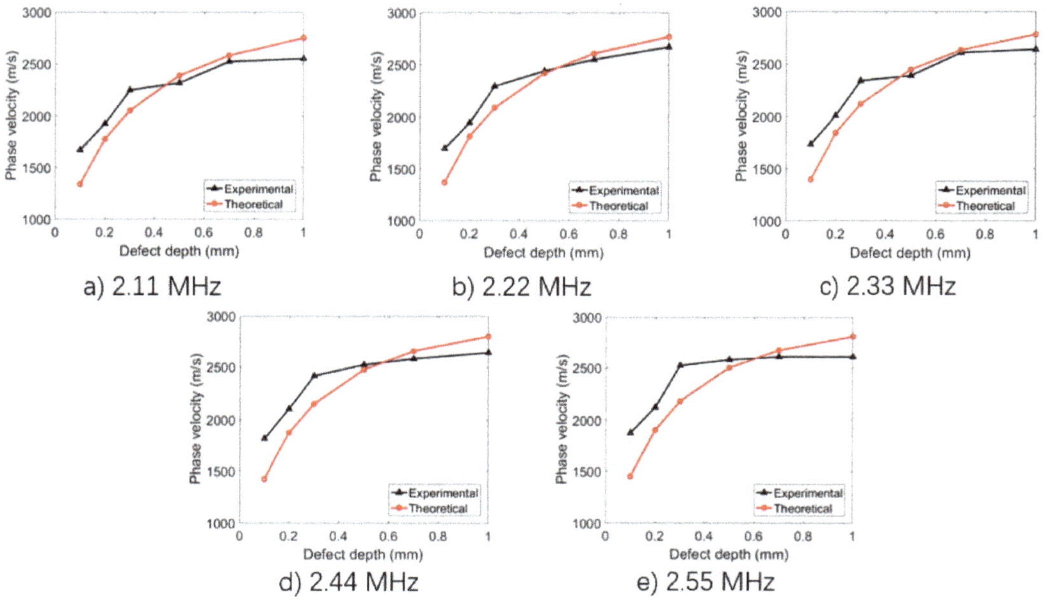

Figure 8. Theoretical and experimental relationship between phase velocity and defect depths at different frequencies: (**a**) 2.11 MHz, (**b**) 2.22 MHz, (**c**) 2.33 MHz, (**d**) 2.44 MHz, (**e**) 2.55 MHz.

Figure 9 shows the depth measurement results of the LGU detection of AM defects mentioned above. The curve correlation coefficient between the designed and measured values is 0.983. The result indicates that it is feasible to measure defect depth based on the dispersion characteristics and wavelet-transform of LGU signals. According to the extracted frequency and sound velocity, the defect can be accurately measured in depth. However, when the defect depth is too large, the main form of the ultrasonic wave is the Rayleigh wave. The energy ratio of the Lamb wave is small, as shown in the time-frequency image (Figure 7f), resulting in an error of 20% when the depth reaches 1 mm. The recommended defect depth range for accurate measurement is suggested to be lower than 0.8 mm, which is enough to meet the inspection layers thickness of AM methods, such as the selective melting method. The accurate position provided by the proposed method in this paper would be helpful for repairing the defective part rapidly and improving the printing efficiency and printing performance of additive/subtractive manufacturing methods.

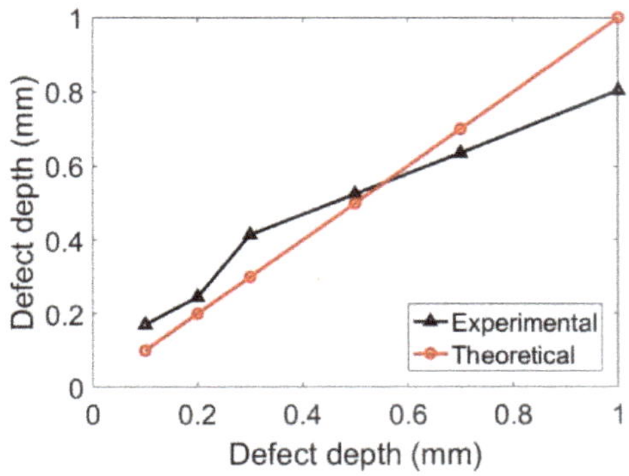

Figure 9. Comparison chart of the experimental and theoretical value of the effect depth.

5. Conclusions

In this paper, a mode-conversion phenomenon from LGU surface waves to Lamb waves caused by subsurface defects at different depths is observed and systematically explored using LGU testing experiments. A novel method to measure the depth of subsurface defects is proposed based on the Lamb waves velocity dispersion analysis by the wavelet-transform. The conclusions are as follows:

(1) The mode-conversion is attributed to the velocity dispersion of the LGU. The central frequency and propagation velocity of the laser-induced surface wave are changed as the depth of the defects change.

(2) The measured result of defect depth is very close to the theoretical value with a fitting coefficient of 0.98. The recommended defect depth range for accurate measurement is suggested to be lower than 0.8 mm, which is enough to meet the inspection layers thickness of AM methods, such as the selective melting method.

In further work, we will consider adding material samples or exciting single-frequency Rayleigh waves for more accurate measuring of the depth of subsurface defects.

Author Contributions: Conceptualization, Z.X.; methodology, W.X.; software, J.Z. and W.X.; validation, Z.X.; formal analysis, Z.X.; investigation, Z.X.; resources, J.Z. and Y.P. and M.W. and B.Y.; data curation, Z.X.; writing—original draft preparation, Z.X.; writing—review and editing, J.Z. and V.P. and B.Y.; visualization, Z.X.; supervision, J.Z. and W.X. and V.P.; project administration, Z.X.; funding acquisition, J.Z. All authors have read and agreed to the published version of the manuscript.

Funding: This work was supported by the National Key R&D Program of China (Grant No. 2018YFB1106100) and the Fast Support Project for Installation and Development (Grant No. 80904010502).

Institutional Review Board Statement: Not applicable.

Informed Consent Statement: Not applicable.

Data Availability Statement: Data Sharing is not applicable for this article.

Conflicts of Interest: The authors declare no conflict of interest.

References

1. Martin, J.H.; Yahata, B.D.; Hundley, J.M.; Mayer, J.; Schaedler, T.A.; Pollock, T.M. 3D printing of high-strength aluminium alloys. *Nature* **2017**, *549*, 365–369. [CrossRef]
2. Ma, Y.; Hu, Z.; Tang, Y.; Ma, S.; Chu, Y.; Li, X.; Luo, W.; Guo, L.; Zeng, X.; Lu, Y. Laser opto-ultrasonic dual detection for simultaneous compositional, structural, and stress analyses for wire + arc additive manufacturing. *Addit. Manuf.* **2020**, *31*, 100956. [CrossRef]
3. Gisario, A.; Kazarian, M.; Martina, F.; Mehrpouya, M. Metal additive manufacturing in the commercial aviation industry: A review. *J. Manuf. Syst.* **2019**, *53*, 124–149. [CrossRef]

4. Du, W.; Bai, Q.; Wang, Y.; Zhang, B. Eddy current detection of subsurface defects for additive/subtractive hybrid manufacturing. *Int. J. Adv. Manuf. Technol.* **2018**, *95*, 3185–3195. [CrossRef]
5. Zhang, J.; Qin, X.; Yuan, J.; Wang, X.; Zeng, Y. The extraction method of laser ultrasonic defect signal based on EEMD. *Opt. Commun.* **2021**, *484*, 126570. [CrossRef]
6. Davis, G.; Nagarajah, R.; Palanisamy, S.; Rashid, R.A.R.; Rajagopal, P.; Balasubramaniam, K. Laser ultrasonic inspection of additive manufactured components. *Int. J. Adv. Manuf. Technol.* **2019**, *102*, 2571–2579. [CrossRef]
7. Lu, J.; Hou, R.; Chen, J.; Shao, H.; Ni, X. A new detection technique for laser-generated Rayleigh wave pulses. *Opt. Commun.* **2001**, *11*, S312–S316. [CrossRef]
8. Zeng, Y.; Wang, X.; Qin, X.; Hua, L.; Xu, M. Laser Ultrasonic inspection of a Wire + Arc Additive Manufactured (WAAM) sample with artificial defects. *Ultrasonics* **2021**, *110*, 106273. [CrossRef]
9. Wang, C.; Sun, A.; Yang, X.; Ju, B.-F.; Pan, Y. Laser-generated Rayleigh wave for width gauging of subsurface lateral rectangular defects. *J. Appl. Phys.* **2018**, *124*, 065104. [CrossRef]
10. Chen, D.; Lv, G.; Guo, S.; Zuo, R.; Liu, Y.; Zhang, K.; Su, Z.; Feng, W. Subsurface defect detection using phase evolution of line laser-generated Rayleigh waves. *Opt. Laser Technol.* **2020**, *131*, 106410. [CrossRef]
11. Zeng, W.; Zou, X.; Yao, F. Finite element simulation of phased array laser-generated surface acoustic wave for identification surface defects. *Optik* **2020**, *224*, 165733. [CrossRef]
12. Gao, F.; Zhou, H.; Huang, C. Defect detection using the phased-array laser ultrasonic crack diffraction enhancement method. *Opt. Commun.* **2020**, *474*, 126070. [CrossRef]
13. Zhou, Z.; Zhang, K.; Zhou, J.; Sun, G.; Wang, J. Application of laser ultrasonic technique for non-contact detection of structural surface-breaking cracks. *Opt. Laser Technol.* **2015**, *73*, 173–178. [CrossRef]
14. Hayashi, Y.; Ogawa, S.; Cho, H.; Takemoto, M. Non-contact estimation of thickness and elastic properties of metallic foils by the wavelet transform of la-ser-generated Lamb waves. *NDT E Int.* **1999**, *32*, 21–27. [CrossRef]
15. Sun, H.; Xu, B.; Xu, G.; Xu, C. Study on laser-generated Lamb waves propagation in viscoelastic and anisotropic plate. *Chin. Opt. Lett.* **2010**, *8*, 776–779. [CrossRef]
16. Benmeddour, F.; Grondel, S.; Assaad, J.; Moulin, E. Study of the fundamental Lamb modes interaction with asymmetrical discontinuities. *NDT E Int.* **2008**, *41*, 330–340. [CrossRef]
17. Benmeddour, F.; Grondel, S.; Assaad, J.; Moulin, E. Study of the fundamental Lamb modes interaction with symmetrical notches. *NDT E Int.* **2008**, *41*, 1–9. [CrossRef]
18. Lee, B.C.; Staszewski, W.J. Lamb wave interaction with structural defects: Modeling and simulations. In Proceedings of the Smart Structures and Materials 2003: Modeling, Signal Processing, and Control, San Diego, CA, USA, 3–6 March 2003; Volume 5049, pp. 146–155. [CrossRef]
19. Wooh, S.C.; Veroy, K. Time-frequency analysis of broadband dispersive waves using the wavelet transform. In Proceedings of the 26th Annual Review of Progress in Quantitative Nondestructive Evaluation, Montreal, QC, Canada, 25–30 July 1999; pp. 831–838.
20. Mostafapour, A.; Davoodi, S. Continuous leakage location in noisy environment using modal and wavelet analysis with one AE sensor. *Ultrasonics* **2015**, *62*, 305–311. [CrossRef]
21. Gur, B.; Niezrecki, C. Autocorrelation based denoising of manatee vocalizations using the undecimated discrete wavelet transform. *J. Acoust. Soc. Am.* **2007**, *122*, 188–199. [CrossRef]
22. Zhang, J.; Lin, G.; Wu, L.; Cheng, Y. Speckle filtering of medical ultrasonic images using wavelet and guided filter. *Ultrasonics* **2016**, *65*, 177–193. [CrossRef]
23. Praveen, A.; Vijayarekha, K.; Abraham, S.T.; Venkatraman, B. Signal quality enhancement using higher order wavelets for ultrasonic TOFD signals from austenitic stainless steel welds. *Ultrasonics* **2013**, *53*, 1288–1292. [CrossRef] [PubMed]
24. Xu, W.; Zhang, J.; Li, X.; Yuan, S.; Ma, G.; Xue, Z.; Jing, X.; Cao, J. Intelligent denoise laser ultrasonic imaging for inspection of selective laser melting components with rough surface. *NDT E Int.* **2022**, *125*, 102548. [CrossRef]
25. Moser, F.; Jacobs, L.J.; Qu, J. Modeling elastic wave propagation in waveguides with the finite element method. *NDT E Int.* **1999**, *32*, 225–234. [CrossRef]

metals

MDPI

Review

Diffraction-Based Residual Stress Characterization in Laser Additive Manufacturing of Metals

Jakob Schröder [1,*], Alexander Evans [1], Tatiana Mishurova [1], Alexander Ulbricht [1], Maximilian Sprengel [1], Itziar Serrano-Munoz [1], Tobias Fritsch [1], Arne Kromm [1], Thomas Kannengießer [1,2] and Giovanni Bruno [1,3,*]

1 Bundesanstalt für Materialforschung und -Prüfung (BAM), Unter den Eichen 87, 12205 Berlin, Germany; Alexander.Evans@bam.de (A.E.); Tatiana.Mishurova@bam.de (T.M.); Alexander.Ulbricht@bam.de (A.U.); Maximilian.Sprengel@bam.de (M.S.); Itziar.Serrano-Munoz@bam.de (I.S.-M.); Tobias.Fritsch@bam.de (T.F.); Arne.Kromm@bam.de (A.K.); Thomas.Kannengiesser@bam.de (T.K.)

2 Institut für Werkstoff- und Fügetechnik, Otto-von-Guericke-Universität Magdeburg, Universitätsplatz 2, 39106 Magdeburg, Germany

3 Institut für Physik und Astronomie, Universität Potsdam, Karl-Liebknecht-Str. 24-25, 14476 Potsdam, Germany

* Correspondence: Jakob.Schroeder@bam.de (J.S.); Giovanni.Bruno@bam.de (G.B.); Tel.: +49-30-8104-3667 (J.S.); +49-30-8104-1850 (G.B.)

Abstract: Laser-based additive manufacturing methods allow the production of complex metal structures within a single manufacturing step. However, the localized heat input and the layer-wise manufacturing manner give rise to large thermal gradients. Therefore, large internal stress (IS) during the process (and consequently residual stress (RS) at the end of production) is generated within the parts. This IS or RS can either lead to distortion or cracking during fabrication or in-service part failure, respectively. With this in view, the knowledge on the magnitude and spatial distribution of RS is important to develop strategies for its mitigation. Specifically, diffraction-based methods allow the spatial resolved determination of RS in a non-destructive fashion. In this review, common diffraction-based methods to determine RS in laser-based additive manufactured parts are presented. In fact, the unique microstructures and textures associated to laser-based additive manufacturing processes pose metrological challenges. Based on the literature review, it is recommended to (a) use mechanically relaxed samples measured in several orientations as appropriate strain-free lattice spacing, instead of powder, (b) consider that an appropriate grain-interaction model to calculate diffraction-elastic constants is both material- and texture-dependent and may differ from the conventionally manufactured variant. Further metrological challenges are critically reviewed and future demands in this research field are discussed.

Keywords: laser-based additive manufacturing; residual stress analysis; X-ray and neutron diffraction; diffraction-elastic constants; strain-free lattice spacing

Citation: Schröder, J.; Evans, A.; Mishurova, T.; Ulbricht, A.; Sprengel, M.; Serrano-Munoz, I.; Fritsch, T.; Kromm, A.; Kannengießer, T.; Bruno, G. Diffraction-Based Residual Stress Characterization in Laser Additive Manufacturing of Metals. *Metals* **2021**, *11*, 1830. https://doi.org/10.3390/met11111830

Academic Editor: Matteo Benedetti

Received: 25 October 2021
Accepted: 9 November 2021
Published: 13 November 2021

1. Introduction

In recent years additive manufacturing (AM) has evolved from a technology for rapid prototyping to a mature production process used in several industries from aerospace to medical applications [1]. In essence, an energy source incrementally manufactures a part in a layer-by-layer process from a wire or powder feedstock [2]. AM processes allow the fabrication of complex structures, which cannot be produced via conventional manufacturing methods [3,4]. This freedom of design enables improvements in component performance and weight reduction of parts [4,5]. In addition, the rapid solidification rates and tailored heat treatment schedules can improve certain material properties, leading to further performance and efficiency gains [6–9]. However, process-related internal stress (IS) may lead to the formation of cracks or delamination [10–13]. IS may severely reduce the applicability of the process to manufacture materials more prone to this type of in-process damage. Moreover, very often IS locks large residual stress (RS) in the parts after production [14].

Therefore, certain materials, which are less susceptible to IS and to related defect formation, are generally preferred to date for the production with laser-based AM methods. These include engineering materials such as stainless steels, titanium-, aluminum-, and nickel-based alloys. In fact, alloys such as 316L, Ti6Al4V, AlSi10Mg, as well as Inconel 625 and 718 are widely used in laser-based AM. It is extremely difficult to monitor IS during production, especially in such complex AM-based processes. Therefore, extensive research has been dedicated to the topics of RS (i.e., the final footprint of IS). The RS determination and mitigation for those alloys are the subjects of this review.

The subjects have a further relevance: In recent years efforts have also been made to extend the laser-based AM production to materials more prone to IS and RS related defects, such as Nickel alloys Inconel 939 [15,16], Inconel 738 [17–20], or martensitic steels [21–23]. In these cases, the control and knowledge of the RS state gains an even greater importance. In fact, investigations have shown that even optimized process parameters (e.g., hatch spacing, laser power, scan speed or scan strategy) can result in high RS magnitudes [24,25]. In general, a careful selection of the process parameters allows the reduction of the RS level and thus increases the overall mechanical performance [26].

Several destructive and non-destructive techniques are available to determine the RS within a material. Due to their non-destructive nature diffraction methods are, naturally, the most widespread for the characterization of RS. The complete stress state within the bulk (by means of neutrons), the subsurface (by means of synchrotron X-rays) and surface (using Lab X-ray) can be characterized.

To allow the precise determination of RS using diffraction-based methods, knowledge about the microstructure, the texture and the processing conditions is required. First, a strain-free lattice spacing (d_0^{hkl}) must be found as a reference to permit the calculation of the strain [27]. The situation is akin to weldments, in which a chemical gradient appears across the weld line, provoking a variation of d_0^{hkl} [27–30]: chemical gradients due to solute-concentration variation are present in AM alloys [31]. This poses a new challenge for the determination of strain and subsequently stress. Secondly, the anisotropic nature of most single crystals requires material specific constants to enable the precise determination of RS by diffraction-based strain measurements [32–34]. The so-called diffraction-elastic constants (DECs) are not only dependent on the alloy, but also rely on the underlying microstructure and texture. In fact, the RS determination by diffraction methods is facilitated if a non-textured polycrystal with relatively small equiaxed grains is measured: in such a case the so-called quasi-isotropic approximation can be used [32,34]. In practice, this assumption is often invalid, as the microstructure can strongly deviate from equiaxed. However, the crystallographic texture and morphology strongly depend on the processing conditions. Rolled or hot-extruded materials, for example, typically exhibit a strong crystallographic texture, which may cause an anisotropic behavior [35,36]. Methods to deal with such process-related peculiar microstructures have been developed in the past for established manufacturing methods [27]. The columnar microstructures, which develop during laser-based AM, typically exhibit a strong crystallographic texture in conjunction with an inhomogeneous grain size along the build direction [37]. Therefore, well established models to determine the DECs in conventional products may fail to predict correct values for AM alloys [38–40].

While detailed reviews on the process parameter dependence and process-specific strategies of RS mitigation can be found in the literature [14,41–44], an extensive review on the methodology of diffraction-based methods with respect to laser-based AM processes is absent.

A first assessment of the critical aspects to account for in the domain of RS determination of AM was provided by Mishurova et al. [45]. Building on this, the present paper showcases an in-depth critical review of the literature in the domain of experimental characterization of RS in laser-based powder AM via non-destructive diffraction methods: An overview of practices and related challenges in diffraction-based RS determination for laser-based AM will be given. Especially, the appropriate choice of the DECs and d_0^{hkl} is paramount to provide accurate absolute RS levels [33,45]. Furthermore, it is indispensable to take the principal stress directions into account, which are for AM materials not necessarily governed by the geometry but instead by the building strategy and, consequently, by the microstructure anisotropy [46]. We will show that significant method development work is still necessary to reliably determine RS by diffraction methods in AM parts.

2. Laser-Based AM Processes

The first laser-based AM process, namely selective laser sintering (SLS), was first developed in 1979, although it took until the 1990's until metal materials were manufacturable [47]. In this process a laser compacts loose powder in a layer-by-layer process to form a green body using a binding polymer [47]. A following infiltration fills the porosity to improve the overall mechanical performance [48]. The development of laser sources allowed EOS (Electro Optical Systems GmbH, Krailing, Germany) to develop a variant of SLS, which no longer needed a binding polymer, as the peripheral region of the powder particles was meltable [47]. The resulted parts were porous but had reasonable mechanical properties [49]. Further development in laser technology finally allowed the manufacturer to fully melt the powder bed [47]. The laser powder bed fusion (LPBF) and laser metal deposition (LMD) processes were then developed. These processes will be introduced in the next paragraphs and are the focus of this review, owing to their propensity to generate high residual stresses. These also are the leading metal AM processes for both new part production and repair engineering [50,51]. Therefore, they have high technological and environmental importance.

2.1. *LPBF (Laser Powder Bed Fusion)*

The usage of increasingly powerful lasers has increased the ability to fully melt the metallic powder [47]. This advance has gradually enabled the production of nearly fully dense (>99.9%) parts, if the process parameters are optimized, with mechanical properties comparable to those of conventionally produced metals [52,53]. Figure 1a illustrates the LPBF process. In a chamber flooded with a protective gas (typically Argon, to prevent oxidation during production), a recoater delivers powder from a reservoir to the build platform. A laser then melts predefined areas within the powder layer. The reservoir and build platform move accordingly to the part design and the steps are repeated in a layer-wise manner until the final part is produced.

The most relevant parameters for process optimization are laser power, scanning speed, layer thickness, hatching distance and, ultimately, the scanning strategy [54,55]. To reduce the temperature gradient during manufacturing, thereby reducing RS and distortions, preheating the baseplate is a typical approach [56]. Nowadays, preheating temperatures up to 1000 °C are realized [57]. The so-called inter layer time, which defines the time passed between deposition of subsequent layers can help to reduce microstructural gradients due to heat accumulation in the part [58]. Not only the process parameters but also the feedstock powder significantly influence the quality of the part. Typically, spherical particles with a size between 10–60 μm are ideal in terms of processability [59]. A comprehensive review on powders for LPBF can be found elsewhere [59]. When the process parameters are carefully controlled, parts with superior properties compared to SLS and direct metal laser sintering (DMLS) are manufacturable [52]. Due to the high heat input and high cooling rates, IS play a major role in those parts, which can lead to distortions and cracking, and remain locked in the part as RS [10,60].

2.2. *LMD (Laser Metal Deposition)*

While in the processes of SLS, DMLS and LPBF a first applied powder layer is selectively melted for part manufacture, during LMD a powder or wire feedstock, is directly fed into the laser beam focus [61]. In a powder-based process a carrier gas drags the powder from the feeder to the nozzle into the melt pool [62]. A second gas is used to prevent oxidation, whereby different gases are available as carrier and shielding [62,63].

Depending on the application different type of nozzles are available; they can influence the efficiency of the process [64,65]. The laser beam then fully melts the feedstock material, and the part is created in an incremental manner (Figure 1b). One of the main advantages of the LMD process is that in contrast to other processes the excess material is minimized, even though material loss can still be a problem due to overspray of the nozzle [65]. In addition, the deposition rates are higher during LMD, but the overall part quality typically suffers compared to LPBF [66]. The most relevant process parameters for process optimization are powder or wire feed rate, laser power, gas flow and scanning velocity [67]. In the LMD process layer thicknesses and particle sizes are commonly larger as compared to the LPBF process.

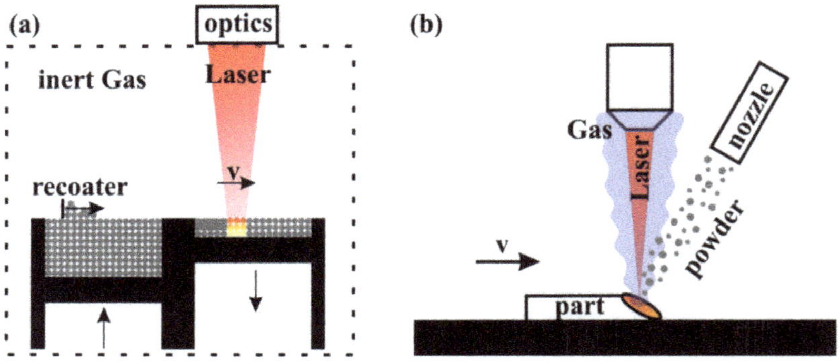

Figure 1. Simplified schematic images of the different laser-based additive manufacturing processes of (**a**) laser-powder bed fusion (adapted from [68]) and (**b**) powder-based laser metal deposition with a lateral injection nozzle (adapted from [69]).

3. Definition of Residual Stress

Residual stress (RS) is stress that exist in a manufactured part without the application of external loads, moments, or thermal gradients [34]. It is very unlikely for manufactured parts to be completely free of RS [70]. Figure 2 visualizes the different types of RS as defined in literature. Depending on the length scale over which the RS self-equilibrate, they can be categorized as the following [71]:

- Type I stresses (σ^{I}) equilibrate over large distances (comparable to the size of the part). This type of RS can be caused e.g., by temperature gradients, machining, and other treatments at the component scale. They depend on the material and its history, as well as on the component geometry.
- Type II or intergranular stresses (σ^{II}) vary over the grain scale and balance over a few grains. They strongly depend on the microstructure, and on the materials history, but weakly on the part geometry. Type II stress is very common in composites and crystallographically anisotropic materials
- Type III stresses (σ^{III}) vary over the atomic scale. Typically, this type is caused by defects of the crystal lattice (e.g., dislocations). They are balanced within each grain and depend on both the microstructure and the materials history.

While the failure of materials can depend on local features, and therefore on Type II and III stresses, in engineering applications usually Type I stress dominates. Indeed, a major contributor to RS in AM polycrystalline parts is Type I RS caused by localized heating, melting, and rapid solidification during the manufacturing process [60].

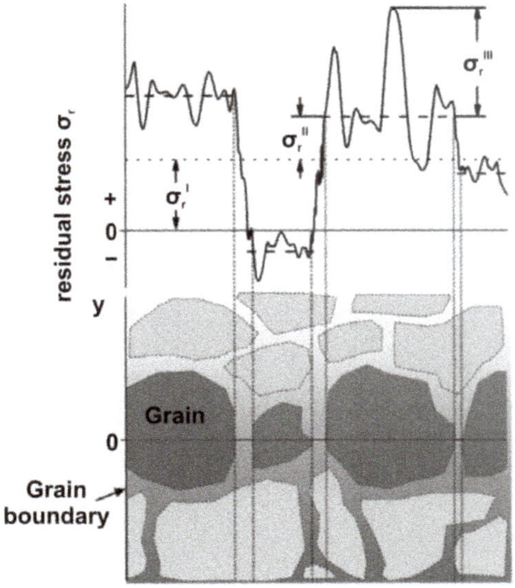

Figure 2. Schematic representation of the different types of RS within a polycrystalline material where σ^{I}, σ^{II}, σ^{III} denote the type I, II and II stresses respectively. Adapted from [72].

4. Residual Stress with Respect to Laser-Based AM

4.1. Origin of Residual Stress

Previous studies showed that RS in AM parts is primarily caused by the thermal gradients in conjunction with the solidification shrinkage that arise due to continuous re-heating, re-melting, and cooling of previously solidified layers [60,73,74]. The local and rapid heating of the upper layer by the laser beam, combined with slow heat conduction (Figure 3a), consequently leads to a steep temperature gradient within the material [60]. However, the already solidified layers restrict the expansion of the uppermost layer, thus leading to the formation of elastic compressive strains [60]. These strains eventually become plastic upon reaching the local temperature dependent yield strength [60]. Therefore, without the presence of mechanical constraints, such plastic strains (ε_{pl}) would lead to bending as indicated in Figure 3a [60]. During cooling, the shrinkage (ε_{th}) of the plastically compressed upper layers leads to an inversion of the bending [60]. The aforementioned is accompanied by the formation of tensile RS in the locally plastically deformed region, balanced by surrounding compression (Figure 3b) [60]. Finally, solidification shrinkage of the molten layer superimposes on the solid-state mechanisms, which leads to tensile RS at the upper most surface balanced by subjacent compression [60]. Extending this phenomenon over multiple layers leads to large thermal gradients particularly along the building direction. Thus, large RS may appear in the final part. The RS itself is influenced by many manufacturing parameters, e.g., the number and the thickness of the layers [60], the geometry, the scanning strategy [38,75–77], and the laser energy density [13]. Optimization of these parameters can significantly reduce RS but also needs to be balanced against the impact on defects and microstructure. The current approach is to optimize some scanning parameters and the scanning strategy, since they highly affect thermal gradients [78]. An alternative approach is the use of stress relieving heat treatments to reduce the magnitude and subsequent impact of RS [41]. These heat treatments must also be balanced against manufacturing cost considerations and both the desired microstructure and the consequent mechanical properties of the alloys.

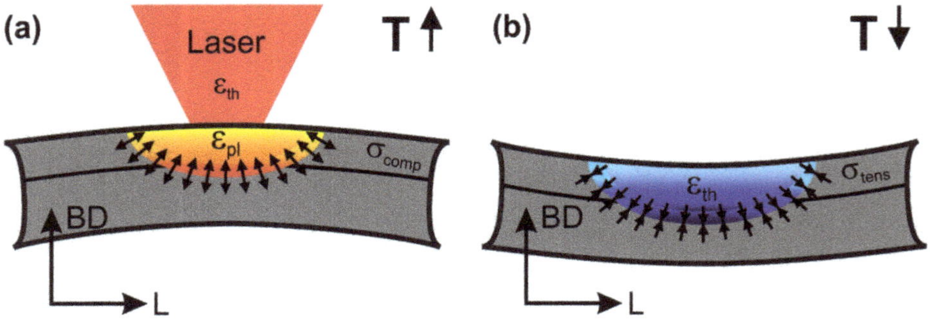

Figure 3. Schematic images showing the effect of the heat input on the stress state during (**a**) heating and (**b**) cooling in LPBF manufacturing (adapted from [60]).

4.2. Distribibition of Residual Stress

An exemplary RS distribution for LPBF of 316L is shown in Figure 4, acquired on 24 mm × 46 mm × 21 mm prisms at middle height [79]. Measurements conducted by X-ray diffraction reveal the presence of high magnitude tensile RS at the surfaces [79]. Bulk neutron diffraction measurements show that stresses invert to compressive RS at an approximate distance of 6 mm from the surfaces, balancing the tensile RS [79]. In fact, it is typical that tensile stresses develop at the surfaces which are balanced by compressive stresses in the bulk [12,60,74,75,79–91]. As mentioned before, the magnitude and distribution of the RS locked in the part is dependent on the manufacturing parameters. However, the general aspects remain unchanged irrespective of the alloy being produced. To characterize the complete stress distribution within a sample, different measurement methods may be required [79]. The methods and the associated challenges to determine the RS from diffraction-based methods in the domain of laser-based AM will be introduced in the following paragraphs.

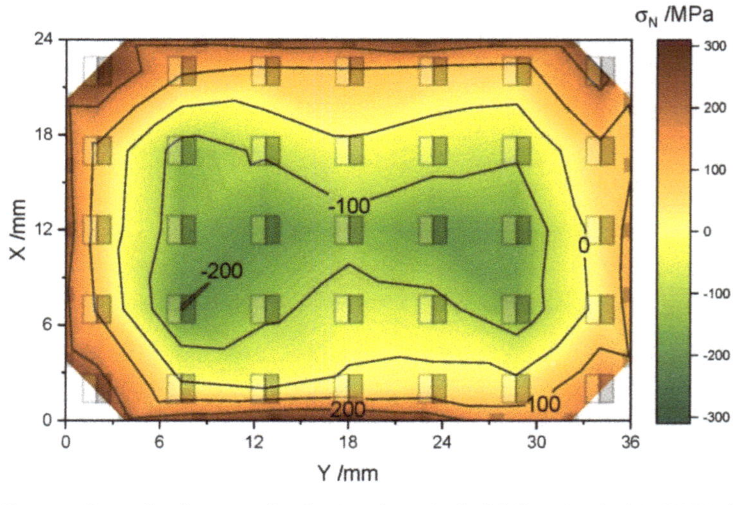

Figure 4. Example of a stress distribution along the build direction (σ_N) in LPBF of 316L prisms measured by ND (bulk) and lab X-ray (surface). Reproduced from [79].

5. Determination of Residual Stresses with Diffraction-Based Methods

The determination of RS can be categorized into destructive (e.g., hole drilling, crack compliance method, hardness testing, etc.) and non-destructive methods (e.g., Bridge curvature method, diffraction, etc.) [92]. However, this paper will solely focus on the methodology of non-destructive diffraction-based methods for RS analysis used for laser-based AM. Therefore, in the following paragraphs the most relevant diffraction techniques will be introduced. Diffraction techniques are well established non-destructive method to evaluate RS in both academia and industry. Determining elastic strains by measuring the variation of lattice spacing provides a powerful method to identify RS at the surface (X-ray diffraction, XRD), at the subsurface (synchrotron energy dispersive diffraction, ED-XRD), as well as in the bulk (synchrotron or neutron diffraction, ND) [13,74,82,83].

5.1. General Aspects of Diffraction-Based Methods

The Bragg equation [93] (Equation (1)) describes the condition for constructive interference of spherical waves emitted by an ordered arrangement of atoms (in lattice planes with distance d^{hkl}), induced by an impinging planar wave of wavelength λ with its order of diffraction n. This law provides the basis for the determination of RS with diffraction-based methods, as the lattice (quantified by the interplanar distance d^{hkl}) can be used as a strain gauge. Consequently, once a material is under the effect of RS the d^{hkl} are altered. Since the beam size in XRD, SXRD or ND measurements is finite, the measured diffraction peak contains a superposition of type I and type II RS within the sampling volume [71]. In all diffraction measurements, the total strain of the lattice is expressed by a shift of the respective diffraction peak (Equation (1)). For the monochromatic case, with a defined wavelength λ, and a known strain-free lattice spacing (d_0^{hkl}), a peak shift to lower scattering angles represents a tensile strain, while a shift to larger scattering angles a compressive one. Type III stresses will mostly contribute to the broadening of the peak or changes in the peak shape [32].

$$2d^{hkl} sin\theta = n\lambda \tag{1}$$

The strain is then calculated as

$$\left\{\varepsilon^{hkl}\right\} = \left\{\frac{d^{hkl} - {d_0}^{hkl}}{{d_0}^{hkl}}\right\} \tag{2}$$

However, to link the determined lattice strains in the laboratory coordinate systems to macroscopic stresses in the sample coordinate systems a few more considerations are necessary. A short description of the fundamentals of RS determination with diffraction-based method is, therefore, presented in the following. For a more detailed description on RS analysis by diffraction-based methods, the reader is referred to the literature [32–34,71,94].

In the general case, RS is derived from lattice strains of a particular set of lattice planes. The measured values are ${d_{\varphi\psi}}^{hkl}$, i.e., interplanar distances at different sample orientations (φ,ψ). For the RS determination, the strains are calculated as in Equation (2) and successively converted to **elastic** stresses via Hooke's law. This yields the general equation for RS determination in the Voigt notation (Equation (3)). Equation (3) connects the elastic lattice strain $\left\{\varepsilon_{\varphi\psi}{}^{hkl}\right\}$ (in all directions (φ,ψ) with the components of the stress tensor in the sample coordinate system by using a transformation matrix (Figure 5). The stress (denoted by $\langle\sigma^S\rangle$) is averaged over all crystallites contained in the gauge volume. The values $\frac{1}{2}{S_2}^{hkl}$ (Equation (4a)) and ${S_1}^{hkl}$ (Equation (4b)) represent the diffraction elastic constants (DECs), which in general depend on the measurement direction in the crystal system. These constants take the elastic anisotropy of the single crystal into account and are discussed in detail later. However, for quasi-isotropic (poly)-crystals they are independent of the sample coordinate system. The DECs serve as proportionality constants, which connect the measured ${d_{\varphi\psi}}^{hkl}$ to a macroscopic RS for the different lattice planes. A further unknown parameter is d_0^{hkl}, which represents the reference value for the determination of the strain. Different strategies are available to determine the d_0^{hkl}, which will be examined later.

$$\left\{\varepsilon_{\varphi\psi}{}^{hkl}\right\} = \left\{\frac{d_{\varphi\psi}^{hkl} - d_0{}^{hkl}}{d_0{}^{hkl}}\right\} = \frac{1}{2}S_2{}^{hkl}\left[sin^2\psi\left(\langle\sigma_{11}^S\rangle cos^2\varphi + \langle\sigma_{22}^S\rangle sin^2\varphi + \langle\sigma_{12}^S\rangle sin2\varphi - \langle\sigma_{33}^S\rangle\right) + \langle\sigma_{33}^S\rangle + 2sin2\psi(\langle\sigma_{13}^S\rangle cos\varphi + \langle\sigma_{23}^S\rangle sin\varphi)\right] + S_1{}^{hkl}\left(\langle\sigma_{11}^S\rangle + \langle\sigma_{22}^S\rangle + \langle\sigma_{33}^S\rangle\right) \quad (3)$$

$$\frac{1}{2}S_2{}^{hkl} = \frac{1+\nu^{hkl}}{E^{hkl}} \quad (4a)$$

$$S_1{}^{hkl} = \frac{-\nu^{hkl}}{E^{hkl}} \quad (4b)$$

Equation (3) represents the most general case, where all stress components are present. If simplifying assumptions can be made, such as the absence of shear stress components (i.e., the fact that the sample coordinate system coincides with the principal stress system), plane stress or plane strain states, or that a particular component vanishes, the equation would simplify. Some cases are developed in more detail below. The same would happen if we can apply simplifications on the DECs, as for instance assume that the material is isotropic.

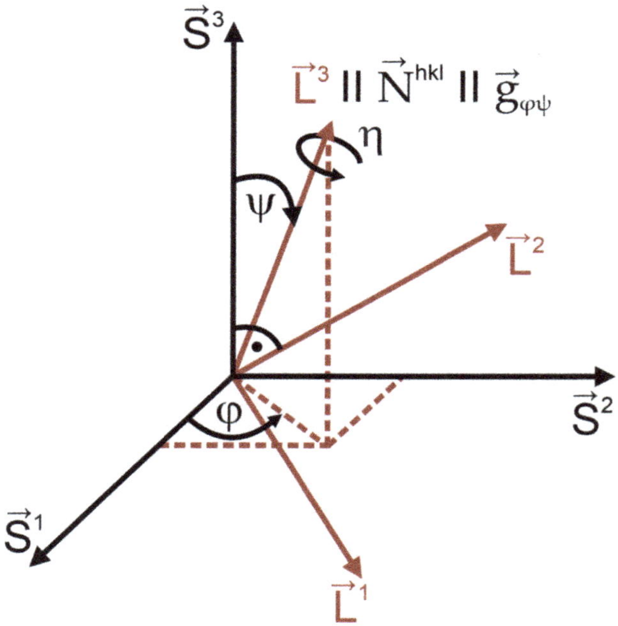

Figure 5. Orientation of the laboratory coordinate system (L) with respect to the sample coordinate system (S), and the associated angles φ and ψ. η denotes the rotation angle around the measurement direction (adapted from [32]).

5.2. X-ray Diffraction

5.2.1. The Monochromatic Case for Surface Analysis

The use of monochromatic X-ray sources for the determination of RS is widely spread. The penetration depth is in the order of a few μm. The general equation for RS determination (Equation (3)) can thus be simplified: The stress components normal to the measurement plane 12 [σ_{i3}= 0 (i = 1, 2, 3)] can be considered zero (Equation (5)).

$$\varepsilon_{\varphi\psi}^{hkl} = \frac{1}{2}S_2\sigma_\varphi sin^2\psi + S_1(\sigma_{11}+\sigma_{22})\text{with } \sigma_\varphi = \sigma_{11}cos^2\varphi + \sigma_{22}sin^2\varphi + \sigma_{12}sin2\varphi \quad (5)$$

As laboratory setups mostly use monochromatic X-rays sources, an appropriate lattice plane representing the bulk material must be chosen. A guideline for this can be found in DIN EN 15305 [95], but will be discussed more in detail in Section 6.4. The main approach used in laboratory X-ray devices is the $sin^2\psi$ method in which the lattice spacing is measured under variation of the ψ angle under a (usually) fixed φ angle (Figure 5). Equation (5) can be considered a linear equation of the form $\varepsilon(sin^2\psi) = a{\cdot}sin^2\psi + b$. The straight line has a slope of $a = \frac{1}{2}S_2\sigma_\varphi$ and intersects the $\varepsilon(sin^2\psi)$ axis at $b = S_1(\sigma_{11}+\sigma_{22})$. From the linear regression of the respective $\varepsilon(sin^2\psi)$—distribution the RS can be determined in the direction φ (Figure 6). In an ideal case, where an elastically isotropic or non-textured material in a homogeneous stress state is sampled, the obtained $\varepsilon(sin^2\psi)$ is truly linear [32]. Even though these requirements are often not fulfilled, the errors are typically of small order and can thus be neglected [32]. However, for strongly textured materials (e.g., rolled, additively manufactured) the deviations can be severe. In the case of present shear stresses (e.g., σ_{13} and σ_{23}) an ellipsoid is observable (different $\varepsilon(sin^2\psi)$ for $\pm\psi$) rather than a linear distribution. By the subtraction of the respective $-\psi$ and $+\psi$ distributions a linear equation is obtained. Finally, from its slope the shear stress component in the direction φ can be determined. Although normal stress components can also be determined within the information depth of the radiation this requires the precise knowledge of ${d_0}^{hkl}$, which is not needed for the determination of shear stresses [32]. Due to the relatively low penetration of lab X-rays into metallic materials, the surface roughness of additively manufactured material impacts the determined stress values [96].

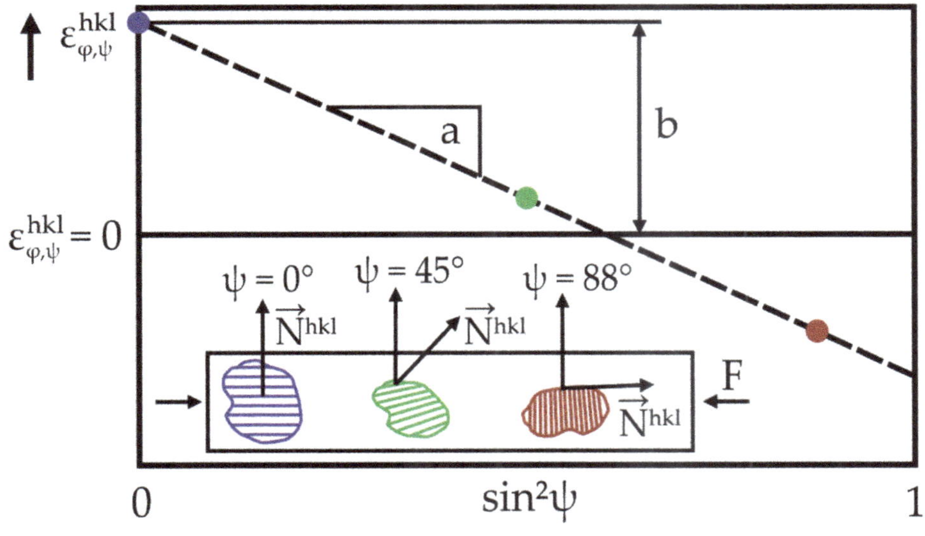

Figure 6. Simplified $\varepsilon(sin^2\psi)$ distribution assuming an elastically isotropic or non-textured material in a homogeneous compression stress state (adapted from [32]).

5.2.2. The Energy Dispersive Case

In addition to the monochromatic (angular dispersive) XRD technique, it is also possible to use polychromatic radiation (white beam) for RS determination. An energy dispersive detector detects the respective energies of the diffracted X-rays at a fixed diffraction angle θ. In such manner the entire diffraction spectrum of the respective material can be obtained for each measurement direction (φ, ψ) [32]. Due to the wide energy range used, the information retrieved arises from different depths of the specimen [97]. The information depth of the respective energy can be calculated using the following equation [97]:

$$\tau_{\eta} = \frac{sin^2\theta \; - \; sin^2\psi + cos^2\theta sin^2\psi sin^2\eta}{2\,u(E^{hkl})sin\theta cos\psi} \tag{6}$$

The information depth is a function of the sample rotation around the diffraction vector η, the diffraction angle θ, the tilt angle ψ and the energy dependent linear absorption coefficient $u(E^{hkl})$. The latter is material dependent. τ_{η} defines the depth below the surface from which 63% of the total diffracted intensity comes from [98].

The Energy of each respective reflection can be directly transferred to the lattice plane spacing by rewriting the Braggs law in terms of photon Energy E^{hkl} [99]:

$$d^{hkl}(\text{Å}) = \frac{\text{h} \cdot \text{C}}{2sin\theta} \cdot \frac{1}{E^{hkl}} \approx \frac{6.199}{sin\theta} \cdot \frac{1}{E^{hkl}} \tag{7}$$

In Equation (7) h is the Planck constant and c the speed of light. The $sin^2\psi$ method is also applicable for the energy dispersive case. The $\varepsilon(sin^2\psi)$ distributions are simply calculated using Equation (7) together with the strain definition (see Equation (2)). The same simplifications (as for lab X-ray) apply whenever measuring in reflection mode or a biaxial stress state can elsewise be justified. In fact, the plane stress assumption might only hold for lower energy ranges with a low penetration depth. This complicates the RS analysis of higher energy reflections, as the triaxial approach could be more suitable. The acquisition of the entire diffraction spectrum allows the stress analysis for each lattice plane observed. Therefore, a depth resolved stress analysis (near surface) is possible up to the maximum information depth (according to Equation (6)). With respect to laser-based AM, authors have extracted RS depth profiles by using the combination of different reflections (under the assumption of vanishing stress component normal to the surface) [38,96]. In addition, a full pattern refinement to obtain an average d^{hkl} can be conducted (e.g., Rietveld refinement) [100]. Recently, Hollmann et al. [101] proposed methods for near surface measurements of materials with cubic symmetry and nearly single crystalline texture (e.g., additively manufactured).

Due to the high energies available in synchrotron facilities even measurements in transmission are possible both in angular (monochromatic) and energy dispersive (polychromatic) modes, depending on the material measured and the sample thickness [32]. In these cases, depending on the geometry, the out-of-plane stress cannot be neglected and hence the triaxial stress analysis approach is required. However, the ability to perform triaxial RS measurements is hampered by the use of elongated lozenge-shaped sampling volume, typical in high energy diffraction measurements (because of the required small diffraction angle [32]). On the one hand, the method therefore allows a very high spatial resolution (in the order of 10–100 µm) in the two in-plane directions, but on the other hand the spatial resolution becomes poor (several millimeters) in the out-of-plane direction. Despite this limitation, energy dispersive techniques are well suited for thick wall geometries, whereby the stress state is closer to the plane strain condition and limit gradients exist though the thickness. Moreover, significant work is reported on the use of transmission set-up for the determination of DECs through in situ tensile testing. This topic is addressed in Section 6.3.3.

5.3. Neutron Diffraction

As neutrons have a high penetration depth in most materials, fully 3D stress states can be probed. The gain in generality of the approach must be paid at a price: the strong dependence of the RS analysis on the reference interplanar spacing, ${d_0}^{hkl}$. Additional complications arise when ${d_0}^{hkl}$ differs over the analyzed region due to chemical gradients over the specimen. These points are discussed in detail in Section 6.1. There are two neutron diffraction methods to determine RS: the monochromatic and the time-of-flight

(TOF) method. The TOF method uses a polychromatic beam and rests on the detection of many diffraction peaks. Thus, the method leverages on the fact that the velocity of the neutrons is inversely proportional to its wavelength. In the monochromatic case, the instrument operates with a fixed wavelength, and most commonly only one peak at a time can be detected. The two methods will be introduced briefly below. For a more detailed description the reader is referred to the literature [94].

5.3.1. The Monochromatic Method

In scattering, a neutron may be described by its wave vector k, of magnitude $2\pi/\lambda$ directed along its velocity part [94]. Due to the wave nature of matter, the de Broglie wavelength of the neutron (λ) is directly linked to the momentum (p) of the respective particle [94]:

$$p = m_n v = \frac{hk}{2\pi} = \hbar k \quad (8)$$

This allows the calculation of an associated wavelength in dependence of the neutron velocity v and mass m_n with the Planck's constant h. In the monochromatic case, neutrons with a given wavelength are used to study the lattice strain within the material [102]. The wavelength of the neutrons is usually selected using a single crystal monochromator from a broader neutron wavelength spectrum [102]. Typically, the wavelength is chosen so that a diffraction angle of around $2\theta \sim 90°$ is used. The condition $2\theta \sim 90°$ allows the definition of a nearly cubic sampling (gauge) volume. Thereby, probing the same region upon any sample rotation. The diffracted signal is usually then detected on a position sensitive detector or a scanning point detector [94]. An example of a typical diffraction peak obtained is shown in Figure 7, which are typically fitted using a symmetric function (e.g., Gaussian).

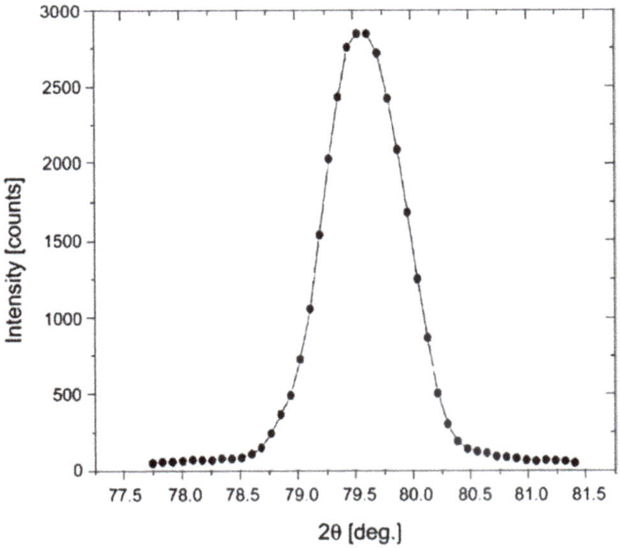

Figure 7. Neutron peak profile. Reproduced from [102] with permission from Elsevier.

The change of diffraction angle with respect to a reference value yields variations of the lattice strain and can be expressed in the angular form as [102]:

$$\varepsilon^{hkl} = \frac{\Delta d^{hkl}}{d^{hkl}} = -\Delta\theta^{hkl} cot\theta^{hkl} \quad (9)$$

Similar to the angular dispersive X-ray methods (see Section 5.2.1), an appropriate lattice-plane must be chosen, which represent the bulk behavior the best. For the stress analysis the same relations remain valid. However, the isotropic form of the Hooke's law typically is applied along three orthogonal principal strain components [94] (see Equation (14)). The consequences for the related assumptions with respect to principal directions are discussed in Section 6.2.

5.3.2. The Time-of-Flight Method

The neutron diffraction (ND) time-of-flight method is the equivalent to polychromatic diffraction in the X-ray case. From the neutron travel time between source and detector, the associated wavelength can be calculated (Equation (10)) [94].

$$\lambda = \frac{ht}{m_n L} \tag{10}$$

As detectors are placed at 2θ~90°, using Bragg's law (Equation (1)) one can directly determine d^{hkl} from the wavelength at which peaks appear in the diffraction spectrum (for a known crystal structure). A typical ND diffraction spectrum is shown in Figure 8. In contrast, to steady state sources (monochromatic), the time-pulsed source instruments (time-of-flight) typically cause an asymmetry due to the moderation process: More complicated fitting functions are typically necessary [94]. Using the TOF methods an average d can be obtained by a full pattern refinement, but also single peak fits are performable [102].

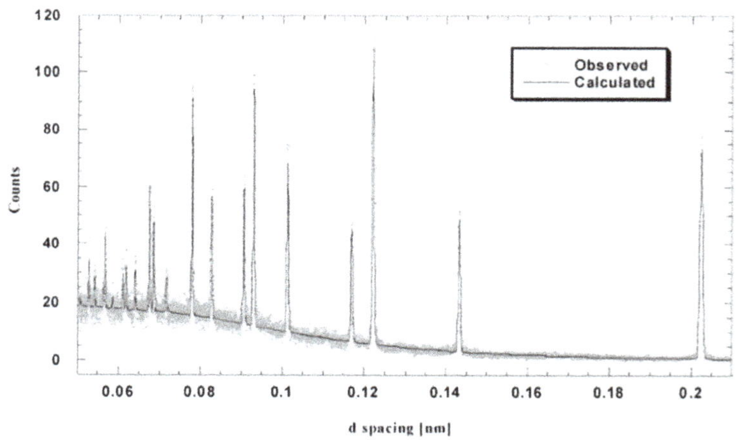

Figure 8. Time-of-flight pattern. Reproduced from [102] with permission from Elsevier.

6. Peculiarities of Diffraction-Based Methods in the Case of AM

6.1. Strain-Free Lattice Spacing ($d_0{}^{hkl}$)

To precisely determine RS in parts using diffraction-based techniques, the knowledge of a $d_0{}^{hkl}$ as a reference is essential (see Equation (2)). A comprehensive description of the methods to obtain a $d_0{}^{hkl}$ value is given by Withers et al. [27]. In the case of laboratory X-ray diffraction measurements ($sin^2\psi$), where out-of-plane stresses can be considered to equal zero (σ_{i3} = 0), a prior knowledge of $d_0{}^{hkl}$ is not required, as it can even be calculated by the combination of several measurements [32]. In addition, the method is relatively insensitive to an inaccuracy in $d_0{}^{hkl}$ up to 10^{-3} nm [32].

For example, one could measure d^{hkl} vs. $sin^2\psi$ for the directions (φ = 90°, ψ) and (φ = 0, ψ) and then determine their average value:

$$\frac{d^{hkl}\ (\varphi = 90^\circ,\ \psi) + d^{hkl}\ (\varphi = 0^\circ,\ \psi)}{2} = (\sigma_{11} + \sigma_{22}) d_0^{hkl} \left[2S_1^{hkl} + \frac{1}{2} S_2^{hkl} sin^2\psi \right] + d_0^{hkl} \tag{11}$$

The right term equals to d_0^{hkl}, when (isotropic, no steep gradient, $\overline{\sigma}_{22} \neq \overline{\sigma}_{11}$) [34]:

$$sin^2\psi = sin^2\psi^* = \frac{-S_1{}^{hkl}}{\frac{1}{2}S_2{}^{hkl}}\left(1+\frac{\overline{\sigma}_{22}}{\overline{\sigma}_{11}}\right) \tag{12}$$

Consequently, the $d_0{}^{hkl}$ can be defined as (e.g., for $\overline{\sigma}_{11} = \overline{\sigma}_{22}$):

$$d_0^{hkl} = \frac{d^{hkl}(\varphi = 0^\circ, \psi*)+d^{hkl}(\varphi = 90^\circ, \psi*)}{2} \text{ with } sin^2\psi^* = \frac{-2S_1{}^{hkl}}{\frac{1}{2}S_2{}^{hkl}} \tag{13}$$

Therefore, in this particular case, the bare elastic constants define the strain-free direction ψ^*, and the half average of d^{hkl} ($\varphi = 90^\circ$, ψ) and d^{hkl} ($\varphi = 0^\circ$, ψ) at the position $sin^2\psi^*$ provides the $d_0{}^{hkl}$ (at the location where the $sin^2\psi$ scan was carried out). A more detailed description and examples for other stress states to derive a $d_0{}^{hkl}$ are given in [34].

Although this method leads to a simplified experimental determination of $d_0{}^{hkl}$ it still bears the problem of DECs values (Equation (12)). As the determination of $d_0{}^{hkl}$ by this method is dependent upon knowledge of DECs (Equation (12)), the reliability of the DECs must be high to determine a correct value for $d_0{}^{hkl}$. The determination of the DECs is a separate topic and will be examined later.

While the method is sensitive to intergranular and interphase stresses [27], a relative comparison of $d_0{}^{hkl}$ near the surface is often still possible. Thiede et al. [82] used this method to determine $d_0{}^{hkl}$ variations across the surface of LPBF manufactured Inconel 718 prisms (assuming $\overline{\sigma}_{11} = \overline{\sigma}_{22}$) (Figure 9a). A small normal stress component σ_n was observed, which was reported to correlate with the scanning strategy. The $sin^2\psi$ method has also been used by other researchers to determine $d_0{}^{hkl}$ in LPBF Ti6Al4V [83,96]. As an alternative, Pant et al. [81] used the d^{hkl} value measured at $\psi = 0^\circ$ as $d_0{}^{hkl}$ for calculating RS values.

For the cases in which the out-of-plane stress cannot be considered to equal zero ($\sigma_{i3} \neq 0$) the precise knowledge of $d_0{}^{hkl}$ remains indispensable. An independent determination of $d_0{}^{hkl}$ can be made by means of the following strategies:

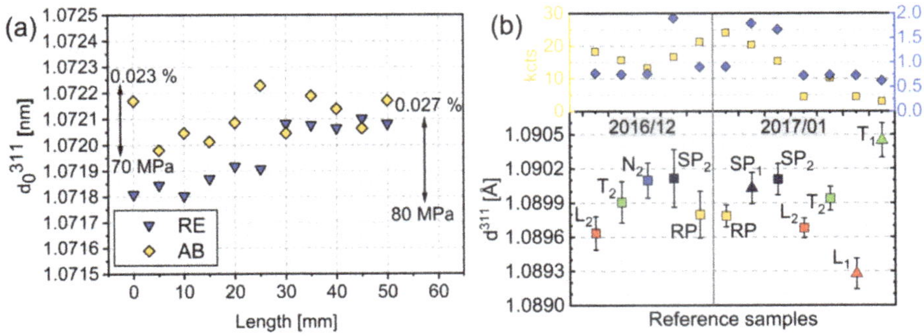

Figure 9. $d_0{}^{311}$ values extracted from d^{311} versus $sin^2\psi$ plots (**a**) and from raw powder (RP), filings (SP) and cuboids (L, T, N) (**b**). Data taken from [82].

6.1.1. Use of Raw Powder

In the case of AM there is also the possibility to obtain $d_0{}^{hkl}$ by measurements on raw powder. This is a method, which does not require a twin specimen and is non-destructive by nature. However, the thermal history of the raw powder, and consequently the local chemical composition and microstructure, may differ significantly from that of the printed part [103]. The macro- and micro-scale differences in chemistry can significantly alter the lattice parameters of the material [104–106]. As Thiede et al. [82] have concluded, a shift due to different local chemical segregation prevented the use of it as a reference (Figure 9b). Similar findings were made by Kolbus et al. [103] and more recently Serrano-Munoz et al. [107].

While there may be examples of low alloyed or commercially pure materials, where the use of the raw powder may be applicable due to the lower amount of segregation, using raw manufacturing powder as ${d_0}^{hkl}$ is generally not recommended in the domain of AM [45].

6.1.2. Use of Mechanical Filings

Mechanical filings from either the specimen itself or taken from a twin specimen can be used. This approach would capture the effect of the thermal cycles on the local chemical segregation and has the advantage that in a powder the macroscopic RS is fully relieved [27]. However, the filing process tends to induce plastic deformation within the powder, leading to strong diffraction peak broadening associated with microscopic stresses (Type III and possibly Type II [103]). In addition, filings certainly contain different intergranular strain than the component, so that they cannot be considered fully stress-free [27].

It was recently shown that using the ${d_0}^{hkl}$ reference value of mechanical filings, a compressive stress was found for all measured points, which contradicts the stress balance condition [107]. This mismatch was attributed to the high accumulation of plastic strain in the as-filed condition, and in fact the FWHM vastly reduced upon heat treatment (indicating a significant plastic recovery). Even if some circumstances lead to the conclusion that the filings from the material are the most suitable ${d_0}^{hkl}$ [82], the applicability of mechanical filings as ${d_0}^{hkl}$ shall be limited to exceptional cases and rarely be considered an appropriate approach in the general case of AM.

6.1.3. Use of Macroscopically Relaxed Samples (Cubes/Combs/Arrays)

In neutron diffraction, it is common to determine the ${d_0}^{hkl}$ with small cubes (or combs) cut from a sister sample. These cubes are assumed to be free of macroscopic stress. Although they appropriately represent the (possible) variation of chemical composition of the specimen, other problems must be considered: The cubes could retain intergranular stresses and are vulnerable to geometrical effects if poorly positioned on the sample manipulator [27]. Thiede et al. [82] measured small 5 mm × 5 mm × 5.5 mm cubes extracted from sister samples of LPBF Inconel 718. However, they found a significant dependence of the d^{hkl} value on the measurement direction (Figure 9b). This suggests that the cubes were not fully macroscopically stress free and thus could not yield a reliable ${d_0}^{hkl}$ [82]. Nevertheless, a similar d^{hkl} dependence on the measurement direction was found by Ulbricht et al. [79] for LPBF manufactured stainless steel 316L, this time using 3 mm × 3 mm × 3 mm coupons.

To obtain a representative ${d_0}^{hkl}$ they averaged the values over all measured directions (which correspond to the main geometrical directions). Kolbus et al. [103] attributed the different ${d_0}^{hkl}$ in different directions of reassembled DMLS Inconel 718 cubes (2.5 mm × 2.5 mm × 2.5 mm and 5 mm × 5 mm × 5 mm) to anisotropic micro stress between the fcc matrix and the precipitation phases. They applied an average obtained from measurements on reference cubes extracted from different positions but did not average over different strain directions. Regardless of the direction being measured, Pant et al. [81] found that the average value of the measured ${d_0}^{hkl}$ on the LPBF manufactured Inconel 718 array (cut by wire electric discharge machining) was position independent. The average value, however, showed to not provide sufficient accuracy concerning the stress balance condition in the cross sections [81]. Other approaches based on relaxing macroscopic stresses by cutting or extracting small geometries from sister samples were conducted by several researchers [31,85,91,108–112]. Although some inconsistencies in defining a representative ${d_0}^{hkl}$ from measurements on macroscopically stress-free samples have been reported, approaches to determine ${d_0}^{hkl}$ using coupons (or small pieces of the printed part) are widespread; to date, this approach is considered the best to produce a reliable measured ${d_0}^{hkl}$.

6.1.4. Stress and Moment Balance

Another method to determine a d_0^{hkl} is based on the continuum mechanics-based requirements that force and moment must balance across selected cross sections or over the whole specimen [27]. Therefore, by mapping the d^{hkl} in the required sample region the reference d_0^{hkl} can be iteratively found by imposing stress and moment balance, even starting from a nominal value [27]. However, great care must be taken to prove the applicability of the method: the experimental data must cover the whole cross section and it must be ensured that a global d_0^{hkl} is appropriate (i.e., the method would not work if d_0^{hkl} varies across the sample) [27]. Serrano-Munoz et al. [107] applied this method to obtain a d_0^{hkl} for different cross sections of LPBF manufactured Inconel 718 prisms. The method produced a similar d_0^{hkl}, indicating no dependance on the scanning strategy and the cross section being analyzed (i.e., there is no spatial variation of d_0^{hkl} along the length of the sample). Therefore, an average value was used for the d_0^{hkl} in the RS calculation [107], which applicability was later shown [80].

In fact, Kolbus et al. [103] proposed the method of stress balance as a possibility to check the measured d_0^{hkl}, as also indicated by Withers et al. [27]. To cross-check the values measured on mechanically relaxed samples, Pant et al. [81] used the stress balance approach and found a significant difference. Such discrepancy was attributed to microstructural variations: the average d_0^{hkl} value obtained by stress balance was used for the final RS calculations. Stress balance is often applied as an alternative approach to obtain a d_0^{hkl} without additional experimental effort [86,113]. However, in order to check the applicability of the hypotheses mentioned above, one should always compare the results (stress fields, d_0^{hkl}) obtained by using the stress balance condition with those obtained using experimental methods [27,80,103].

Indeed, the applicability of the stress balance approach for AM materials, which possibly exhibit 2D or 3D chemical variations due to the differential cooling rates, still requires further experimentation to test the robustness of the approach. Although this method would aid to make the RS determination by diffraction fully non-destructive, great care must be taken to avoid large errors in the RS values. In fact, Wang et al. [31] showed a LMD manufactured Inconel 625 wall displayed local variations of d_0^{hkl} due to the chemical and microstructural heterogeneity of the builds (Figure 10). This fact impeded the applicability of the stress balance condition.

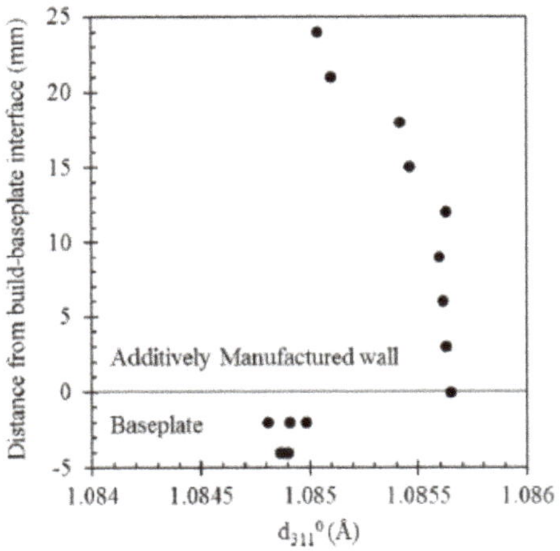

Figure 10. Reference d_0^{311} as a function of distance from the build-baseplate interface measured by neutron diffraction in the 40 s dwell time stress-free reference samples without heat treatment. Reproduced from [31] with permission from Elsevier.

The following table (Table 1) summarizes available methods to obtain ${d_0}^{hkl}$. It also reports the references in which each method was applied (in the case of laser-based AM parts).

Table 1. Overview of the different methods to obtain a ${d_0}^{hkl}$ in the domain of laser-based AM.

Method	Advantages	Disadvantages	References
Cubes/ matchsticks	• Can capture spatial variations when taken from several regions • Represent thermal history • Contain same chemical segregation if taken from representative twin	• Need a sister sample • Require precise sample alignment • Contain possible type I and II stresses • Display possible anisotropy of ${d_0}^{hkl}$ depending on direction • Are destructive for twin or actual specimen	[79,82,103, 109,110,112]
Comb/arrays	• Can capture spatial variations • Are easier to align (relative to cubes) • Are easier to extract with respect to cubes taken from distinct positions	Same as Cubes/matchsticks and: • Are time consuming to measure • Require twin or destruction of specimen	[31,46,80,81, 91,111]
Stress balance	• Does not need twin sample • Is fully non-destructive	• Cannot capture spatial variations of ${d_0}^{hkl}$ • Only works with a high density of measurement points over a cross section • Needs to be experimentally verified	[80,81,86,103, 107,113]
Feedstock powder	• Is easily obtained for powder-based AM	• Does not represent the thermal history of the AM process • Has different segregation from the AM part	[82,103,107]
Powder filings	• Represent thermal history of the sample	• Are plastically deformed • Need a twin specimen (or partially destroy the specimen) • Can yield spatially resolved ${d_0}^{hkl}$ only if extracted from distinct locations	[82,107]
***Sin*$^2\psi$**	• Is fully non-destructive	• Can only be applied for plane stress states • Relies on the DECs	[38,82]

6.2. Principal Stress Directions

A simplification of Equation (3) with the hypothesis of isotropic elastic constants (i.e., with the use of E, Young's modulus, and ν Poisson's ratio) would read for normal stress and strain components ($ii = xx, yy, zz$ in cartesian coordinates) [94]:

$$\sigma_{\text{ii}} = \frac{E^{hkl}}{\left(1+\nu^{hkl}\right)}\left[\varepsilon_{\text{ii}} + \frac{\nu^{hkl}}{\left(1-2\nu^{hkl}\right)}\left(\varepsilon_{xx}+\varepsilon_{yy}+\varepsilon_{zz}\right)\right] \tag{14}$$

For isotropic materials, Equation (14) is valid also in presence of shear stress [114]. However, without the knowledge about the principal stress directions, such a determination would not necessarily capture the maximum stress values. A common assumption to reduce the experimental effort is, that the principal stress directions coincide with the sample geometrical axes (e.g., see [82,83,103,110,115]). If the principal stress directions are known, Equation (14) can be used to calculate the principal stress components. This would reduce the number of measurements needed down to 3, if d_0^{hkl} is known. In the general case, where the principal stress axes are not known particular attention (and effort) needs to be dedicated to this aspect. In AM parts, the determination of the principal stress directions goes through the knowledge of the manufacturing process and of its impact on the principal stress directions.

Although several process parameters largely influence the magnitude of the RS, such as layer thickness, scanning speed, beam power, and vector length, the major influencing parameter on the stress distribution (and principal axes) is the relative orientation of the scanning pattern to the corresponding geometry [107].

Investigations about the principial stress direction in AM-Material started on simple geometries such as prisms [108]. Six strain directions were used for the calculation of the principal stress direction, which was not found to coincide with the sample geometrical ones. In contrast, complex structures were investigated by Fritsch et al. [116] using ND. It was shown, that for LPBF manufactured IN 625 lattice structures measurements along 6 independent strain directions are not sufficient to determine the principal stress directions and magnitudes. The authors found that at least seven independent directions are required to experimentally determine the direction of principal stress and even 8 directions are needed if the correct RS magnitude needs to be determined. In that case the calculated directions become insensitive to the choice of the measurement directions. Furthermore, it was proven that the RS tensor ellipsoid axes align well with the orientation of the struts within the lattice structure. [116].

Gloaguen et al. [46] showed, for example, that when assuming the principal stresses along the geometrical specimen axes for LPBF manufactured Ti6Al4V, the RS is affected by significant errors. This can be attributed to the fact that the principal stress axes deviate from the sample axes. This observation was made even though a simple bidirectional scanning strategy along the geometry with a 90° interlayer rotation was applied.

In fact, Vrancken [117] found for LPBF manufactured Ti6Al4V produced by a comparable scanning strategy, that the principal stress directions coincide with the direction of the scanning tracks. In other studies researchers found the principal stress directions to align with the sample geometrical axes [76,82], if the scanning strategy is more complicated (e.g., rotation between subsequent layers etc.).

These results emphasize that an increasing part complexity requires advanced measurement techniques and strategies to reach the desired precision for a reliable assessment of RS states. Again, given the complexity of laser-based additive manufacturing processes, the general assumption that the principal stress directions are governed by the sample geometry must be used carefully [46]. Therefore, for the alignment of the specimen in the laboratory coordinate system it is recommended not to make any assumption about the principal direction of stress and measure at least eight independent directions at all locations.

6.3. Diffraction-Elastic Constants (DECs)

To obtain stress values, the DECs act as proportionality constants to link the measured microscopic (i.e., lattice) strains to macroscopic stresses (see Equation (3)). Their precise knowledge is important, because the magnitude of the resulting RS depends on the values of the DECs (see Equation (3)). RS are thus highly vulnerable to errors if reliable values of the DEC are not used.

Two methods are available to obtain the DECs: They can be calculated from the single crystal elastic constants (SCEC) using different theoretical schemes (for instance a grain interaction model for the polycrystalline aggregate). This method is to be preferred if the SCECs are reliably known (note that much work needs still to be made for AM materials). The presence of a strong crystallographic texture in conjunction with crystal anisotropy can hamper the determination of the DECs by theoretical calculations, as one must properly take the texture into account. Alternatively, one can directly determine them in an in situ deformation test during diffraction. In this case, the microscopic response is monitored during a macroscopic deformation, and the proportionality constant between applied stress and lattice strain is the plane-specific Young's modulus E^{hkl}. A guideline for this is given in DIN EN 15305 [95]. The latter method, however, is connected to a relatively high experimental effort.

6.3.1. The Anisotropy of Single Crystals

The anisotropy of the single crystal can be expressed by the differences of the different elements of the compliance tensor. For cubic materials Zener [118] proposed the following coefficient, written in the Voigt notation, to calculate the anisotropy of the single crystal:

$$A^Z = \frac{2 \cdot C_{44}}{C_{11} - C_{12}} \tag{15}$$

In this definition, full isotropy is expressed by a value of $A^Z = 1$. Any deviation from $A^Z = 1$ signifies a certain degree of crystal anisotropy. However, as the Zener ratio only remains valid in the cubic case, researchers were motivated to formulate a more general anisotropy index, which would be valid for an arbitrary crystal structure. Such an index (A^U) was derived by Rangathan and Ostoja-Starzewski [119]. It is based on the fractional difference between the upper (Voigt) and lower (Reuss) bounds on the bulk (κ^V, κ^R) and shear (μ^V, μ^R) moduli. The values can be determined by the following equation (Equation (16)).

$$A^U = \frac{\kappa^v}{\kappa^R} + 5\frac{u^v}{u^R} - 6 \tag{16}$$

The main advantage of this formulation is its applicability to any type of crystal symmetry. However, it remains a relative measure of anisotropy. In fact, it has not been proven, that a crystal with twice an A^U also is twice as anisotropic. Therefore, Kube [120] provided an alternative definition, the anisotropy index A^L (Equation (17)), whereby the value of $A^L = 0$ expresses isotropy.

$$A^L(C^v, C^R) = \sqrt{\left[\ln\left(\frac{\kappa^v}{\kappa^R}\right)\right]^2 + 5\left[\ln\left(\frac{u^v}{u^R}\right)\right]^2} \tag{17}$$

There are also different approaches such as the Ledbetter and Migliori ratio [121] or the method proposed by Chung and Buessem [122]. However, we will use A^L in the following to compare the single crystal anisotropy of the commonly materials used in laser-based AM. One last important remark must be made: the applicability of all DEC calculation schemes heavily rests on the availability of reliable SCEC. A compiled list with the single crystal elastic constants (SCEC) of important alloys for laser based additive manufacturing is given in Table 2. The significant difference in the elastic anisotropy of the different single crystals is evident. The data shown are mainly inferred from measurements on conventionally produced polycrystalline materials or represent measurements on the respective single crystals. Data directly related to additively manufactured materials are still lacking. This may have an impact on the determination of the DECs and of RS. This is because the calculation of DECs is made under the assumption, that tabulated SCECs are still suitable for additively manufactured materials. Nevertheless, some authors have already tackled the problem of the determination of SCEC from experimental data on textured polycrystalline alloys [123].

Table 2. Single crystal elastic constants (SCEC) of several engineering alloys in GPa, with their dimensionless calculated Zener (A^Z) and universal (A^L) anisotropy ratios. For the calculation of A^L the Matlab script provided by Kube [120] was used.

Material	Crystal Structure	C_{11}	C_{12}	C_{44}	C_{33}	C_{13}	Ref.	A^Z	A^L [$\cdot 10^{-2}$]
Aluminium	FCC	108.2	61.3	28.5	-	-	[124]	1.2	2.04
		107.9	60.4	28.6	-	-	[125]	1.2	1.85
		106.8	60.7	28.2	-	-	[126]	1.2	2.18
		112.4	66.3	27.7	-	-	[127]	1.2	1.81
		108.2	62.2	28.4	-	-	[128]	1.2	2.38
		105.6	63.9	28.5	-	-	[129]	1.4	5.22
		107.3	60.9	28.3	-	-	[130]	1.2	2.12
Average		108.1	62.2	28.3	-	-	-	1.2	2.35
Ti6Al4V	HCP	150	83	42	137	53	[123]	-	5.67

Table 2. *Cont.*

Material	Crystal Structure	C_{11}	C_{12}	C_{44}	C_{33}	C_{13}	Ref.	A^Z	A^L [$\cdot 10^{-2}$]
Inconel 625	FCC	243.3	156.7	117.8	-	-	[131]	2.7	51.88
Inconel 718	FCC	240.9	140.5	105.7	-	-	[132]	2.1	29.17
		259.6	179	109.6	-	-	[133]	2.7	51.85
		231.2	145.1	117.2	-	-	[134]	2.7	51.95
Average		243.9	154.9	110.8	-	-	-	2.5	43.36
316L	FCC	191.2	117.9	138.6	-	-	[135]	3.8	89.33
		215.9	144.6	128.9	-	-	[136]	3.6	83.72
		198	125	122	-	-	[137]	3.3	71.38
Average		204.4	131.8	128.8	-	-	-	3.6	81.41

6.3.2. Grain Interaction Models for the Calculation of DECs

Several models have been developed to calculate DECs from SCEC. The first model developed by Voigt [138] (Figure 11a) assumes that adjacent grains undergo the same strain during deformation. However, this assumption violates the equilibrium of forces at the interfaces. On the other hand, Reuss [139] later proposed a model where the equilibrium of forces is fulfilled as a homogenous stress state is assumed (Figure 11b). This leads to the problem, that the different crystals undergo different strains, which would not satisfy the compatibility conditions [114]. To solve these problems Kröner developed a model based on Eshelby's theory [140], which fulfills the interface and the compatibility conditions (Figure 11c). Such scheme considers a spherical particle of arbitrary anisotropy embedded in an isotropic material. With the assumption of spherical particles and isotropic matrix, Kröner derived a closed (analytical) solution to the problem [141]. If the surrounding matrix is not texture free (e.g., as in the case of AM materials), numerical approaches must be considered [141]. In general, the Voigt model is the least applicable, as it results in elastic properties, such as E^{hkl}, that are independent on the plane {hkl}. This does not apply for most crystals. In contrast, the Kröner model has been shown to well match to experimentally determined values in an excellent manner for non-textured microstructures [142–144]. Interestingly, if a strong texture is present, as it has been observed in certain cases (including AM materials), the Reuss model displays better agreement with experimental data [32]. In fact, for columnar structures (the case of AM microstructures) the assumption that each crystal undergoes the same stress could be a good approximation.

From the discussion above, it is clear that for the application of each model, the microstructure and texture present in the material must be considered to determine appropriate values for the DECs. Indeed, many modifications and developments of the three schemes mentioned above have been made over the past years, to encompass the microstructure in the calculation of DECs. Initially, Dölle and Hauk [145] introduced the so-called stress factors to account for the texture using the crystallographic orientation distribution function (ODF). Several authors (e.g., Slim et al. [146], Brakman et al. [147], Welzel et al. [148–150], Gnäupel-Herold et al. [151]) proposed alternative approaches to embed the ODF in the determination of the DECS. More recently Mishurova et al. [40] have shown, that the use of Wu's tensor [152] is equivalent to using Kröner´s approach. In addition, they showcased the applicability of the procedure to LPBF Ti6Al4V. They concluded that, since hexagonal polycrystals possess transverse isotropy and LPBF Ti6Al4V had a fiber texture, the calculated DECs (using the best-fit isotropy assumption) reasonably agreed with experimentally determined values.

6.3.3. Experimental Determination of Diffraction Elastic Constants

The main method for the experimental determination of DECs are in situ mechanical tests, i.e., during high-energy X-ray or neutron diffraction experiments. The response of each lattice plane is monitored as a function of applied stress. It is important to mention that this approach rests on the hypothesis that a statistically significant ensemble of grains with the normal to the planes {*hkl*} is oriented along the load axis. From these datasets, the

DECs for each monitored plane then can be derived (see Table 3). For LPBF Ti6Al4V and IN718 a comparison of the model prediction with experimentally obtained values is given in Figure 12.

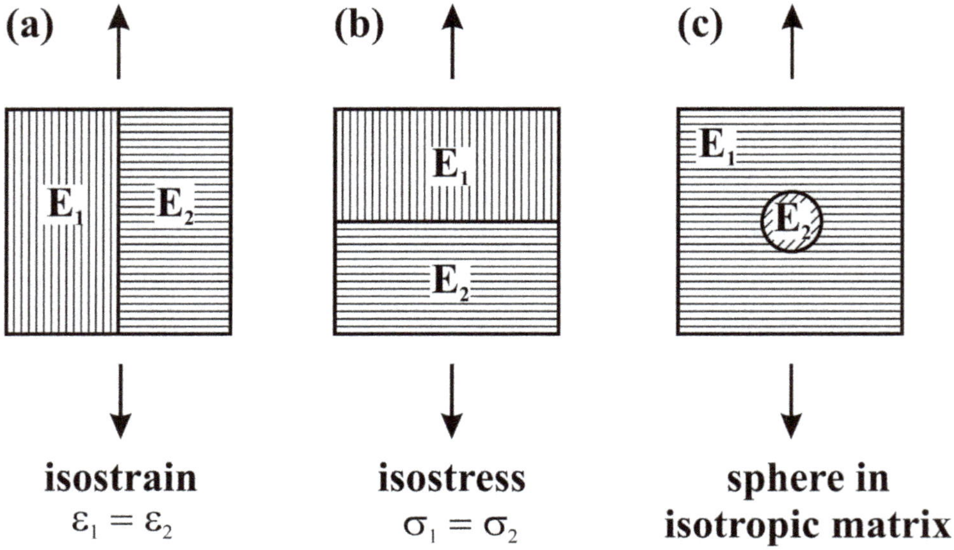

Figure 11. Overview of different model approaches for the calculation of the diffraction elastic constants. (**a**) Voigt model [138] (**b**) Reuss model [139] (**c**) Eshelby–Kröner model [141].

Table 3. Experimentally determined diffraction elastic constants by the means of diffraction methods. The plane-specific elastic moduli (E^{hkl}) are given in GPa.

Material	Process	Condition	E^{200}	E^{311}	E^{420}	E^{220}	E^{331}	E^{111}		Ref.
AlSi10Mg	LPBF	As built tension	66	68	-	71	-	73		[153]
IN625	LMD	As built compression	123	156	169	210	219	278		[131]
IN718	LPBF	FHT * tension	194	196	231	-	230	-		[39]
IN718	LPBF	DA ** tension	152	173	173	199	227	197		[39]
316L	LPBF	As built tension	139	180	-	219	-	246		[154]
Ti6Al4V	LPBF			$E^{10\bar{1}0}$	$E^{10\bar{1}1}$	$E^{11\bar{2}2}$	E^{0002}	$E^{10\bar{1}2}$	$E^{10\bar{1}3}$	
		As built tension		110	106	117	-	107	117	[155]
		HT-730 tension		106	116	126	134	128	125	[155]
		HT-900 tension		111	114	113	132	118	127	[155]
			$E^{21\bar{3}0}$	$E^{11\bar{2}0}$	$E^{10\bar{1}1}$	$E^{11\bar{2}2}$	$E^{20\bar{2}3}$	$E^{10\bar{1}2}$	$E^{10\bar{1}3}$	
		As built tension	108	110	115	115	116	120	125	[156]
		As built compression	-	115	-	117	123	125	126	[156]

* FHT (°C/h/MPa): 1066/1.5 + 1150/3/105 + 982/1 + 720/8 + 620/10, ** DA (°C/h): 1066/1.5 + 720/8.

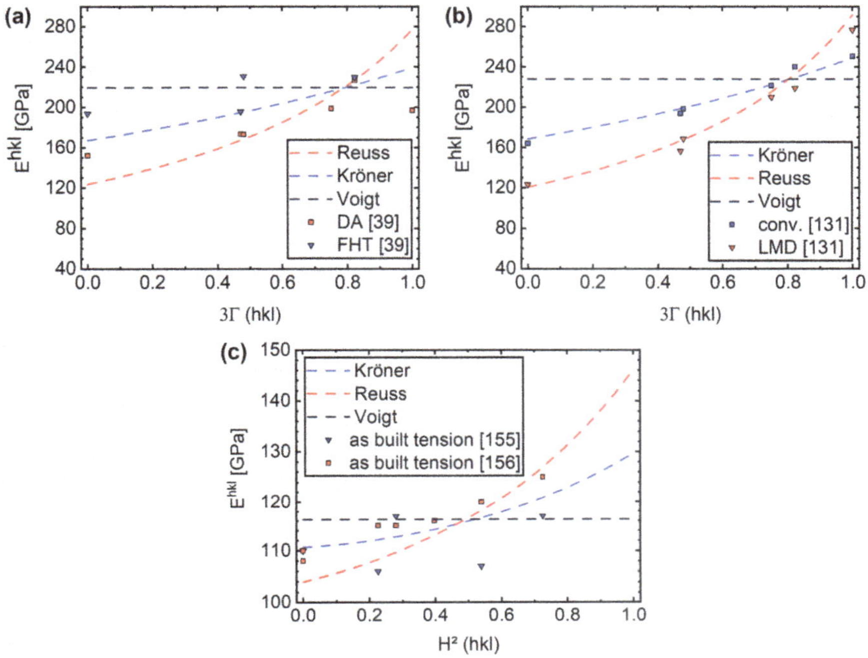

Figure 12. Comparison of the model predictions of Reuss, Voigt and Kröner for Ni-based alloys Inconel 718 (**a**), Inconel 625 (**b**) [39] and Ti6Al4V [155,156] (**c**).

For the alloys 316L and AlSi10Mg such a comparative figure is not necessary, since the model prediction of Kröner nearly perfectly matches the experimental values of the elastic moduli [153,154]. This is different for additively manufactured Ti6Al4V, Inconel 718 and Inconel 625. For a recrystallized and, thus, untextured microstructure (FHT) the Kröner model best matches the experimental values (Figure 12a) [39]. When, on the contrary, the columnar as-built microstructure (exhibiting relatively strong crystallographic texture) was retained, the model prediction of Reuss best fit the experimental data for Inconel 718 and 625 (Figure 12a,b) [39,131]. The AM Ti6Al4V alloy seems to deviate from this behavior: Mishurova et al. [156] showed, that for low H^2 (prismatic planes) the model predictions by Kröner agreed better with experimental data than other schemes (Figure 12c). In contrast, for higher H^2 (basal planes) the model prediction of Reuss better matched with the data [156]. This can be explained by the transverse isotropy of the single crystal elastic tensor, exhibiting an isotropic behavior in the basal directions but a strong anisotropy along its c-axis. It has been shown that when considering the transversal isotropy in the material model a reasonable agreement between model and experimental data can be obtained [40]. In presence of crystallographic texture, materials with a higher anisotropy factor (Table 2) tend to be better described by the predictions of the Reuss model than by those of Kröner's scheme. The exception to this trend is given by the alloy 316, for which the model approach of Kröner yields a good prediction of the polycrystal behavior, although the single crystal itself is highly anisotropic [154]. Such an agreement can be explained by the rather weak crystallographic texture along the loading direction in conjunction with the relatively small grain size [154].

As Mishurova et al. [45,156] argue, it is mandatory to report the DECs used to obtain stress values if one wants to compare own data with literature. Severe differences in the RS magnitude are the consequence if different model predictions considered for the determination of DEC. This was shown by Serrano-Munoz et al. [38] for LPBF manufactured Inconel 718: Applying the Kröner model led to a spiky stress depth profile. Stress values of 1200 MPa were reached, which far exceed the yield strength of the non-heat-treated material (630–800 MPa). Much more realistic stress values (up to 870 MPa) were obtained by applying the Reuss model to the experimental data. Also, the spikes of the stress depth profile were smoothed. This indicates the ability of the Reuss model to reasonably describe the intergranular behavior of LPBF Inconel 718 [157]. This was also supported by the findings of Pant et al. [81], who found stresses up to 1000 MPa in their study of as built LPBF manufactured Inconel 718. They used DECs measured for conventionally manufactured Inconel 718 and nearly equal to the Kröner model calculations [158].

Currently, a lack of consistency is observable in the open literature, as summarized in Table 4. It is to remark that, so far, DECs of additively manufactured microstructures have been mainly determined for loading along the building direction. The micromechanical behavior of the microstructure perpendicular to the build direction is, to the best of the authors' knowledge, not yet reported.

Table 4. Origins of applied diffraction elastic constants used for stress calculation in additively manufactured specimens.

Origin of DECs	References
Not given	[75,76,85,88,109,113,159–163]
unknown origin	[10,103,110,164]
Experimental values (conventional)	[81,86,165,166]
Reuss Model	[38,107,167]
Eshelby–Kröner Model	[13,26,74,79,82,83,96,108,112,168–170]
Voigt-Reuss-Hill	[171]
Experimental values (AM)	[31,115]

To summarize, in order to reliably calculate the DECs the proper SCECs of the material and both the microstructure and texture of the specimen should be considered. The crystal and macroscopic anisotropy provide guidance for the choice of the model to consider.

One last remark must be made: To obtain the plane specific Poisson ratio (ν^{hkl}) one would have to track the same set of grains during the deformation in both, the transverse and axial direction (i.e., along the tensile axis) [156]. However, this is practically impossible. In the case of nearly texture-free conventionally manufactured materials with small grain size the calculation of the ν^{hkl} using measurements in two perpendicular sample direction is a good approximation, as the gauge volume contains a sufficient amount of randomly oriented crystals. This approximation, in contrast, cannot be made for strongly textured AM materials. In this case, the use of suitable model schemes is recommended.

6.4. Choice of the Appropriate Lattice Planes

In the angular dispersive (monochromatic) case, one uses specific grains with specific lattice orientations as strain monitors [172]. It is assumed that the statistical ensemble is representative of the material. However, because of their particular elastic and plastic response, these grains are not necessarily representative of the overall stress state [172]. Consequently, the choice of a suitable reflection, whose grains represent the macroscopic RS in a body, is of utter importance [172,173]. Thereby, three main aspects need to be addressed:

- Insensitivity to intergranular stress accumulation (material dependent)
- Crystal symmetry
- Texture of the material

Whenever a sample is under stresses, a superposition of macroscopic (Type I) and intergranular stress (Type II) occurs [172]. If our goal is to determine the macroscopic stress state, a lattice plane, which exhibits a low tendency to accumulate intergranular stresses during deformation should be chosen. This tendency can be tested during in-situ loading experiments. Increasing non-linearity of the lattice plane response to a macroscopic load is an indication for the intergranular strain accumulation [174,175]. In fact, if the lattice strain vs. applied stress curve is non-linear residual strain is retained upon unloading. Such residual strain increases with increasing macroscopic plastic deformation.

The accumulation of intergranular stresses is critically dependent on the elastic and plastic anisotropy of the material [175]. In general, it is advisable to choose the lattice planes with the lowest Miller indices possible, as a high multiplicity of the lattice plane helps to reduce the required measurement time.

Besides these general considerations, one should take the underlying texture into account [176]. For example, for a (cubic) material with a strong cube texture, one should use the 200 reflection in spite of its typically high sensitivity to intergranular strains [172]. In fact, the 200 reflection represents most of the grains in such particular case [172]. For conventionally manufactured materials a general guideline on the selection of an appropriate lattice plane is given in ISO-21432 [177]. However, for AM the situation might be different, as strong textures typically prevail.

Very little studies on the topic of the accumulation of intergranular strains in laser-based AM materials are available in the open literature. Table 5 shows the lattice planes typically considered for RS analysis of different laser-based AM materials and outlines their suitability compared to their conventionally manufactured counterparts. In the case of fcc materials the 311 reflection is almost exclusively used [26,74,82,83,103,108]. However, it has been shown by Choo et al. [178] for LPBF 316L that the {311} oriented grains accumulate more intergranular strain than the {111} and {220} grains, which is in contrast to conventional rolled 316L [179]. In fact, considering the <220> texture along the building direction [178], the 220 reflection is more easily detected than others, and yields less data scatter. Likewise, Wang et al. [180] observed a strong nonlinear micromechanical response during initial loading of LPBF manufactured 316L, which has been attributed to anisotropic residual strains within the as-built samples. Consequently, the hierarchical heterogenous microstructures of AM 316L may give rise to significant differences in the buildup of intergranular stresses and should be accounted for.

For hexagonally closed packaged (hcp) materials the pyramidal planes {102} and {103} are considered to exhibit low intergranular stress (for conventionally processed materials) [181,182]. However, Cho et al. [183] showed for conventionally manufactured α-Ti-834 that the first eight diffraction peaks (i.e., those with the lowest Miller indices) accumulate significant intergranular strains. In fact, studies on the topic of intergranular strain accumulation are absent for additively manufactured hcp materials. Although Zhang et al. [155] showed that a high dislocation density is present within the α' Phase of as-built LPBF Ti6Al4V, the micromechanical response show anomalies to conventional Ti6Al4V: sometimes it remains linear well beyond the early stages of loading, sometimes it shows footprints of twinning [184,185]. Thus, the question of the accumulation of intergranular strains is yet far from fully elucidated in AM hcp materials.

In general, more research needs to be dedicated to the topic of intergranular stresses within the domain of laser-based AM materials. One must carefully evaluate whether the requirements are fulfilled for a certain lattice plane to represent the bulk behavior. The approach of using the full pattern refinement of the lattice parameter minimizes (actually, averages) the possible contributions of high intergranular stress to the determined macroscopic type I RS.

Table 5. Material dependent selection of Lattice planes for RS determination in laser-based AM.

Material	AM	Conventional	Conclusion
Inconel 718 (fcc)	• DA: <200> texture along BD max. 3 m.r.d (corresponds to loading direction) [39] • 311 only slightly non-linear upon loading [39] • As built: 311 remains linear upon loading (<220> textured along BD) [186] • Heat-treated: 311 shows increased non-linearity (untextured) [186] • 311, 111 linear upon loading in as built condition, with increased non-linearity of (311) in heat-treated condition (possibly γ shedding load to δ) [187]	• Low texture (max. 1.5 m.r.d) [143] • 311, 111 remain linear upon loading [143] • 220 displays non-linearity at larger strains [143] • 222 and 311 accumulate low intergranular strain [143]	• Heat-treatments affect the micromechanical behavior upon loading • 311 appears suitable for as built samples • Texture up to 3 m.r.d has little influence on the choice of the lattice plane
316L (fcc)	• <220> texture along BD (<200> perpendicular which corresponds to LD max. 4 m.r.d) [154] • 311: remains linear upon loading in the loading direction [154] • 200, 111, 220 display increased non-linearity upon loading [154] • <110> texture along BD (max. 2.8 m.r.d) 311, 200 accumulates significant intergranular strain; 220, 111 accumulate low intergranular strain [178] • Nearly untextured ($\approx$ 1 m.r.d) lower magnitude of total intergranular strain accumulated [178]	• Weak cube texture (max. 2 m.r.d) [179] • 311 slightly non-linear at larger strains [179] • 311, 111 accumulate low residual strain [179]	• 311 accumulates significant intergranular strain, but remains linear upon loading • 220 lowest intergranular strain accumulation and representative for the bulk (along BD; dependent on processing conditions)
304 (fcc)	• Austenite <200> textured perpendicular to BD (max. 3 m.r.d corresponds to loading direction) [188] • 311: Slight increase in non-linearity upon loading [188] • Different behavior in compression/tension for 200, 111, 220 [188] • 200 carries load shed from 220 and 111 [188]	• Low texture (max. 1 m.r.d) [188] • 311: linear behavior [188] • No difference in compression/tension [188] • 200 carries load shed from 220 and 111 [188]	→ Comparison to full pattern fit required to draw a general conclusion → Texture must be taken into consideration

Table 5. *Cont.*

Material	AM	Conventional	Conclusion
AlSiMg10 (fcc)	• Texture with max. 3.4 m.r.d [153] • Interphase and intergranular strains between Al matrix and Si particles [153] • Strong non-linearity in plastic region for every reflection [153] • <100> texture along BD with max. 4.5 m.r.d [189] • Al phase accumulates significant residual strains for 220 < 111 < 311 < 200 (parallel to loading direction) [189] • Si phase accumulates significant residual strains for 111 and 311 [189] • Rietveld refinement reasonably agrees with the trend for both Al and Si 311 [189]	-	• Significant residual strains accumulate in 311 • Still 311 displays best agreement with average obtained by Rietveld refinement (represents bulk behavior) • 220 accumulates least residual strain
Ti6Al4V (hcp)	• <002> texture along BD (max. 6 m.r.d) in as built specimen [155] • Increased non-linearity for 100, 101, 102 at higher deformations in as built condition [155] • <002> texture along BD [184] • Early deviation from linearity for 002 and 110 (elastic region) [184] • 102 and 103 start to deviate at higher strains [184] • Strong non-linearity for every reflection upon compressive loading for different build orientations (horizontal, 45°, vertical) [185] • Strong twinning formation in vertically build specimen (driven by initial texture) [185]	Unidirectionally rolled plate (UD) (loading along rolling direction (RD)) [181]: • <002> texture along transverse direction (max. 3.8 m.r.d) • 103, 002 accumulate lowest residual strain • Largest in 101, 201 • Cross rolled plate [181]: • <002> texture along RD (max. 3.5 m.r.d) • 103, 002 accumulate lowest residual strain • Largest in 101, 201 • Bar vs. UD plate [182]: • UD: Strong texture up to max. 6 m.r.d • Bar: texture up to max. 2 m.r.d • Low accumulation of residual lattice strains for 102 and 002 in both cases • Larger accumulation of residual lattice strains for 100 and 110 in UD due to stronger texture	• Comparable texture intensity of AM and conventional • Still significant change of micromechanical behavior between additively manufactured and conventional Ti6Al4V • Full pattern fit recommended • For AM materials (along BD) 002 may represent the bulk behavior the best → Question not resolved: Studies on accumulation of residual strains required

7. Summary & Outlook

Additive manufacturing (AM) methods allow the fabrication of complex structures within a single manufacturing step. Still the heterogeneity of the process often leads to mechanically anisotropic, columnar, and textured microstructures. While one of the biggest challenges in AM is to develop mitigation strategies for the large residual stress that inevitably appears after production, the precise determination of such residual stress remains challenging. Diffraction-based methods provide a powerful tool to non-destructively determine the residual stress. However, the peculiar microstructures of AM materials pose challenges for the characterization of residual stress. Therefore, assumptions and measurement conditions must be chosen with great care:

- First, one must evaluate if the assumption of a biaxial stress state can be justified (e.g., surface measurements with $sin^2\psi$ method) or a triaxial stress state must be considered. In the latter case, neutron diffraction should be preferred to other techniques and precise knowledge about the strain-free lattice spacing (${d_0}^{hkl}$) is required. To obtain such a reference, measurements on mechanically relaxed samples are recommended. The stress balance method is recommended as a validation method. If the requirements for the correct application of stress balance conditions (no spatial variation of composition with large number of points) are known to be fulfilled, the stress balance method can be used to obtain a global ${d_0}^{hkl}$. Still, the strategy to determine ${d_0}^{hkl}$ needs to be tailored for each case.
- Secondly, the principal stress directions should be known in advance if one wants to determine the maximum stress values. For conventional processes such as forging or rolling these are often known (they coincide with the main geometrical sample axes). In the case of AM, the complexity of the process conditions hinders the prior knowledge of the principal stress directions. Although research indicates the principal directions to be determined by the scanning strategy (i.e., the main stress axes follow the scanning vector) it is recommended to run experimental checks. Ideally the full stress tensor should be characterized.
- Thirdly, the microstructure and texture of the sample should be well characterized. Texture is one of the driving factors for the determination of the diffraction elastic constants (DECs). Furthermore, the DECs are material-dependent, dictated by the single crystal properties. Therefore, choosing the appropriate modeling scheme for the calculation of DECs from single crystal elastic constants is challenging. At best the DECs should be experimentally determined. If that is not possible, it is indispensable to take the microstructure and the texture into account in the selection of the grain-interaction model.
- Lastly, an appropriate lattice plane must be chosen in the case of a monochromatic measurement technique (Laboratory XRD or steady state Neutron sources), as stresses are derived from one single lattice plane. Such plane should be insensitive to accumulation of intergranular strain and possess a high multiplicity, to represent the macroscopic behavior of the sample.

The amount of research dedicated to the methodology of diffraction-based methods in the domain of AM is increasing but still limited. In particular, the understanding of the influence of the microstructure and texture on the DECs should be addressed for all metal AM processes. This would aid to provide a general strategy to determine the DECs for an additively manufactured material. Further research is needed to develop a uniform strategy to determine an appropriate ${d_0}^{hkl}$; this would increase the comparability of results. It is also worthwhile to dedicate research to gain a better understanding of intergranular stress accumulation for the hierarchical structures occurring in laser-based AM.

Author Contributions: Conceptualization, J.S., A.E. and M.S.; methodology, J.S. and A.K.; validation, J.S., T.M. and I.S.-M.; formal analysis, T.M.; investigation, J.S. and A.U.; resources, G.B.; writing—original draft preparation, J.S., A.E. and A.U.; writing—review and editing, J.S., A.E., T.M., A.U., M.S., I.S.-M., T.F., A.K., T.K. and G.B.; visualization, J.S., A.K. and T.F.; supervision, A.E., G.B. and T.K.; project administration, G.B. and T.K. All authors have read and agreed to the published version of the manuscript.

Funding: This research received no external funding.

Institutional Review Board Statement: Not applicable.

Informed Consent Statement: Not applicable.

Data Availability Statement: Not applicable.

Acknowledgments: The authors kindly acknowledge the fruitful scientific discussions on the topic with Michael Hofmann (TU Munich), Winfried Petry (TU Munich), Christoph Genzel (HZB, Berlin), and Manuela Klaus (HZB, Berlin). This work was supported by the internal BAM focus area materials project AGIL "Microstructure development in additively manufactured metallic components: from powder to mechanical failure" and the internally funded project MIT1-2019-45.

Conflicts of Interest: The authors declare no conflict of interest.

References

1. Appleyard, D. Powering up on powder technology. *Met. Powder Rep.* **2015**, *70*, 285–289. [CrossRef]
2. Emmelmann, C.; Kranz, J.; Herzog, D.; Wycisk, E. Laser Additive Manufacturing of Metals. *Biol. Med. Phys. Biomed.* **2013**, 143–191. [CrossRef]
3. Kranz, J.; Herzog, D.; Emmelmann, C. Design guidelines for laser additive manufacturing of lightweight structures in TiAl6V4. *J. Laser Appl.* **2015**, *27*, S14001. [CrossRef]
4. Khorasani, M.; Ghasemi, A.; Rolfe, B.; Gibson, I. Additive manufacturing a powerful tool for the aerospace industry. *Rapid Prototyp. J.* **2021**, ahead-of-Print. [CrossRef]
5. Herzog, D.; Seyda, V.; Wycisk, E.; Emmelmann, C. Additive manufacturing of metals. *Acta Mater.* **2016**, *117*, 371–392. [CrossRef]
6. Gallmeyer, T.G.; Moorthy, S.; Kappes, B.B.; Mills, M.J.; Amin-Ahmadi, B.; Stebner, A.P. Knowledge of process-structure-property relationships to engineer better heat treatments for laser powder bed fusion additive manufactured Inconel 718. *Addit. Manuf.* **2020**, *31*, 100977. [CrossRef]
7. Voisin, T.; Forien, J.B.; Perron, A.; Aubry, S.; Bertin, N.; Samanta, A.; Baker, A.; Wang, Y.M. New insights on cellular structures strengthening mechanisms and thermal stability of an austenitic stainless steel fabricated by laser powder-bed-fusion. *Acta Mater.* **2021**, *203*, 116476. [CrossRef]
8. Pröbstle, M.; Neumeier, S.; Hopfenmuller, J.; Freund, L.P.; Niendorf, T.; Schwarze, D.; Goken, M. Superior creep strength of a nickel-based superalloy produced by selective laser melting. *Mater. Sci. Eng. A* **2016**, *674*, 299–307. [CrossRef]
9. Vrancken, B.; Thijs, L.; Kruth, J.P.; Van Humbeeck, J. Heat treatment of Ti6Al4V produced by Selective Laser Melting: Microstructure and mechanical properties. *J. Alloy. Compd.* **2012**, *541*, 177–185. [CrossRef]
10. Yadroitsev, I.; Yadroitsava, I. Evaluation of residual stress in stainless steel 316L and Ti6Al4V samples produced by selective laser melting. *Virtual. Phys. Prototyp.* **2015**, *10*, 67–76. [CrossRef]
11. Kruth, J.P.; Froyen, L.; Van Vaerenbergh, J.; Mercelis, P.; Rombouts, M.; Lauwers, B. Selective laser melting of iron-based powder. *J. Mater. Process. Technol.* **2004**, *149*, 616–622. [CrossRef]
12. Mukherjee, T.; Zhang, W.; DebRoy, T. An improved prediction of residual stresses and distortion in additive manufacturing. *Comput. Mater. Sci.* **2017**, *126*, 360–372. [CrossRef]
13. Mishurova, T.; Cabeza, S.; Artzt, K.; Haubrich, J.; Klaus, M.; Genzel, C.; Requena, G.; Bruno, G. An Assessment of Subsurface Residual Stress Analysis in SLM Ti-6Al-4V. *Materials* **2017**, *10*, 348. [CrossRef]
14. Carpenter, K.; Tabei, A. On Residual Stress Development, Prevention, and Compensation in Metal Additive Manufacturing. *Materials* **2020**, *13*, 255. [CrossRef]
15. Marchese, G.; Parizia, S.; Saboori, A.; Manfredi, D.; Lombardi, M.; Fino, P.; Ugues, D.; Biamino, S. The Influence of the Process Parameters on the Densification and Microstructure Development of Laser Powder Bed Fused Inconel 939. *Metals* **2020**, *10*, 882. [CrossRef]
16. Kanagarajah, P.; Brenne, F.; Niendorf, T.; Maier, H.J. Inconel 939 processed by selective laser melting: Effect of microstructure and temperature on the mechanical properties under static and cyclic loading. *Mater. Sci. Eng. A* **2013**, *588*, 188–195. [CrossRef]
17. Zhang, X.Q.; Chen, H.B.; Xu, L.M.; Xu, J.J.; Ren, X.K.; Chen, X.Q. Cracking mechanism and susceptibility of laser melting deposited Inconel 738 superalloy. *Mater. Des.* **2019**, *183*, 108105. [CrossRef]
18. Muñiz-Lerma, J.A.; Tian, Y.; Wang, X.; Gauvin, R.; Brochu, M. Microstructure evolution of Inconel 738 fabricated by pulsed laser powder bed fusion. *Prog. Addit. Manuf.* **2019**, *4*, 97–107. [CrossRef]
19. Ramakrishnan, A.; Dinda, G.P. Direct laser metal deposition of Inconel 738. *Mater. Sci. Eng. A* **2019**, *740*, 1–13. [CrossRef]

20. Sotov, A.V.; Agapovichev, A.V.; Smelov, V.G.; Kokareva, V.V.; Dmitrieva, M.O.; Melnikov, A.A.; Golanov, S.P.; Anurov, Y.M. Investigation of the IN-738 superalloy microstructure and mechanical properties for the manufacturing of gas turbine engine nozzle guide vane by selective laser melting. *Int. J. Adv. Manuf. Tech.* **2020**, *107*, 2525–2535. [CrossRef]
21. Seede, R.; Shoukr, D.; Zhang, B.; Whitt, A.; Gibbons, S.; Flater, P.; Elwany, A.; Arroyave, R.; Karaman, I. An ultra-high strength martensitic steel fabricated using selective laser melting additive manufacturing: Densification, microstructure, and mechanical properties. *Acta Mater.* **2020**, *186*, 199–214. [CrossRef]
22. Boes, J.; Rottger, A.; Theisen, W.; Cui, C.; Uhlenwinkel, V.; Schulz, A.; Zoch, H.W.; Stern, F.; Tenkamp, J.; Walther, F. Gas atomization and laser additive manufacturing of nitrogen-alloyed martensitic stainless steel. *Addit. Manuf.* **2020**, *34*, 101379. [CrossRef]
23. Saeidi, K.; Zapata, D.L.; Lofaj, F.; Kvetkova, L.; Olsen, J.; Shen, Z.J.; Akhtar, F. Ultra-high strength martensitic 420 stainless steel with high ductility. *Addit. Manuf.* **2019**, *29*, 100803. [CrossRef]
24. Lu, Y.; Wu, S.; Gan, Y.; Huang, T.; Yang, C.; Junjie, L.; Lin, J. Study on the microstructure, mechanical property and residual stress of SLM Inconel-718 alloy manufactured by differing island scanning strategy. *Opt. Laser Technol.* **2015**, *75*, 197–206. [CrossRef]
25. Valdez, M.; Kozuch, C.; Faierson, E.J.; Jasiuk, I. Induced porosity in Super Alloy 718 through the laser additive manufacturing process: Microstructure and mechanical properties. *J. Alloys Compd.* **2017**, *725*, 757–764. [CrossRef]
26. Nadammal, N.; Mishurova, T.; Fritsch, T.; Serrano-Munoz, I.; Kromm, A.; Haberland, C.; Portella, P.D.; Bruno, G. Critical role of scan strategies on the development of microstructure, texture, and residual stresses during laser powder bed fusion additive manufacturing. *Addit. Manuf.* **2021**, *38*, 101792. [CrossRef]
27. Withers, P.J.; Preuss, M.; Steuwer, A.; Pang, J.W.L. Methods for obtaining the strain-free lattice parameter when using diffraction to determine residual stress. *J. Appl. Crystallogr.* **2007**, *40*, 891–904. [CrossRef]
28. Dixneit, J.; Kromm, A.; Boin, M.; Wimpory, R.C.; Kannengiesser, T.; Gibmeier, J.; Schroepfer, D. Residual stresses of LTT welds in large-scale components. *Weld. World* **2017**, *61*, 1089–1097. [CrossRef]
29. Kromm, A. Evaluation of weld filler alloying concepts for residual stress engineering by means of Neutron and X-ray diffraction. *Adv. Mater. Res.* **2014**, *996*, 469–474. [CrossRef]
30. Altenkirch, J.; Gibmeier, J.; Kromm, A.; Kannengiesser, T.; Nitschke-Pagel, T.; Hofmann, M. In situ study of structural integrity of low transformation temperature (LTT)-welds. *Mater. Sci. Eng. A* **2011**, *528*, 5566–5575. [CrossRef]
31. Wang, Z.; Denlinger, E.; Michaleris, P.; Stoica, A.D.; Ma, D.; Beese, A.M. Residual stress mapping in Inconel 625 fabricated through additive manufacturing: Method for neutron diffraction measurements to validate thermomechanical model predictions. *Mater. Des.* **2017**, *113*, 169–177. [CrossRef]
32. Spieß, L.; Teichert, G.; Schwarzer, R.; Behnken, H.; Genzel, C. *Moderne Röntgenbeugung*, 2nd ed.; Vieweg+Teubner Verlag: Wiesbaden, Germany, 2009. [CrossRef]
33. Noyan, I.C.; Cohen, J.B. *Residual Stress: Measurement by Diffraction and Interpretation*; Springer: New York, NY, USA, 1987; p. x. 276p.
34. Hauk, V.; Behnken, H. *Structural and Residual Stress Analysis by Nondestructive Methods: Evaluation, Application, Assessment*; Elsevier: Amsterdam, The Netherlands, 1997; p. xiv. 640p.
35. Ye, T.; Li, L.X.; Liu, X.; Liu, W.H.; Guo, P.C.; Tang, X. Anisotropic deformation behavior of as-extruded 6063-T4 alloy under dynamic impact loading. *Mater. Sci. Eng. A* **2016**, *666*, 149–155. [CrossRef]
36. You, Z.; Fu, H.; Qu, S.; Bao, W.; Lu, L. Revisiting anisotropy in the tensile and fracture behavior of cold-rolled 316L stainless steel with heterogeneous nano-lamellar structures. *Nano Mater. Sci.* **2020**, *2*, 72–79. [CrossRef]
37. Charmi, A.; Falkenberg, R.; Avila, L.; Mohr, G.; Sommer, K.; Ulbricht, A.; Sprengel, M.; Neumann, R.S.; Skrotzki, B.; Evans, A. Mechanical anisotropy of additively manufactured stainless steel 316L: An experimental and numerical study. *Mater. Sci. Eng. A* **2021**, *799*, 140154. [CrossRef]
38. Serrano-Munoz, I.; Fritsch, T.; Mishurova, T.; Trofimov, A.; Apel, D.; Ulbricht, A.; Kromm, A.; Hesse, R.; Evans, A.; Bruno, G. On the interplay of microstructure and residual stress in LPBF IN718. *J. Mater. Sci.* **2021**, *56*, 5845–5867. [CrossRef]
39. Schröder, J.; Mishurova, T.; Fritsch, T.; Serrano-Munoz, I.; Evans, A.; Sprengel, M.; Klaus, M.; Genzel, C.; Schneider, J.; Bruno, G. On the influence of heat treatment on microstructure and mechanical behavior of laser powder bed fused Inconel 718. *Mater. Sci. Eng. A* **2021**, *805*, 140555. [CrossRef]
40. Mishurova, T.; Bruno, G.; Evsevleev, S.; Sevostianov, I. Determination of macroscopic stress from diffraction experiments: A critical discussion. *J. Appl. Phys.* **2020**, *128*, 025103. [CrossRef]
41. Bartlett, J.L.; Li, X. An overview of residual stresses in metal powder bed fusion. *Addit. Manuf.* **2019**, *27*, 131–149. [CrossRef]
42. DebRoy, T.; Wei, H.L.; Zuback, J.S.; Mukherjee, T.; Elmer, J.W.; Milewski, J.O.; Beese, A.M.; Wilson-Heid, A.; De, A.; Zhang, W. Additive manufacturing of metallic components—Process, structure and properties. *Prog. Mater. Sci.* **2018**, *92*, 112–224. [CrossRef]
43. Acevedo, R.B.O.; Kantarowska, K.; Santos, E.C.; Fredel, M.C. Residual stress measurement techniques for Ti6Al4V parts fabricated using selective laser melting: State of the art review. *Rapid Prototyp. J.* **2020**. [CrossRef]
44. Fang, Z.C.; Wu, Z.L.; Huang, C.G.; Wu, C.W. Review on residual stress in selective laser melting additive manufacturing of alloy parts. *Opt. Laser Technol.* **2020**, *129*, 106283. [CrossRef]
45. Mishurova, T.; Serrano-Munoz, I.; Fritsch, T.; Ulbricht, A.; Sprengel, M.; Evans, A.; Kromm, A.; Madia, M.; Bruno, G. A Critical Discussion on the Diffraction-Based Experimental Determination of Residual Stress in AM Parts. In *Structural Integrity of Additive Manufactured Materials and Parts*; Shamsaei, N., Seifi, M., Eds.; ASTM International: West Conshohocken, PA, USA, 2020; pp. 122–138. [CrossRef]

46. Gloaguen, D.; Girault, B.; Courant, B.; Dubos, P.A.; Moya, M.J.; Edy, F.; Kornmeier, J.R. Study of Residual Stresses in Additively Manufactured Ti-6Al-4V by Neutron Diffraction Measurements. *Metall. Mater. Trans. A* **2020**, *51*, 951–961. [CrossRef]
47. Vayre, B.; Vignat, F.; Villeneuve, F. Metallic additive manufacturing: State-of-the-art review and prospects. *Mech. Ind.* **2012**, *13*, 89–96. [CrossRef]
48. Kumar, S.; Kruth, J.P. Effect of bronze infiltration into laser sintered metallic parts. *Mater. Des.* **2007**, *28*, 400–407. [CrossRef]
49. Kruth, J.P.; Vandenbroucke, B.; Van Vaerenbergh, J.; Mercelis, P. Benchmarking of different SLS/SLM processes as rapid manufacturing techniques. In Proceedings of the International Conference Polymers & Moulds Innovations PMI, Gent, Belgium, 20–24 April 2005.
50. Saboori, A.; Aversa, A.; Marchese, G.; Biamino, S.; Lombardi, M.; Fino, P. Application of Directed Energy Deposition-Based Additive Manufacturing in Repair. *Appl. Sci.* **2019**, *9*, 3316. [CrossRef]
51. Onuike, B.; Bandyopadhyay, A. Additive manufacturing in repair: Influence of processing parameters on properties of Inconel 718. *Mater. Lett.* **2019**, *252*, 256–259. [CrossRef]
52. Kruth, J.P.; Mercelis, P.; Van Vaerenbergh, J.; Froyen, L.; Rombouts, M. Binding mechanisms in selective laser sintering and selective laser melting. *Rapid Prototyp. J.* **2005**, *11*, 26–36. [CrossRef]
53. Poprawe, R.; Loosen, P.; Hoffmann, H.-D. The future of high power laser techniques. In *XVI International Symposium on Gas Flow, Chemical Lasers, and High-Power Lasers, Pts 1 and 2*; International Society for Optics and Photonics: Bellingham, WA, USA, 2007; Volume 6346, p. 34602. [CrossRef]
54. Aboulkhair, N.T.; Everitt, N.M.; Ashcroft, I.; Tuck, C. Reducing porosity in AlSi10Mg parts processed by selective laser melting. *Addit. Manuf.* **2014**, *1–4*, 77–86. [CrossRef]
55. Oliveira, J.P.; LaLonde, A.D.; Ma, J. Processing parameters in laser powder bed fusion metal additive manufacturing. *Mater. Des.* **2020**, *193*, 108762. [CrossRef]
56. Buchbinder, D.; Meiners, W.; Pirch, N.; Wissenbach, K.; Schrage, J. Investigation on reducing distortion by preheating during manufacture of aluminum components using selective laser melting. *J. Laser Appl.* **2014**, *26*, 012004. [CrossRef]
57. von Müller, A.; Schlick, G.; Neu, R.; Anstatt, C.; Klimkait, T.; Lee, J.; Pascher, B.; Schmitt, M.; Seidel, C. Additive manufacturing of pure tungsten by means of selective laser beam melting with substrate preheating temperatures up to 1000 degrees C. *Nucl. Mater. Energy.* **2019**, *19*, 184–188. [CrossRef]
58. Mohr, G.; Altenburg, S.J.; Hilgenberg, K. Effects of inter layer time and build height on resulting properties of 316L stainless steel processed by laser powder bed fusion. *Addit. Manuf.* **2020**, *32*, 101080. [CrossRef]
59. Vock, S.; Klöden, B.; Kirchner, A.; Weißgräber, T.; Kieback, B. Powders for powder bed fusion: A review. *Prog. Addit. Manuf.* **2019**, *4*, 383–397. [CrossRef]
60. Mercelis, P.; Kruth, J.P. Residual stresses in selective laser sintering and selective laser melting. *Rapid Prototyp. J.* **2006**, *12*, 254–265. [CrossRef]
61. Gu, D.D.; Meiners, W.; Wissenbach, K.; Poprawe, R. Laser additive manufacturing of metallic components: Materials, processes and mechanisms. *Int. Mater. Rev.* **2012**, *57*, 133–164. [CrossRef]
62. Ruiz, J.E.; Cortina, M.; Arrizubieta, J.I.; Lamikiz, A. Study of the Influence of Shielding Gases on Laser Metal Deposition of Inconel 718 Superalloy. *Materials* **2018**, *11*, 1388. [CrossRef]
63. Cortina, M.; Arrizubieta, J.I.; Ruiz, J.E.; Lamikiz, A.; Ukar, E. Design and manufacturing of a protective nozzle for highly reactive materials processing via Laser Material Deposition. *Procedia CIRP* **2018**, *68*, 387–392. [CrossRef]
64. Arrizubieta, J.I.; Ruiz, J.E.; Martinez, S.; Ukar, E.; Lamikiz, A. Intelligent nozzle design for the Laser Metal Deposition process in the Industry 4.0. *Procedia Manuf.* **2017**, *13*, 1237–1244. [CrossRef]
65. Bernhard, R.; Neef, P.; Eismann, T.; Wiche, H.; Hoff, C.; Hermsdorf, J.; Kaierle, S.; Wesling, V. Additive manufacturing of LMD nozzles for multi-material processing. *Procedia CIRP* **2020**, *94*, 336–340. [CrossRef]
66. Gruber, S.; Grunert, C.; Riede, M.; Lopez, E.; Marquardt, A.; Brueckner, F.; Leyens, C. Comparison of dimensional accuracy and tolerances of powder bed based and nozzle based additive manufacturing processes. *J. Laser Appl.* **2020**, *32*, 032016. [CrossRef]
67. Mahamood, R.M. Processing Parameters in Laser Metal Deposition Process. In *Laser Metal Deposition Process of Metals, Alloys, and Composite Materials. Engineering Materials and Processes*, 1st ed.; Springer: Cham, Switzerland, 2018. [CrossRef]
68. Frazier, W.E. Metal Additive Manufacturing: A Review. *J. Mater. Eng. Perform.* **2014**, *23*, 1917–1928. [CrossRef]
69. Gibson, I.; Rosen, D.; Stucker, B. Directed Energy Deposition Processes. In *Additive Manufacturing Technologies: 3D Printing, Rapid Prototyping, and Direct Digital Manufacturing*; Springer: New York, NY, USA, 2015; pp. 245–268. [CrossRef]
70. Withers, P.J.; Bhadeshia, H.K.D.H. Overview—Residual stress part 2—Nature and origins. *Mater. Sci. Technol.* **2001**, *17*, 366–375. [CrossRef]
71. Withers, P.J.; Bhadeshia, H.K.D.H. Overview—Residual stress part 1—Measurement techniques. *Mater. Sci. Technol.* **2001**, *17*, 355–365. [CrossRef]
72. Mo, F.J.; Sun, G.G.; Li, J.; Zhang, C.S.; Wang, H.; Chen, Y.; Liu, Z.; Yang, Z.K.; Li, H.J.; Yang, Z.L.; et al. Recent Progress of Residual Stress Distribution and Structural Evolution in Materials and Components by Neutron Diffraction Measurement at RSND. *Quantum Beam Sci.* **2018**, *2*, 15. [CrossRef]
73. Patterson, A.E.; Messimer, S.L.; Farrington, P.A. Overhanging Features and the SLM/DMLS Residual Stresses Problem: Review and Future Research Need. *Technologies* **2017**, *5*, 15. [CrossRef]

74. Nadammal, N.; Cabeza, S.; Mishurova, T.; Thiede, T.; Kromm, A.; Seyfert, C.; Farahbod, L.; Haberland, C.; Schneider, J.A.; Portella, P.D.; et al. Effect of hatch length on the development of microstructure, texture and residual stresses in selective laser melted superalloy Inconel 718. *Mater. Des.* **2017**, *134*, 139–150. [CrossRef]
75. Wu, A.S.; Brown, D.W.; Kumar, M.; Gallegos, G.F.; King, W.E. An Experimental Investigation into Additive Manufacturing-Induced Residual Stresses in 316L Stainless Steel. *Metall. Mater. Trans. A* **2014**, *45*, 6260–6270. [CrossRef]
76. Bagg, S.D.; Sochalski-Kolbus, L.M.; Bunn, J.R. The Effect of Laser Scan Strategy on Distortion and Residual Stresses of Arches Made With Selective Laser Melting. In Proceedings of the American Society for Precision Engineering (ASPE), Raleigh, NC, USA, 27–30 June 2016.
77. Schmidt, M.; Merklein, M.; Bourell, D.; Dimitrov, D.; Hausotte, T.; Wegener, K.; Overmeyer, L.; Vollertsen, F.; Levy, G.N. Laser based additive manufacturing in industry and academia. *CIRP Ann.-Manuf. Technol.* **2017**, *66*, 561–583. [CrossRef]
78. Vastola, G.; Zhang, G.; Pei, Q.X.; Zhang, Y.W. Controlling of residual stress in additive manufacturing of Ti6Al4V by finite element modeling. *Addit. Manuf.* **2016**, *12*, 231–239. [CrossRef]
79. Ulbricht, A.; Altenburg, S.J.; Sprengel, M.; Sommer, K.; Mohr, G.; Fritsch, T.; Mishurova, T.; Serrano-Munoz, I.; Evans, A.; Hofmann, M.; et al. Separation of the Formation Mechanisms of Residual Stresses in LPBF 316L. *Metals* **2020**, *10*, 1234. [CrossRef]
80. Serrano-Munoz, I.; Evans, A.; Mishurova, T.; Sprengel, M.; Pirling, T.; Kromm, A.; Bruno, G. The Importance of Subsurface Residual Stress in Laser Powder Bed Fusion IN718. *Adv. Eng. Mater.* **2021**, 2100895. [CrossRef]
81. Pant, P.; Proper, S.; Luzin, V.; Sjostrom, S.; Simonsson, K.; Moverare, J.; Hosseini, S.; Pacheco, V.; Peng, R.L. Mapping of residual stresses in as-built Inconel 718 fabricated by laser powder bed fusion: A neutron diffraction study of build orientation influence on residual stresses. *Addit. Manuf.* **2020**, *36*, 101501. [CrossRef]
82. Thiede, T.; Cabeza, S.; Mishurova, T.; Nadammal, N.; Kromm, A.; Bode, J.; Haberland, C.; Bruno, G. Residual stress in selective laser melted Inconel 718: Influence of the removal from base plate and deposition hatch length. *Mater. Perform. Charact.* **2018**, *7*, 717–735. [CrossRef]
83. Mishurova, T.; Cabeza, S.; Thiede, T.; Nadammal, N.; Kromm, A.; Klaus, M.; Genzel, C.; Haberland, C.; Bruno, G. The influence of the support structure on residual stress and distortion in SLM Inconel 718 parts. *Metall. Mater. Trans. A* **2018**, *49*, 3038–3046. [CrossRef]
84. Casavola, C.; Campanelli, S.L.; Pappalettere, C. Preliminary investigation on distribution of residual stress generated by the selective laser melting process. *J. Strain Anal. Eng. Des.* **2009**, *44*, 93–104. [CrossRef]
85. Zaeh, M.F.; Branner, G. Investigations on residual stresses and deformations in selective laser melting. *Prod. Eng.* **2010**, *4*, 35–45. [CrossRef]
86. Moat, R.J.; Pinkerton, A.J.; Li, L.; Withers, P.J.; Preuss, M. Residual stresses in laser direct metal deposited Waspaloy. *Mater. Sci. Eng. A* **2011**, *528*, 2288–2298. [CrossRef]
87. Kruth, J.-P.; Deckers, J.; Yasa, E.; Wauthlé, R. Assessing and comparing influencing factors of residual stresses in selective laser melting using a novel analysis method. *Proc. Inst. Mech. Eng. Part B J. Eng. Manuf.* **2012**, *226*, 980–991. [CrossRef]
88. Liu, Y.; Yang, Y.; Wang, D. A study on the residual stress during selective laser melting (SLM) of metallic powder. *Int. J. Adv. Manuf. Technol.* **2016**, *87*, 647–656. [CrossRef]
89. Cheng, B.; Shrestha, S.; Chou, K.V. Stress and deformation evaluations of scanning strategy effect in selective laser melting. *Addit. Manuf.* **2016**, *12*, 240–251. [CrossRef]
90. Attallah, M.M.; Jennings, R.; Wang, X.; Carter, L.N. Additive manufacturing of Ni-based superalloys: The outstanding issues. *MRS Bull.* **2016**, *41*, 758–764. [CrossRef]
91. Ghasri-Khouzani, M.; Peng, H.; Rogge, R.; Attardo, R.; Ostiguy, P.; Neidig, J.; Billo, R.; Hoelzle, D.; Shankar, M.R. Experimental measurement of residual stress and distortion in additively manufactured stainless steel components with various dimensions. *Mater. Sci. Eng. A* **2017**, *707*, 689–700. [CrossRef]
92. Lu, J.; Society for Experimental Mechanics (U.S.). *Handbook of Measurement of Residual Stresses*; Fairmont Press: Lilburn, GA, USA; Prentice Hall PTR: Upper Saddle River, NJ, USA, 1996; p. xv. 238p.
93. Bragg, W.H.; Bragg, W.L. The reflection of X-rays by crystals. *Proc. R. Soc. Lond. A* **1913**, *88*, 428–438. [CrossRef]
94. Hutchings, M.T.; Withers, P.J.; Holden, T.M.; Lorentzen, T. *Introduction to the Characterization of Residual Stress by Neutron Diffraction*; CRC Press Taylor & Francis Group: Boca Raton, FL, USA, 2005. [CrossRef]
95. DIN-EN-15305. *Non-Destructive Testing—Test Method for Residual Stress Analysis by X-ray Diffraction*; German Institute for Standardization: Berlin, Germany, 2019. [CrossRef]
96. Mishurova, T.; Artzt, K.; Haubrich, J.; Requena, G.; Bruno, G. Exploring the Correlation between Subsurface Residual Stresses and Manufacturing Parameters in Laser Powder Bed Fused Ti-6Al-4V. *Metals* **2019**, *9*, 261. [CrossRef]
97. Genzel, C. Formalism for the Evaluation of Strongly Nonlinear Surface Stress-Fields by X-Ray-Diffraction Performed in the Scattering Vector Mode. *Phys. Status Solidi A* **1994**, *146*, 629–637. [CrossRef]
98. Genzel, C.; Denks, I.A.; Gibmeler, J.; Klaus, M.; Wagener, G. The materials science synchrotron beamline EDDI for energy-dispersive diffraction analysis. *Nucl. Instrum. Meth. A* **2007**, *578*, 23–33. [CrossRef]
99. Giessen, B.C.; Gordon, G.E. X-ray Diffraction—New High-Speed Technique Based on X-ray Spectrography. *Science* **1968**, *159*, 973–975. [CrossRef]
100. Apel, D.; Klaus, M.; Genzel, M.; Genzel, C. Rietveld-based energy-dispersive residual stress evaluation: Analysis of complex stress fields $\sigma_{ij}(z)$. *J. Appl. Crystallogr.* **2014**, *47*, 511–526. [CrossRef]

101. Hollmann, A.; Meixner, M.; Klaus, M.; Genzel, C. Concepts for nondestructive and depth-resolved X-ray residual stress analysis in the near-surface region of nearly single crystalline materials with mosaic structure. *J. Appl. Crystallogr.* **2021**, *54*, 22–31. [CrossRef]
102. Webster, G.A.; Wimpory, R.C. Non-destructive measurement of residual stress by neutron diffraction. *J. Mater. Process. Technol.* **2001**, *117*, 395–399. [CrossRef]
103. Kolbus, L.; Payzant, E.; Cornwell, P.; Watkins, T.; Babu, S.; Dehoff, R.; Lorenz, M.; Ovchinnikova, O.; Duty, C. Comparison of Residual Stresses in Inconel 718 Simple Parts Made by Electron Beam Melting and Direct Laser Metal Sintering. *Metall. Mater. Trans. A* **2015**, *46*, 1419–1432. [CrossRef]
104. Liu, W.C.; Yao, M.; Chen, Z.L.; Wang, S.G. Niobium segregation in Inconel 718. *J. Mater. Sci.* **1999**, *34*, 2583–2586. [CrossRef]
105. Bobbio, L.D.; Bocklund, B.; Otis, R.; Borgonia, J.P.; Dillon, R.P.; Shapiro, A.A.; McEnerney, B.; Liu, Z.K.; Beese, A.M. Characterization of a functionally graded material of Ti-6Al-4V to 304L stainless steel with an intermediate V section. *J. Alloys Compd.* **2018**, *742*, 1031–1036. [CrossRef]
106. Carroll, B.E.; Otis, R.A.; Borgonia, J.P.; Suh, J.O.; Dillon, R.P.; Shapiro, A.A.; Hofmann, D.C.; Liu, Z.K.; Beese, A.M. Functionally graded material of 304L stainless steel and inconel 625 fabricated by directed energy deposition: Characterization and thermodynamic modeling. *Acta Mater.* **2016**, *108*, 46–54. [CrossRef]
107. Serrano-Munoz, I.; Ulbricht, A.; Fritsch, T.; Mishurova, T.; Kromm, A.; Hofmann, M.; Wimpory, R.C.; Evans, A.; Bruno, G. Scanning Manufacturing Parameters Determining the Residual Stress State in LPBF IN718 Small Parts. *Adv. Eng. Mater.* **2021**, *23*, 2100158. [CrossRef]
108. Bayerlein, F.; Bodensteiner, F.; Zeller, C.; Hofmann, M.; Zaeh, M.F. Transient development of residual stresses in laser beam melting—A neutron diffraction study. *Addit. Manuf.* **2018**, *24*, 587–594. [CrossRef]
109. Syed, A.K.; Ahmad, B.; Guo, H.; Machry, T.; Eatock, D.; Meyer, J.; Fitzpatrick, M.E.; Zhang, X. An experimental study of residual stress and direction-dependence of fatigue crack growth behaviour in as-built and stress-relieved selective-laser-melted Ti6Al4V. *Mater. Sci. Eng. A* **2019**, *755*, 246–257. [CrossRef]
110. Brown, D.W.; Bernardin, J.D.; Carpenter, J.S.; Clausen, B.; Spernjak, D.; Thompson, J.M. Neutron diffraction measurements of residual stress in additively manufactured stainless steel. *Mater. Sci. Eng. A* **2016**, *678*, 291–298. [CrossRef]
111. Woo, W.; Kim, D.-K.; Kingston, E.J.; Luzin, V.; Salvemini, F.; Hill, M.R. Effect of interlayers and scanning strategies on through-thickness residual stress distributions in additive manufactured ferritic-austenitic steel structure. *Mater. Sci. Eng. A* **2019**, *744*, 618–629. [CrossRef]
112. Pratt, P.; Felicelli, S.D.; Wang, L.; Hubbard, C.R. Residual Stress Measurement of Laser-Engineered Net Shaping AISI 410 Thin Plates Using Neutron Diffraction. *Metall. Mater. Trans. A* **2008**, *39*, 3155–3163. [CrossRef]
113. Cottam, R.; Wang, J.; Luzin, V. Characterization of microstructure and residual stress in a 3D H13 tool steel component produced by additive manufacturing. *J. Mater. Res.* **2014**, *29*, 1978–1986. [CrossRef]
114. Hosford, W.F. *Mechanical Behavior of Materials*; Cambridge University Press: Cambridge, UK; New York, NY, USA, 2005; p. xx. 425p.
115. An, K.; Yuan, L.; Dial, L.; Spinelli, I.; Stoica, A.D.; Gao, Y. Neutron residual stress measurement and numerical modeling in a curved thin-walled structure by laser powder bed fusion additive manufacturing. *Mater. Des.* **2017**, *135*, 122–132. [CrossRef]
116. Fritsch, T.; Sprengel, M.; Evans, A.; Farahbod-Sternahl, L.; Saliwan-Neumann, R.; Hofmann, M.; Bruno, G. On the determination of residual stresses in additively manufactured lattice structures. *J. Appl Crystallogr.* **2021**, *54*, 228–236. [CrossRef] [PubMed]
117. Vrancken, B. *Study of Residual Stresses in Selective Laser Melting*; KU Leuven: Leuven, Belgium, 2016.
118. Zener, C.M. *Elasticity and Anelasticity of Metals*; University of Chicago Press: Chicago, IL, USA, 1948; p. x. 170p.
119. Ranganathan, S.I.; Ostoja-Starzewski, M. Universal elastic anisotropy index. *Phys. Rev. Lett.* **2008**, *101*, 055504. [CrossRef]
120. Kube, C.M. Elastic anisotropy of crystals. *AIP Adv.* **2016**, *6*, 095209. [CrossRef]
121. Ledbetter, H.; Migliori, A. A general elastic-anisotropy measure. *J. Appl. Phys.* **2006**, *100*, 063516. [CrossRef]
122. Chung, D.H.; Buessem, W.R. Elastic Anisotropy of Crystals. *J. Appl. Phys.* **1967**, *38*, 2010–2012. [CrossRef]
123. Heldmann, A.; Hoelzel, M.; Hofmann, M.; Gan, W.M.; Schmahl, W.W.; Griesshaber, E.; Hansen, T.; Schell, N.; Petry, W. Diffraction-based determination of single-crystal elastic constants of polycrystalline titanium alloys. *J. Appl. Crystallogr.* **2019**, *52*, 1144–1156. [CrossRef]
124. Dieter, G.E. *Mechanical Metallurgy*, 3rd ed.; McGraw-Hill: New York, NY, USA, 1986; p. xxiii. 751p.
125. Tallon, J.L.; Wolfenden, A. Temperature-Dependence of the Elastic-Constants of Aluminum. *J. Phys. Chem. Solids* **1979**, *40*, 831–837. [CrossRef]
126. Kamm, G.N.; Alers, G.A. Low-Temperature Elastic Moduli of Aluminum. *J. Appl. Phys.* **1964**, *35*, 327–330. [CrossRef]
127. Sutton, P.M. The Variation of the Elastic Constants of Crystalline Aluminum with Temperature between 63-Degrees-K and 773-Degrees-K. *Phys. Rev.* **1953**, *91*, 816–821. [CrossRef]
128. Goens, E. The main resilience constants of monocrystal of cooper, gold and lead. *Phys. Z.* **1936**, *37*, 321–326.
129. Lazarus, D. The Variation of the Adiabatic Elastic Constants of Kcl, Nacl, Cuzn, Cu, and Al with Pressure to 10,000 Bars. *Phys. Rev.* **1949**, *76*, 545–553. [CrossRef]
130. Schmunk, R.E.; Smith, C.S. Pressure Derivatives of the Elastic Constants of Aluminum and Magnesium. *J. Phys. Chem. Solids* **1959**, *9*, 100–112. [CrossRef]
131. Wang, Z.; Stoica, A.D.; Ma, D.; Beese, A.M. Diffraction and single-crystal elastic constants of Inconel 625 at room and elevated temperatures determined by neutron diffraction. *Mater. Sci. Eng. A* **2016**, *674*, 406–412. [CrossRef]

132. Haldipur, P.; Margetan, F.J.; Thompson, R.B. Estimation of single-crystal elastic constants from ultrasonic measurements on polycrystalline specimens. *AIP* **2004**, *700*, 1061–1068. [CrossRef]
133. Martin, G.; Ochoa, N.; Sai, K.; Herve-Luanco, E.; Cailletaud, G. A multiscale model for the elastoviscoplastic behavior of Directionally Solidified alloys: Application to FE structural computations. *Int. J. Solids Struct.* **2014**, *51*, 1175–1187. [CrossRef]
134. Jothi, S.; Merzlikin, S.V.; Croft, T.N.; Andersson, J.; Brown, S.G.R. An investigation of micro-mechanisms in hydrogen induced cracking in nickel-based superalloy 718. *J. Alloys Compd.* **2016**, *664*, 664–681. [CrossRef]
135. Kikuchi, M. Elastic Anisotropy and Its Temperature Dependence of Single-Crystals and Polycrystal of 18-12 Type Stainless-Steel. *Trans. Jpn. Inst. Met.* **1971**, *12*, 417–421. [CrossRef]
136. Bradfield, G. Comparison of Elastic Anisotropy of 2 Austenitic Steels. *J. Iron Steel Inst.* **1964**, *202*, 616.
137. Mangalick, M.C.; Fiore, N.F. Orientation Dependence of Dislocation Damping and Elastic Constants in Fe-18cr-Ni Single Crystals. *Trans. Metall. Soc. AIME* **1968**, *242*, 2363.
138. Voigt, W. Ueber die Beziehung zwischen den beiden Elasticitätsconstanten isotroper Körper. *Ann. Phys.* **1889**, *274*, 573–587. [CrossRef]
139. Reuss, A. Account of the liquid limit of mixed crystals on the basis of the plasticity condition for single crystal. *Z. Angew. Math. Mech.* **1929**, *9*, 49–58. [CrossRef]
140. Eshelby, J.D. The Determination of the Elastic Field of an Ellipsoidal Inclusion, and Related Problems. *Proc. R. Soc. Lond. Ser. A-Math. Phys. Sci.* **1957**, *241*, 376–396. [CrossRef]
141. Kröner, E. Berechnung Der Elastischen Konstanten Des Vielkristalls Aus Den Konstanten Des Einkristalls. *Z. Fur Phys.* **1958**, *151*, 504–518. [CrossRef]
142. Holden, T.M.; Holt, R.A.; Clarke, A.P. Intergranular strains in Inconel-600 and the impact on interpreting stress fields in beat steam-generator tubing. *Mater. Sci. Eng. A* **1998**, *246*, 180–198. [CrossRef]
143. Wagner, J.N.; Hofmann, M.; Wimpory, R.; Krempaszky, C.; Stockinger, M. Microstructure and temperature dependence of intergranular strains on diffractometric macroscopic residual stress analysis. *Mater. Sci. Eng. A* **2014**, *618*, 271–279. [CrossRef]
144. Daymond, M.R.; Bouchard, P.J. Elastoplastic deformation of 316 stainless steel under tensile loading at elevated temperatures. *Metall. Mater. Trans. A-Phys. Metall. Mater. Sci.* **2006**, *37*, 1863–1873. [CrossRef]
145. Dölle, H.; Hauk, V. Influence of Mechanical Anisotropy of Polycrystal (Texture) Upon Stress Evaluation by Means of X-rays. *Z. Fur Met.* **1978**, *69*, 410–417.
146. Slim, M.F.; Alhussein, A.; Zgheib, E.; Francois, M. Determination of single-crystal elasticity constants of the beta phase in a multiphase tungsten thin film using impulse excitation technique, X-ray diffraction and micro-mechanical modeling. *Acta Mater.* **2019**, *175*, 348–360. [CrossRef]
147. Brakman, C.M. Diffraction Elastic-Constants of Textured Cubic Materials—The Voigt Model Case. *Philos. Mag. A-Phys. Condens. Matter Struct. Defects Mech. Prop.* **1987**, *55*, 39–58. [CrossRef]
148. Welzel, U.; Mittemeijer, E.J. Diffraction stress analysis of macroscopically elastically anisotropic specimens: On the concepts of diffraction elastic constants and stress factors. *J. Appl. Phys.* **2003**, *93*, 9001–9011. [CrossRef]
149. Welzel, U.; Ligot, J.; Lamparter, P.; Vermeulen, A.C.; Mittemeijer, E.J. Stress analysis of polycrystalline thin films and surface regions by X-ray diffraction. *J. Appl. Crystallogr.* **2005**, *38*, 1–29. [CrossRef]
150. Welzel, U.; Freour, S.; Kumar, A.; Mittemeijer, E.J. Diffraction stress analysis using direction dependent grain-interaction models. *Mater. Sci. Forum* **2005**, *490–491*, 7–12. [CrossRef]
151. Gnäupel-Herold, T.; Creuziger, A.A.; Iadicola, M. A model for calculating diffraction elastic constants. *J. Appl. Crystallogr.* **2012**, *45*, 197–206. [CrossRef]
152. Wu, T.T. The effect of inclusion shape on the elastic moduli of a two-phase material. *Int. J. Solids Scrut.* **1966**, *2*, 1–8. [CrossRef]
153. Zhang, X.X.; Lutz, A.; Andrä, H.; Lahres, M.; Gan, W.M.; Maawad, E.; Emmelmann, C. Evolution of microscopic strains, stresses, and dislocation density during in-situ tensile loading od additively manufactured AlSi10Mg. *Int. J. Plast.* **2021**, *139*, 102946. [CrossRef]
154. Chen, W.; Voisin, T.; Zhang, Y.; Florien, J.B.; Spadaccini, C.M.; McDowell, D.L.; Zhu, T.; Wang, Y.M. Microscale residual stresses in additively manufactured stainless steel. *Nat. Commun.* **2019**, *10*, 4338. [CrossRef]
155. Zhang, D.C.; Wang, L.Y.; Zhang, H.; Maldar, A.; Zhu, G.M.; Chen, W.; Park, J.S.; Wang, J.; Zeng, X.Q. Effect of heat treatment on the tensile behavior of selective laser melted Ti-6Al-4V by in situ X-ray characterization. *Acta Mater.* **2020**, *189*, 93–104. [CrossRef]
156. Mishurova, T.; Artzt, K.; Haubrich, J.; Evsevleev, S.; Evans, A.; Meixner, M.; Munoz, I.S.; Sevostianov, I.; Requena, G.; Bruno, G. Connecting Diffraction-Based Strain with Macroscopic Stresses in Laser Powder Bed Fused Ti-6Al-4V. *Metall. Mater. Trans. A* **2020**, *51*, 3194–3204. [CrossRef]
157. Klaus, M.; Genzel, C. Reassessment of evaluation methods for the analysis of near-surface residual stress fields using energy-dispersive diffraction. *J. Appl. Crystallogr.* **2019**, *52*, 94–105. [CrossRef]
158. Aba-Perea, P.E.; Pirling, T.; Withers, P.J.; Kelleher, J.; Kabra, S.; Preuss, M. Determination of the high temperature elastic properties and diffraction elastic constants of Ni-base superalloys. *Mater. Des.* **2016**, *89*, 856–863. [CrossRef]
159. Song, J.; Wu, W.; Zhang, L.; He, B.; Lu, L.; Ni, X.; Long, Q.; Zhu, G. Role of scanning strategy on residual stress distribution in Ti-6Al-4V alloy prepared by selective laser melting. *Optik* **2018**, *170*, 342–352. [CrossRef]
160. Anderson, L.S.; Venter, A.M.; Vrancken, B.; Marais, D.; van Humbeeck, J.; Becker, T.H. Investigating the Residual Stress Distribution in Selective Laser Melting Produced Ti-6Al-4V using Neutron Diffraction. *Mater. Res. Proc.* **2018**, *4*, 73–78. [CrossRef]

161. Wang, L.; Felicelli, S.D.; Pratt, P. Residual stresses in LENS-deposited AISI 410 stainless steel plates. *Mater. Sci. Eng. A* **2008**, *496*, 234–241. [CrossRef]
162. Lesyk, D.A.; Martinez, S.; Mordyuk, B.N.; Dzhemelinskyi, V.V.; Lamikiz, A.; Prokopenko, G.I. Post-processing of the Inconel 718 alloy parts fabricated by selective laser melting: Effects of mechanical surface treatments on surface topography, porosity, hardness and residual stress. *Surf. Coat. Tech.* **2020**, *381*, 125136. [CrossRef]
163. Li, L.; Pan, T.; Zhang, X.C.; Chen, Y.T.; Cui, W.Y.; Yan, L.; Liou, F. Deformations and stresses prediction of cantilever structures fabricated by selective laser melting process. *Rapid Prototyp. J.* **2021**, *27*, 453–464. [CrossRef]
164. van Zyl, I.; Yadroitsava, I.; Yadroitsev, I. Residual Stress in Ti6al4v Objects Produced by Direct Metal Laser Sintering. *S. Afr. J. Ind. Eng.* **2016**, *27*, 134–141. [CrossRef]
165. Shah, K.; Haq, I.U.; Shah, S.A.; Khan, F.U.; Khan, M.T.; Khan, S. Experimental Study of Direct Laser Deposition of Ti-6Al-4V and Inconel 718 by Using Pulsed Parameters. *Sci. World J.* **2014**, *2014*, 841549. [CrossRef]
166. Levkulich, N.C.; Semiatin, S.L.; Gockel, J.E.; Middendorf, J.R.; DeWald, A.T.; Klingbeil, N.W. The effect of process parameters on residual stress evolution and distortion in the laser powder bed fusion of Ti-6Al-4V. *Addit. Manuf.* **2019**, *28*, 475–484. [CrossRef]
167. Serrano-Munoz, I.; Mishurova, T.; Thiede, T.; Sprengel, M.; Kromm, A.; Nadammal, N.; Nolze, G.; Saliwan-Neumann, R.; Evans, A.; Bruno, G. The residual stress in as-built Laser Powder Bed Fusion IN718 alloy as a consequence of the scanning strategy induced microstructure. *Sci. Rep.* **2020**, *10*, 14645. [CrossRef]
168. Mishurova, T.; Artzt, K.; Haubrich, J.; Requena, G.; Bruno, G. New aspects about the search for the most relevant parameters optimizing SLM materials. *Addit. Manuf.* **2019**, *25*, 325–334. [CrossRef]
169. Rangaswamy, P.; Griffith, M.L.; Prime, M.B.; Holden, T.M.; Rogge, R.B.; Edwards, J.M.; Sebring, R.J. Residual stresses in LENS® components using neutron diffraction and contour method. *Mater. Sci. Eng. A* **2005**, *399*, 72–83. [CrossRef]
170. Rangaswamy, P.; Holden, T.M.; Rogge, R.B.; Griffith, M.L. Residual stresses in components formed by the laserengineered net shaping (LENS®) process. *J. Strain Anal. Eng. Des.* **2003**, *38*, 519–527. [CrossRef]
171. Vrancken, B.; Wauthle, R.; Kruth, J.-P.; Humbeeck, J. Study of the influence of material properties on residual stress in selective laser melting. In Proceedings of the 24th International SFF Symposium: An Additive Manufacturing Conference, Austin, TX, USA, 12–14 August 2013; pp. 393–407.
172. Clausen, B.; Leffers, T.; Lorentzen, T. On the proper selection of reflections for the measurement of bulk residual stresses by diffraction methods. *Acta Mater.* **2003**, *51*, 6181–6188. [CrossRef]
173. Daymond, M.R.; Bourke, M.A.M.; VonDreele, R.B.; Clausen, B.; Lorentzen, T. Use of Rietveld refinement for elastic macrostrain determination and for evaluation of plastic strain history from diffraction spectra. *J. Appl. Phys.* **1997**, *82*, 1554–1562. [CrossRef]
174. Dye, D.; Stone, H.J.; Reed, R.C. Intergranular and interphase microstresses. *Curr. Opin. Solid State Mater Sci.* **2001**, *5*, 31–37. [CrossRef]
175. Clausen, B.; Lorentzen, T.; Leffers, T. Self-consistent modelling of the plastic deformation of FCC polycrystals and its implications for diffraction measurements of internal stresses. *Acta Mater.* **1998**, *46*, 3087–3098. [CrossRef]
176. Daymond, M.R.; Tome, C.N.; Bourke, M.A.M. Measured and predicted intergranular strains in textured austenitic steel. *Acta Mater.* **2000**, *48*, 553–564. [CrossRef]
177. ISO-21432. *Non-Destructive Tesing—Standard Test Method for Determining Residual Stresses by Neutron Diffraction*; International Organization for Standardization: Geneva, Switzerland, 2019. [CrossRef]
178. Choo, H.; Koehler, M.R.; White, L.P.; Ren, Y.; Morin, D.; Garlea, E. Influence of defect characteristics on tensile deformation of an additively manufactured stainless steel: Evolutions of texture and intergranular strain. *Mater. Sci. Eng. A* **2020**, *791*, 139637. [CrossRef]
179. Clausen, B.; Lorentzen, T.; Bourke, M.A.M.; Daymond, M.R. Lattice strain evolution during uniaxial tensile loading of stainless steel. *Mater. Sci. Eng. A* **1999**, *259*, 17–24. [CrossRef]
180. Wang, Y.M.; Voisin, T.; McKeown, J.T.; Ye, J.C.; Calta, N.P.; Li, Z.; Zeng, Z.; Zhang, Y.; Chen, W.; Roehling, T.T.; et al. Additively manufactured hierarchical stainless steels with high strength and ductility. *Nat. Mater.* **2018**, *17*, 63–71. [CrossRef]
181. Warwick, J.L.W.; Coakley, J.; Raghunathan, S.L.; Tailing, R.J.; Dye, D. Effect of texture on load partitioning in Ti-6Al-4V. *Acta Mater.* **2012**, *60*, 4117–4127. [CrossRef]
182. Stapleton, A.M.; Raghunathan, S.L.; Bantounas, I.; Stone, H.J.; Lindley, T.C.; Dye, D. Evolution of lattice strain in Ti-6Al-4V during tensile loading at room temperature. *Acta Mater.* **2008**, *56*, 6186–6196. [CrossRef]
183. Cho, J.R.; Dye, D.; Conlon, K.T.; Daymond, M.R.; Reed, R.C. Intergranular strain accumulation in a near-alpha titanium alloy during plastic deformation. *Acta Mater.* **2002**, *50*, 4847–4864. [CrossRef]
184. Voisin, T.; Calta, N.P.; Khairallah, S.A.; Forien, J.B.; Balogh, L.; Cunningham, R.W.; Rollett, A.D.; Wang, Y.M. Defects-dictated tensile properties of selective laser melted Ti-6Al-4V. *Mater. Des.* **2018**, *158*, 113–126. [CrossRef]
185. Vallejos, J.M.; Barriobero-Vila, P.; Gussone, J.; Haubrich, J.; Kelm, K.; Stark, A.; Schell, N.; Requena, G. In Situ High-Energy Synchrotron X-Ray Diffraction Reveals the Role of Texture on the Activation of Slip and Twinning during Deformation of Laser Powder Bed Fusion Ti-6Al-4V. *Adv. Eng. Mater.* **2021**, 2001556. [CrossRef]
186. Sangid, M.D.; Book, T.A.; Naragani, D.; Rotella, J.; Ravi, P.; Finch, A.; Kenesei, P.; Park, J.-S.; Sharma, H.; Almer, J.; et al. Role of heat treatment and build orientation in the microstructure sensitive deformation characteristics of IN718 produced via SLM additive manufacturing. *Addit. Manuf.* **2018**, *22*, 479–496. [CrossRef]

187. Capek, J.; Polatidis, E.; Knapek, M.; Lyphout, C.; Casati, N.; Pederson, R.; Strobl, M. The Effect of γ'' and δ Phase Precipitation on the Mechanical Properties of Inconel 718 Manufactured by Selective Laser Melting: An In Situ Neutron Diffraction and Acoustic Emission Study. *JOM* **2021**, *73*, 223–232. [CrossRef]
188. Brown, D.W.; Adams, D.P.; Balogh, L.; Carpenter, J.S.; Clausen, B.; King, G.; Reedlunn, B.; Palmer, T.A.; Maguire, M.C.; Vogel, S.C. In Situ Neutron Diffraction Study of the Influence of Microstructure on the Mechanical Response of Additively Manufactured 304L Stainless Steel. *Metall. Mater. Trans. A* **2017**, *48*, 6055–6069. [CrossRef]
189. Zhang, X.X.; Andra, H.; Harjo, S.; Gong, W.; Kawasaki, T.; Lutz, A.; Lahres, M. Quantifying internal strains, stresses, and dislocation density in additively manufactured AlSi10Mg during loading-unloading-reloading deformation. *Mater. Des.* **2021**, *198*, 109339. [CrossRef]

metals

MDPI

Article

Relative Density Measurement of PBF-Manufactured 316L and AlSi10Mg Samples via Eddy Current Testing

Marvin Aaron Spurek [1,2,*], Viet Hiep Luong [2], Adriaan Bernardus Spierings [1], Marc Lany [3], Gilles Santi [3], Bernard Revaz [3] and Konrad Wegener [2]

1 Inspire AG, Innovation Center for Additive Manufacturing Switzerland (ICAMS), Fürstenlandstrasse 122, 9014 St. Gallen, Switzerland; spierings@inspire.ethz.ch

2 Swiss Federal Institute of Technology, ETH Zurich, Institute of Machine Tools and Manufacturing (IWF), Leonhardstrasse 21, 8092 Zurich, Switzerland; viet.hiep.luong@alumni.ethz.ch (V.H.L.); wegener@iwf.mavt.ethz.ch (K.W.)

3 Sensima Inspection SARL, 2 Route Cité Ouest, 1196 Gland, Switzerland; marc.lany@sensima.ch (M.L.); gilles.santi@sensima.ch (G.S.); bernard.revaz@sensima.ch (B.R.)

* Correspondence: spurek@inspire.ethz.ch

Citation: Spurek, M.A.; Luong, V.H.; Spierings, A.B.; Lany, M.; Santi, G.; Revaz, B.; Wegener, K. Relative Density Measurement of PBF-Manufactured 316L and AlSi10Mg Samples via Eddy Current Testing. *Metals* **2021**, *11*, 1376. https://doi.org/10.3390/met11091376

Academic Editor: Giovanni Bruno

Received: 2 August 2021
Accepted: 26 August 2021
Published: 31 August 2021

Abstract: Powder bed fusion (PBF) is the most commonly used additive manufacturing process for fabricating complex metal parts via the layer-wise melting of powder. Despite the tremendous recent technological development of PBF, manufactured parts still lack consistent quality in terms of part properties such as dimensional accuracy, surface roughness, or relative density. In addition to process-inherent variability, this is mainly owing to a knowledge gap in the understanding of process influences and the inability to adequately control them during part production. Eddy current testing (ECT) is a well-established nondestructive testing technique primarily used to detect near-surface defects and measure material properties such as electrical conductivity in metal parts. Hence, it is an appropriate technology for the layer-wise measuring of the material properties of the fused material in PBF. This study evaluates ECT's potential as a novel in situ monitoring technology for relative part density in PBF. Parts made from SS316L and AlSi10Mg with different densities are manufactured on a PBF machine. These parts are subsequently measured using ECT, as well as the resulting signals correlated with the relative part density. The results indicate a statistically significant and strong correlation (316L: $r(8) = 0.998$, $p < 0.001$, AlSi10Mg: $r(8) = 0.992$, $p < 0.001$) between relative part density and the ECT signal component, which is mainly affected by the electrical conductivity of the part. The results indicate that ECT has the potential to evolve into an effective technology for the layer-wise measuring of relative part density during the PBF process.

Keywords: powder bed fusion (PBF); eddy current testing (ECT); part quality; in situ relative part density measurement; quality management

1. Introduction

In the last decade, additive manufacturing technologies have evolved from rapid prototyping to established manufacturing technologies that are increasingly used in industrial production. Powder bed fusion (PBF) is the most commonly used additive manufacturing process for fabricating metal parts, and has evolved to a state-of-the-art technology adopted in various industrial fields such as aerospace, medical, defence, as well as tool, and mould making [1,2]. The PBF process is characterized by the layer-wise melting of a powder bed using a laser beam; hence it enables the direct manufacturing of complex-shaped parts. Despite the tremendous recent technological development of PBF, Debroy et al. [3] identified the lack of consistent part quality as a major challenge impeding the wider commercial adoption of PBF. Part quality is mainly characterized by mechanical properties such as Young's modulus, and part properties such as dimensional accuracy, surface roughness, and relative density [4]. Process-inherent variability, a remaining knowledge gap in understanding the influences on part quality, and the inability to adequately control these influences during part production are the main reasons for inconsistent part quality [3]. In situ process monitoring technologies are expected to play a major role in obtaining consistent part quality in the future by improving the repeatability of the process as stated by Debroy et al. [3]. The majority of in situ monitoring technologies available today focus on monitoring the melt pool shape, size, or temperature [4–11]. Melt pool monitoring could be a reasonable choice for real-time controlling process parameters, such as laser power, which can help to avoid pore formation owing to local overheating of the melt pool [12]. However, owing to the nature of the PBF process, each solidified layer is remolten at least once. Hence, initially present defects can be healed, as determined by Ulbricht et al. [13], and new ones introduced during the layer remelting. Therefore, only the material integrity of the layer after remelting is relevant for the quality of the part. Although, there have been attempts to spatially map defects via melt pool monitoring [6,14], the detection of defects, which were not artificially introduced, lacks reliability. Owing to the inability to reliably monitor the part quality during production, the part certification for industrial use still requires a brute force approach, which involves time-consuming and expensive material tests such as CT scanning [3].

Eddy current testing (ECT) is a standardized nondestructive testing technique [15] that is adopted in various industries to control and certify the quality of electrically conductive parts. ECT can detect surface and near-surface defects such as cracks [16], as well as measure material properties such as electrical conductivity [17]. The industrial application of ECT to PBF remains limited to the quality control of parts during post-processing [18]. This limits the testable area of the part to the region near the surface. In contrast, integrating ECT into the PBF process cycle enables the layer-wise measurement of previously fused layers that provides quality information on the entire part volume after finishing the build cycle. Therefore, the ability of ECT to measure surface and near-surface defects fits well into the layer-wise build process in PBF. Available studies on ECT as a monitoring technology for PBF are summarized as follows: Todorov et al. [19] developed and patented a sensor array and method [20] and integrated it into a laboratory PBF machine. The system was tested in situ on parts with artificially introduced defects such as notches and regions of unfused material, being able to detect these defects. The authors claim that the regions of unfused material with sizes of $10 \times 3 \times 0.12$ mm^3 and $10 \times 3 \times 0.044$ mm^3 are good representations of lack of fusion in PBF. However, research on defect formation in PBF suggests that lack of fusion pores are often smaller than 100 µm in diameter [3]. Ehlers et al. [21] developed a sensor array using giant magnetoresistance (GMR) sensors and studied the capability of the system to detect artificially introduced surface defects in a wrought sample and in a PBF-manufactured sample both made from 316L stainless steel. However, the majority of defects in PBF-manufactued parts are located below the surface [22]. Both of the study presented by Todorov et al. [19] and Ehlers et al. [21] used parts with artificially introduced defects to demonstrate the capability of their ECT systems falling short of providing evidence that real defects caused during the PBF process can be detected reliably. In contrast to detecting individual defects, measuring the porosity of a certain part volume, including regions beyond the remelting zone, is an alternative approach to monitoring the process and part quality. Eisenbarth et al. [23] adopted ECT to identify unique keys designed by introducing porosity in 316L samples by adjusting the process parameters on an industrial PBF machine. However, the authors did not achieve the required sensitivity to distinguish between all different PBF samples sufficiently well and did not compare the ECT results to the relative sample density. Obatron et al. [24] studied ECT's ability to distinguish between PBF-manufactured lattice structures with different overall densities by measuring the electrical conductivity of the lattice. Hippert [25] used a similar approach correlating the electrical conductivity of the sample with its relative density. However, Hippert used samples with large artificially introduced holes and could only distinguish between samples with a density difference of 4% relative density.

Nevertheless, the approach of Hippert and Obaton et al. is promising if sufficiently small relative part density differences in PBF-manufactured parts can be measured.

In this study, the feasibility of ECT to measure relative part density variations owing to defects caused by the PBF process is evaluated. The prerequisite is that the electrical conductivity which can be measured by ECT is sufficiently correlated with the relative part density. To investigate this, parts made from AlSi10Mg and 316L stainless steel are manufactured on a PBF machine with varied process parameters to create different densities caused by lack of fusion and keyhole. The parts are subsequently measured using ECT and the results correlated with the relative part density.

2. Materials and Methods

2.1. Part Production

One cuboid part per scan speed with a size of 25 × 30 × 10 mm^3 was manufactured using a Concept Laser M2 PBF machine (Concept Laser GmbH, Lichtenfels, Germany) with the process parameters presented in Table 1. An additional cube per scan speed with a size of 10 × 10 × 10 mm^3 was fabricated in close proximity to the cuboid parts to study the pore distribution and the melt pool shape. Gas atomized PBF powders by the Carpenter Technology Corporation were used with particle size distributions of 15–45 µm (316L) and 10–60 µm (AlSi10Mg). All parts were fabricated on a 1-mm thick support structure onto 245 × 245-mm^2 build plates and afterwards removed by wire cutting.

Table 1. Powder bed fusion (PBF) process parameters.

Parameter		316L	AlSi10Mg
Scan speed	(mm/s)	500, 750, 1100, 1200 1250, 1300, 1400, 1750, 2250	650, 750, 850, 950, 1000 1050, 1150, 1250, 1500, 2000
Hatch distance	(µm)	75	100
Laser power	(W)	180	
Layer thickness	(µm)	30	
Laser spot diameter	(µm)	105	
Scan pattern	-	90° alternating	
Recoater type	-	Silicone reinforced brush	
Shielding gas	-	Nitrogen	

2.2. Part Characterization

The relative part density of the cuboid parts was measured via the Archimedes method, utilizing an AE200 balance with the measuring unit AB33360 (Mettler Toledo Inc., Columbus, OH, USA). The cubes were cut perpendicular to the scan direction of the top layer, embedded in epoxy resin, ground using SiC grinding paper (320, 600 and 1200 grit sizes), and polished up to 0.5 µm using SiO_2 suspension. Images showing the porosity in the cross sections were taken at 50× magnification using a DM6 optical microscope (Leica Microsystems GmbH, Wetzlar, Germany). To reveal the melt pool boundaries, the 316L parts were etched in V2A etchant at 60 °C for 60 s, and the AlSi10Mg parts were etched in NaOH at room temperature for 12 s. The melt pools were subsequently characterized using the aforementioned optical microscope.

2.3. Eddy Current Testing

2.3.1. Measurement Principle

The physical theory of ECT is explained by Maxwell's equations [26], which are not comprehensively discussed here. The simplified principle is explained using Figure 1a as follows.

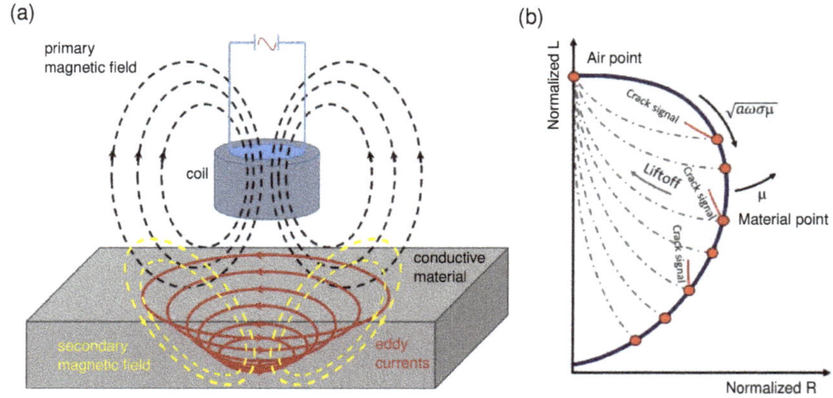

Figure 1. (**a**) Eddy current testing (ECT) measurement principle. (**b**) Impedance plane representation. Adapted from Hippert [25].

A coil is excited with an alternating current with predefined amplitude and frequency f, which generates a time-varying primary magnetic field around it. In proximity to conductive material, the primary magnetic field induces eddy currents in the material. The eddy currents create a secondary magnetic field, which opposes the primary magnetic field that alters the impedance of the coil. Discontinuities of the electrical conductivity σ and magnetic permeability μ in the material triggered by defects such as cracks affect the eddy currents and the secondary field. The impedance change of the coil owing to the secondary magnetic field is measured and evaluated to characterize the material discontinuity. The impedance Z of the coil is calculated as:

$$Z = \frac{V}{I} = R + iX = R + i\omega L \tag{1}$$

where V, I, R, X, ω, and L represent the voltage across the coil, current in the coil, coil resistance, reactance, angular frequency and coil inductance, respectively [25]. The impedance is typically visualized in a normalized form relative to the impedance in air, as presented in Figure 1b. Reducing the distance between the coil and the conductive material leads to a signal transition from the air point to the material point along the lift-off direction. The location of the material point in the impedance plane changes as a function of $\sqrt{a\omega\sigma\mu}$, where a, ω, σ, and μ denote the coil radius, angular frequency, electrical conductivity, and magnetic permeability of the material, respectively. The red dots moving from the air point along the blue curve presented in Figure 1b represent material points of different alloys. Signal responses to defects such as cracks or changes in electrical conductivity are identified by characteristic phase angles in the impedance plane, which differ from the phase angle of a lift-off variation, as presented in Figure 1b. By rotating the signals, i.e., adjusting the phase angle in the impedance plane for a given material, undesired alterations in lift-off during ECT can be shifted to the x-component of the signal in the impedance plane. Therefore, the y-component of the signal in the impedance plane represents the lift-off independent signal response owing to the property of interest. The principle is explained comprehensively in the book of Udpa et al. [27].

2.3.2. Measurement Equipment and Experimental Setup

A standard UPEC tester made by the Sensima Inspection SARL was adopted for the ECT, as presented in Figure 2a. The selected sensor was a ferrite rod coil sensor with a 3-mm diameter (L = 47 μH) operated in absolute mode. Absolute mode means that the measured signal is simply the value of the coil impedance itself [26]. The sensor was connected to the UPEC tester in a bridge configuration with an identical coil for balancing, including two 50-Ω resistors. The standard penetration depth δ is defined as the depth at which the eddy current density decreases to approximately 37% of its surface value [16]. It is calculated as:

$$\delta = \frac{1}{\sqrt{\pi f \sigma \mu}} \tag{2}$$

where f is the frequency of the excitation current, σ the electrical conductivity and $\mu = \mu_0\mu_r$ the magnetic permeability of the material, with μ_r being the relative magnetic permeability and $\mu_0 = 1.256 \times 10^6$ H/m being the magnetic permeability in vacuum [26]. An excitation frequency of $f = 201.6$ kHz was set to ensure a sufficiently high penetration depth for both materials used. The penetration depths for 316L and AlSi10Mg, which were calculated according to Equation (2) are presented in Table 2.

Table 2. Calculated standard penetration depths for 316L and AlSi10Mg. * assumed because Aluminum alloys are paramagnetic [26].

Parameter			316L	AlSi10Mg
Excitation frequency	f	(kHz)	201.6	201.6
Electrical conductivity	σ	(MS/m)	1.38 [26]	12.85 [27]
Relative magnetic permeability	μ_r	(-)	1.02 [26]	1.00 *
Standard penetration depth	δ	(µm)	945	313

The system was mounted onto a X-Y table laboratory test bench illustrated in Figure 2b. Linear encoders with a resolution of 0.1 mm were mounted on the axes and connected to the UPEC tester to map measured data to the sensor position during an acquisition. The cuboid parts were clamped onto the test bench as presented in Figure 2b to ensure a constant lift-off of 0.5 mm, which is the distance between the sensor and part surface. The as-built parts were orientated with the final layer created in the PBF process facing the sensor. Two-dimensional images were obtained by performing a raster scan with a pitch in the y-direction of 0.5 mm on the test bench guiding the sensor over the parts. Prior to the experiments, the lift-off phase angle in the impedance plane was determined for both materials by measuring one of the respective parts with different lift-offs. Accordingly, counter clockwise phase rotation angles of 69° and 79° for 316L and AlSi10Mg were obtained, respectively. The respective phase rotation filter was applied to the ECT data, and the absolute value of the lift-off independent signal component was analyzed, which is named rotated signal in the following chapters.

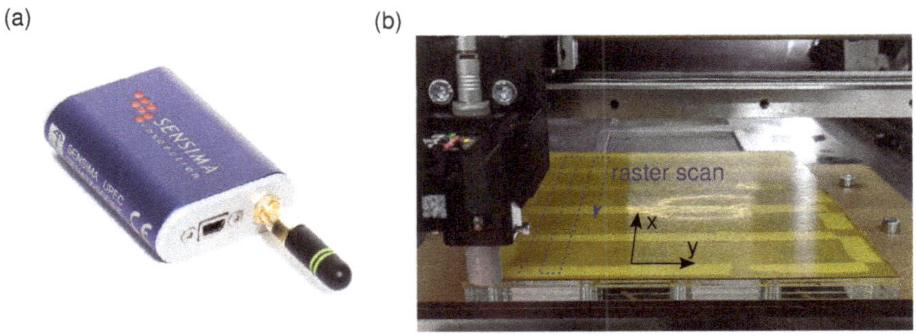

Figure 2. Measurement setup used for ECT. (**a**) Standard UPEC tester of Sensima Inspection SARL. (**b**) X-Y table laboratory test bench.

3. Results

3.1. Relative Part Density

To assess the feasibility of ECT in measuring the relative density of PBF-manufactued parts, process parameters (Table 1) were chosen to cover a wide range of part densities and to include the primary causes of porosity in the PBF process. Figure 3 presents the relative density of the cuboid parts manufactured from AlSi10Mg and 316L, as a function of the scan speed.

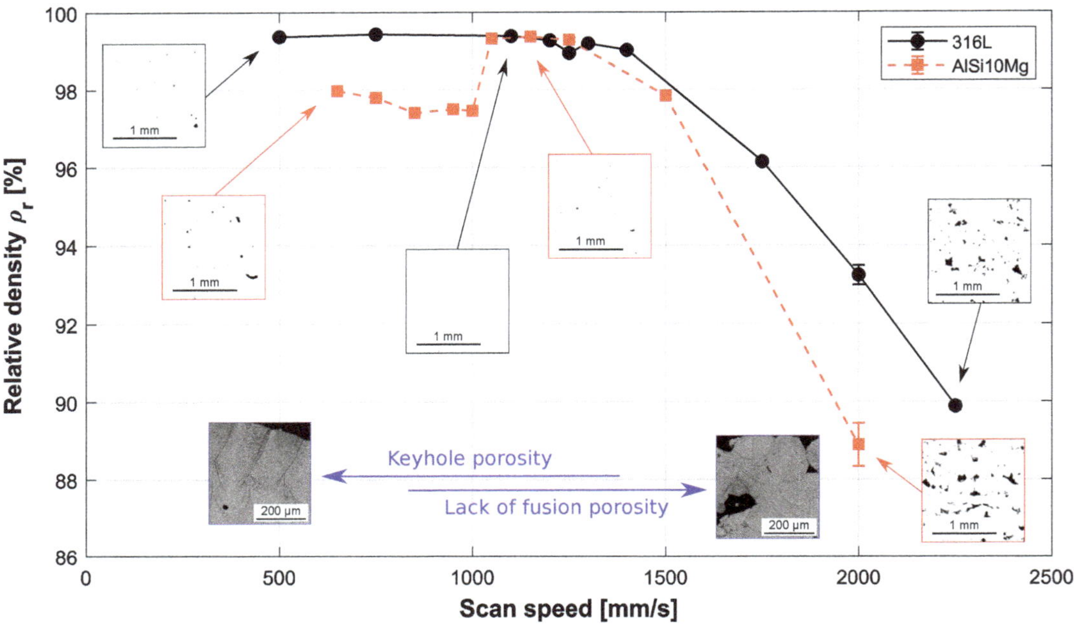

Figure 3. Relative part density of parts manufactured from AlSi10Mg and 316L powder using varied scan speeds. The error bars depict the standard deviations of three density measurements per part. The micrograph and melt pool images were taken from x-z cross sections of cubes ($10 \times 10 \times 10$ mm^3) manufactured in close proximity to the parts on the build plate, using the same process parameters.

The variation of the scan speed yields relative part densities covering a range of 89–99.5%. According to the widely used definition of the volumetric energy density by Stoffregen et al. [28], the energy input into the melt pool is inversely related to the scan speed. By adjusting the scan speed, the resulting relative density of the part can be controlled. In the process region, where insufficient energy input triggers lack of fusion, increasing the scan speed decreases the relative part density, which is consistent with the data presented in Figure 3. The micrograph images presented at the highest scan speeds in Figure 3 exhibit several irregularly shaped pores. These pores are characteristic for lack of fusion as studied extensively in the literature [22,29–32]. The lack of fusion is primarily triggered by the insufficient penetration of the melt pool into the previous layer as determined in different studies [33–35]. At the other end of the process window, low scan speeds yield an increased energy input into the melt pool. Consequently, the temperature of the melt pool at the center of the laser beam can reach the boiling temperature of the material, which substantially increases the amount of material evaporation and triggers the formation of a keyhole-shaped melt pool, as demonstrated by King et al. [36]. In the keyhole region of the process window, the collapse of the keyhole can lead to the formation of entrapped vapor, which causes keyhole porosity in the solidified material [36]. The micrograph images presented at the lowest scan speed in Figure 3 exhibit an increased amount of keyhole porosity, which was confirmed by melt pool shape analysis. Exemplaric images of the melt pool shapes in the different regions of the process window are presented in Figure 3. For most industrial use cases, PBF process windows are experimentally obtained to ensure dense parts; hence, the keyhole formation and lack of fusion are circumvented via the appropriate process parameter selection. Nevertheless, geometrical features of the parts, such as overhang areas, can lead to keyhole porosity owing to the local overheating of the material [37] while process influences, such as smoke obscuring the laser beam, can cause lack of fusion porosity [38]. Hence, relative part density monitoring systems for PBF must at minimum be able to detect porosity coming from these causes.

3.2. Eddy Current Testing

The 2D image obtained from measuring the 10 cuboid parts made from 316L using ECT is illustrated in Figure 4. The positions represented on the x- and y-axes reflect the position of the sensor during the raster scan. The signal weakening that is visible at the edge of the parts is owing to the edge effect, which has been extensively studied in the literature on ECT [39–41]. The size of the edge effect can be reduced, e.g., by adjusting the sensor design [41]; however, this is not the focus of this work.

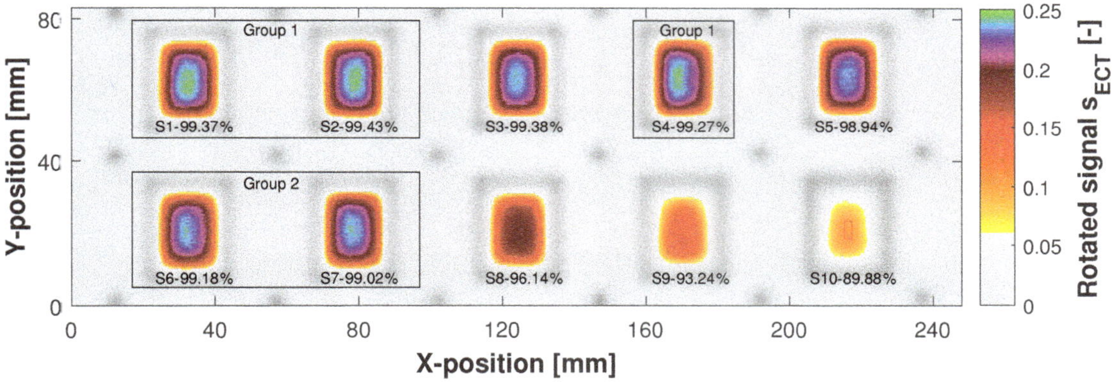

Figure 4. Image obtained from the 2D scan of the 316L parts. The data contain the rotated signal mapped to the X and Y positions of the sensor during the raster scan, as described in Section 2.3. Below each part, the part description and relative density are presented. Non-significant differences between mean rotated signals of samples are indicated by grouping the respective parts. One sample used for the statistical analysis obtained from the central region of the part 2×5 mm^2 (X × Y) is highlighted with a red rectangle on part S10.

Therefore, the following analyses are focused on the values of the rotated signal in the central region of the parts, where there is no influence exerted by the edges. This was ensured by measuring the size of the edge effect for the specific materials and making the parts sufficiently large. Samples of $n = 209$ individual measurements per part were selected from an area of 2×5 mm^2 (X × Y) at the center of each part. The corresponding area is highlighted with a red rectangle on part S10 in Figure 4. To determine statistically significant differences between sample means, Welch's analysis of variance and the Games–Howell post hoc test were applied.

The results of the statistical analysis are summarized in Table 3, which presents the p-values of the pairwise multiple comparisons between samples S1–S10 using the Games–Howell post hoc test. $p < 0.05$ is considered statistically significant. Most of the pairwise comparisons are statistically significant, which means that the respective parts can be statistically significantly distinguished from each other by the rotated signal. The relative density ρ_r of the samples is reported alongside the sample description in Table 3. Even parts with small differences in relative density, such as S2 and S3, with a difference in relative density of 0.05%, can be statistically significantly distinguished by the rotated signal ($p < 0.01$). Non-significant sample differences, i.e., pairwise comparisons of samples with $p \geq 0.05$, are indicated by grouping the respective parts in Figure 4. The parts of each group have differences in relative density smaller than 0.2%. Although S3 has a similar relative density as the parts in Group 1, the rotated signal is significantly smaller in the central region. This is probably attributed to the limitations of the Archimedes density measurement, which only provides the relative density of the entire part, but not of the near-surface volume fraction in the center of the part, which is measured by ECT. The relative density of the entire part is not necessarily equal to the relative density in the near-surface region owing to a potential inhomogeneous pore distribution within PBF-manufactured parts, as determined by Carlton et al. [22].

Table 3. Results of the pairwise multiple comparisons using the Games–Howell post hoc test based on Welch's analysis of variance ($F(9,\ 844) = 59847$, $p < 0.01$). The samples with sample size $n = 209$ were obtained from a 2 × 5-mm^2 (X × Y) region in the center of each of the 10 parts (316L). ** $p < 0.01$, * $p < 0.05$, ^{ns} not significant.

Sample		**S2**	**S3**	**S4**	**S5**	**S6**	**S7**	**S8**	**S9**	**S10**
	ρ_r (%)	**99.43**	**99.38**	**99.27**	**98.94**	**99.18**	**99.02**	**96.14**	**93.24**	**89.88**
S1	99.37	0.12 ^{ns}	<0.01 **	0.16 ^{ns}	<0.01 **	<0.01 **	<0.01 **	<0.01 **	<0.01 **	<0.01 **
S2	99.43		<0.01 **	<0.01 **	<0.01 **	<0.01 **	<0.01 **	<0.01 **	<0.01 **	<0.01 **
S3	99.38			<0.01 **	<0.01 **	<0.01 **	<0.01 **	<0.01 **	<0.01 **	<0.01 **
S4	99.27				<0.01 **	<0.01 **	<0.01 **	<0.01 **	<0.01 **	<0.01 **
S5	98.94					<0.01 **	<0.01 **	<0.01 **	<0.01 **	<0.01 **
S6	99.18						0.87 ^{ns}	<0.01 **	<0.01 **	<0.01 **
S7	99.02							<0.01 **	<0.01 **	<0.01 **
S8	96.14								<0.01 **	<0.01 **
S9	93.24									<0.01 **

The 2D image obtained from measuring the 10 cuboid AlSi10Mg parts using ECT is presented in Figure 5. The same statistical analysis used for the 316L parts was conducted on the AlSi10Mg parts, and the results are summarized in Table 4. Similar to the 316L parts, most of the AlSi10Mg parts can be statistically significantly distinguished from each other by the rotated signal. The parts in each group have relative density variations smaller than 0.1%. However, the differences in relative density between part A3, A4, and A5 are also smaller than 0.1%, and these parts can be statistically significantly distinguished from each other by the rotated signal. As previously discussed, this is probably attributed to the different fractions of the part volume measured by the Archimedes density measurement and ECT. Hence, it is inferred that parts with small differences in relative density within the near-surface region can be statistically significantly distinguished by the rotated signal obtained from ECT.

The samples containing $n = 209$ individual measurements from the aforementioned area of 2 × 5 mm^2 (X × Y) at the center of each part were used to analyze the correlation between the rotated signal and the relative density of the 316L and AlSi10Mg parts. In Figure 6, the correlation plots for 316L (a) and AlSi10Mg (b) are presented. The correlation is strong and significant for both materials with Pearson correlation coefficients of $r(8) = 0.998$, $p < 0.001$ (316L) and $r(8) = 0.992$, $p < 0.001$ (AlSi10Mg), respectively. Based on the theory of Dodd et al. [42], the impedance of the coil solely depends on the lift-off and electrical conductivity of the material for a fixed magnetic permeability, coil size, and excitation frequency, as given in the experiments. The influence of lift-off was eliminated from the data by phase rotation, such that the presented rotated signal primarily depends on the electrical conductivity of the material. Therefore, by measuring the electrical conductivity of a part using ECT, its relative density can be determined using alloy-specific correlation curves, equal to those presented in Figure 6. These correlation curves serve the calibration, i.e., converting the respective rotated signal to relative density.

The correlation between the electrical conductivity and relative part density has already be demonstrated in the literature for metal foams [43,44], as well as the PBF-manufactured parts with internal cavities [25,45]. However, the correlations presented in this work are based on PBF-manufactured parts with relative density variations caused by introducing porosity owing to the two primary causes in PBF, which are lack of fusion and keyhole [3]. Hence, the data represent more realistic conditions for the PBF process.

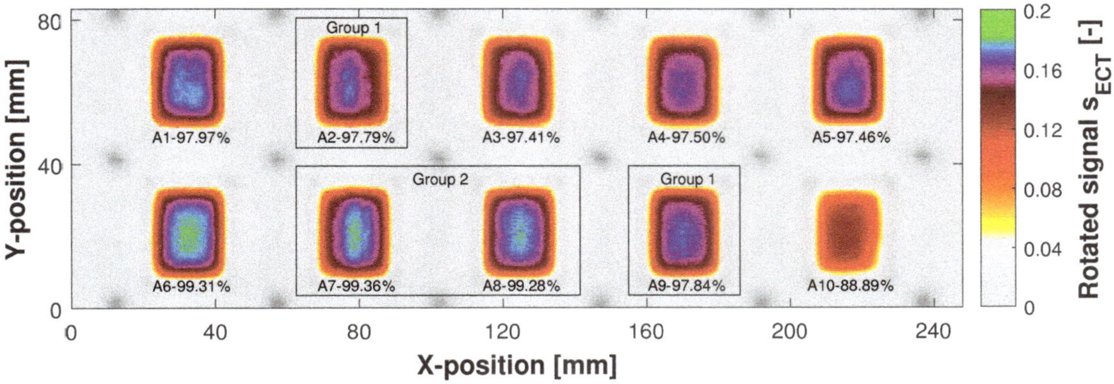

Figure 5. Image obtained from the 2D scan of the AlSi10Mg parts. The data contain the rotated signal mapped to the X and Y position of the sensor during the raster scan as described in Section 2.3. Below each part, the part description and relative density are presented. Non-significant differences between mean rotated signals of samples are indicated by grouping the respective parts.

Table 4. Results of the pairwise multiple comparisons via the Games–Howell post hoc test based on Welch's analysis of variance ($F(9, 845) = 18760$, $p < 0.01$). The samples with sample size $n = 209$ were obtained from a $2 \times 5\text{-mm}^2$ (X × Y) region in the center of each of the 10 parts (AlSi10Mg). ** $p < 0.01$, * $p < 0.05$, ^{ns} not significant.

Sample		**A2**	**A3**	**A4**	**A5**	**A6**	**A7**	**A8**	**A9**	**A10**
	ρ_r (%)	**97.79**	**97.41**	**97.50**	**97.46**	**99.31**	**99.36**	**99.28**	**97.84**	**88.89**
A1	97.97	<0.01 **	<0.01 **	<0.01 **	<0.01 **	<0.01 **	<0.01 **	<0.01 **	<0.01 **	<0.01 **
A2	97.79		<0.01 **	<0.01 **	<0.01 **	<0.01 **	<0.01 **	<0.01 **	0.05 ^{ns}	<0.01 **
A3	97.41			<0.01 **	<0.01 **	<0.01 **	<0.01 **	<0.01 **	<0.01 **	<0.01 **
A4	97.50				<0.01 **	<0.01 **	<0.01 **	<0.01 **	<0.01 **	<0.01 **
A5	99.46					<0.01 **	<0.01 **	<0.01 **	<0.01 **	<0.01 **
A6	99.31						<0.01 **	<0.01 **	<0.01 **	<0.01 **
A7	99.36							1.00 ^{ns}	<0.01 **	<0.01 **
A8	99.28								<0.01 **	<0.01 **
A9	99.84									<0.01 **

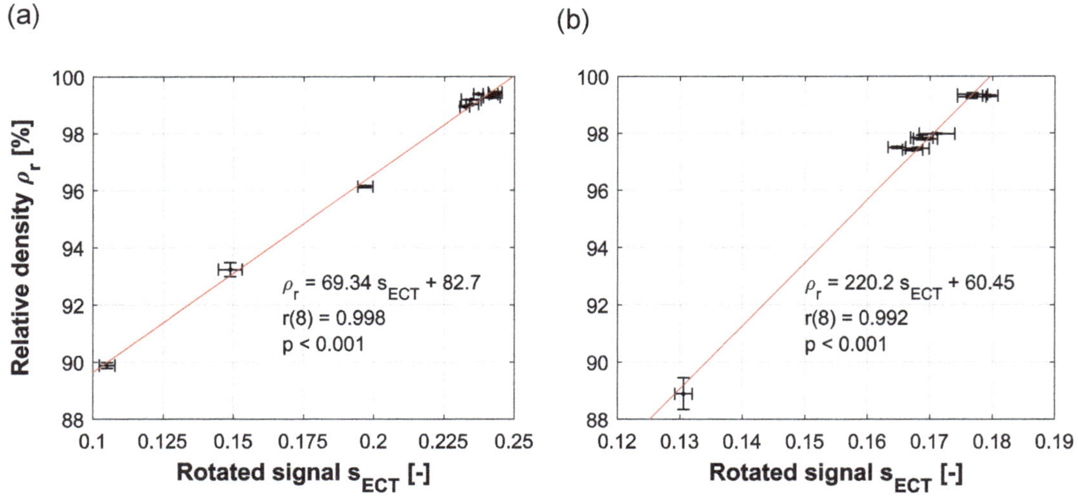

Figure 6. (**a**) Correlation of the rotated signal and the relative part density of 316L parts. (**b**) Correlation of the rotated signal and the relative part density of AlSi10Mg parts. The ECT data presented is extracted from a 2 × 5-mm^2 (X × Y) region in the center of each part to exclude the edge effect. The error bars depict the standard deviation of the rotated signal across the extracted area (horizontal) and the standard deviation of three density measurements per part (vertical). The curve was fitted to the given data via the least squares method.

4. Discussion

In this study, part densities were measured using the Archimedes principle, which provides the relative part density averaged across the entire volume of the part. ECT measurements were performed on the last layers of the parts, and only the central region was considered for the correlation analysis to eliminate the influence of the edge effect. The pore distribution within PBF-manufactured parts is not homogeneous [22]. Therefore, the relative density of the entire part measured by the Archimedes principle does not necessarily accurately represent the actual relative density in the near-surface volume fraction measured by ECT. Hence, it can be assumed that a large fraction of the signal variance indicated by the standard deviations of the rotated signal presented in Figure 6 is owing to actual small relative density differences in the parts. Accordingly, only a small fraction of this signal variance is caused by limitations of the ECT system itself, which could solely be verified by comparing relative density and ECT data obtained from the same fraction of the part volume.

To assess the limitations of the ECT system, the uncertainty of the determined relative density via ECT due to the instrument noise U_I is calculated based on the principle of the expanded uncertainty explained in the Guide to the Expression of Uncertainty in Measurement (GUM) [46] as:

$$U_I = ku_I \tag{3}$$

where k is the coverage factor and u_I is the standard uncertainty. A coverage factor $k = 3$ is selected, which corresponds to a confidence level of 99.7% given the sampling distribution of the sample mean is normal. The sampling distribution of the mean is the distribution of the mean as a random variable derived from random samples with the size n. According to the central limit theorem [47], normality of the sampling distribution of the mean can be assumed, because of the sufficiently large sample size $n = 209$ per part measured. The standard uncertainty u_I owing to the instrument noise is calculated as:

$$u_I = \frac{\sigma_I}{\sqrt{n}} \tag{4}$$

where σ_I is the standard deviation of the instrument noise and n is the sample size, which is $n = 209$ in this study. Equations (3) and (4) are combined, and the unit of U_I is converted to % relative density by multiplying it with sensitivity b, which is the slope of the respective least-square fitted curve presented in Figure 6, as:

$$U_I = kb\frac{\sigma_I}{\sqrt{n}} \tag{5}$$

The results of the uncertainty calculation are presented in Table 5. The uncertainty in the determined relative density via ECT due to the instrument noise U_I is 0.02% (316L) and 0.06% (AlSi10Mg), which indicates that small differences in average relative part density can be measured at high confidence levels.

Table 5. Estimation of the uncertainty of the determined relative density due to the instrument noise U_I.

Parameter			316L	AlSi10Mg
Standard deviation of the instrument noise	σ_n	(-)	0.0013	0.0013
Sensitivity	b	(%)	220.20	69.34
Uncertainty of the determined relative density via ECT	U_I	(%)	0.02%	0.06%

Note that the error bars in Figure 6 are significantly larger than these estimated uncertainties as they represent the standard deviations of the observed signals, which include fluctuations due to inhomogeneities, e.g., local density differences, in the material itself. Furthermore, the aforementioned standard deviations simply describe the data variability within the actual sample, whereas the calculated uncertainty U_I refers to the accuracy with which the mean value of each sample can be determined. The following remarks have to be considered while interpreting the results of this study. Standard instrument settings were adopted, that means that by optimizing parameters, such as gains, the sensitivity can be further increased. Moreover, the sensitivity can be improved by reducing the lift-off and applying more sophisticated signal processing methods, as well as an improved sensor design. Hence, this study demonstrates that such an ECT system mounted onto the recoater of a PBF-machine has the potential to evolve into a effective technology for layer-wise measuring the relative density of PBF-manufactured parts.

5. Conclusions

This study investigates the feasibility of measuring the relative part density of PBF-manufactured parts by ECT. Parts made from AlSi10Mg and 316L were manufactured with different process parameters yielding different densities. The relative part density differences were triggered by the two primary reasons for porosity in PBF, which are lack of fusion and keyhole. The parts were measured by ECT, and the results were correlated with the relative part density.

The ECT signal component, which mainly contains the electrical conductivity of the parts is strongly and significantly correlated with the relative part density for both 316L ($r(8) = 0.998$, $p < 0.001$) and AlSi10Mg ($r(8) = 0.992$, $p < 0.001$). Considering that the measured relative density is an averaged relative density across the entire part volume, and that the ECT data were obtained on the final layers, the correlation between the relative part density and the ECT signal is excellent. The sensitivity of the system can be further increased by reducing the lift-off, applying advanced signal processing methods, and adopting an improved sensor design.

This study presents a pathway for directly layer-wise measuring relative part density during the PBF process using an ECT system mounted on the recoater of a PBF-machine. Because the adopted ECT system is a compliant nondestructive testing instrument, it can furthermore serve the direct qualification and certification of PBF-manufactured parts. By measuring some additionally introduced test geometries to a build job, the system can also be used to assess the process window stability or to monitor the machine condition. Compared to monitoring techniques such as melt pool monitoring, where it is challenging to translate the large amount of generated data to relevant part properties, ECT can provide relevant and compliant part and process information obtained from direct measurements during the PBF process.

Author Contributions: Conceptualization, M.A.S. and A.B.S.; methodology, M.A.S., A.B.S., M.L., G.S., B.R.; formal analysis, M.A.S.; investigation, M.A.S., V.H.L.; resources, K.W.; data curation, M.A.S.; writing—original draft preparation, M.A.S.; writing—review and editing, M.A.S., A.B.S., G.S., K.W.; visualization, M.A.S.; supervision, K.W.; project administration, M.A.S.; funding acquisition, A.B.S., B.R. All authors have read and agreed to the published version of the manuscript.

Funding: Open Access funding provided by ETH Zurich. The authors like to thank Innosuisse—Swiss Innovation Agency for co-financing the investigations within the frame of innovation project 33657.1.

Institutional Review Board Statement: Not applicable.

Informed Consent Statement: Not applicable.

Data Availability Statement: The data presented in this study are available from the corresponding author on reasonable request.

Conflicts of Interest: The authors declare no conflict of interest.

References

1. Milewski, J.O. *Additive Manufacturing of Metals*; Springer Series in Materials Science; Springer International Publishing: Cham, Switzerland, 2017; Volume 258. [CrossRef]
2. Wohlers, T.; Campbell, R.I.; Diegel, O.; Huff, R.; Kowen, J. *Wohlers Report 2020: 3D Printing and Additive Manufacturing State of the Industry*; Wohlers Associates: Fort Collins, CO, USA, 2020.
3. DebRoy, T.; Wei, H.L.; Zuback, J.S.; Mukherjee, T.; Elmer, J.W.; Milewski, J.O.; Beese, A.M.; Wilson-Heid, A.; De, A.; Zhang, W. Additive manufacturing of metallic components–Process, structure and properties. *Prog. Mater. Sci.* **2018**, *92*, 112–224. [CrossRef]
4. Grasso, M.; Colosimo, B.M. Process defects and in situ monitoring methods in metal powder bed fusion: A review. *Meas. Sci. Technol.* **2017**, *28*, 044005. [CrossRef]
5. Yan, Z.; Liu, W.; Tang, Z.; Liu, X.; Zhang, N.; Li, M.; Zhang, H. Review on thermal analysis in laser-based additive manufacturing. *Opt. Laser Technol.* **2018**, *106*, 427–441. [CrossRef]
6. Mitchell, J.A.; Ivanoff, T.A.; Dagel, D.; Madison, J.D.; Jared, B. Linking pyrometry to porosity in additively manufactured metals. *Addit. Manuf.* **2020**, *31*, 100946. [CrossRef]
7. Renken, V.; Lübbert, L.; Blom, H.; Von Freyberg, A.; Fischer, A. Model assisted closed-loop control strategy for selective laser melting. *Procedia CIRP* **2018**, *74*, 659–663. [CrossRef]
8. Fisher, B.A.; Lane, B.; Yeung, H.; Beuth, J. Toward determining melt pool quality metrics via coaxial monitoring in laser powder bed fusion. *Manuf. Lett.* **2018**, *15*, 119–121. [CrossRef]
9. Colosimo, B.M.; Grossi, E.; Caltanissetta, F.; Grasso, M. Penelope: A Novel Prototype for In Situ Defect Removal in LPBF. *JOM* **2020**, *72*, 1332–1339. [CrossRef]
10. Özel, T.; Shaurya, A.; Altay, A.; Yang, L. Process monitoring of meltpool and spatter for temporal-spatial modeling of laser powder bed fusion process. *Procedia CIRP* **2018**, *74*, 102–106. [CrossRef]
11. Mohr, G.; Altenburg, S.J.; Ulbricht, A.; Heinrich, P.; Baum, D.; Maierhofer, C.; Hilgenberg, K. In-situ defect detection in laser powder bed fusion by using thermography and optical tomography—comparison to computed tomography. *Metals* **2020**, *10*, 103. [CrossRef]
12. Martin, A.A.; Calta, N.P.; Khairallah, S.A.; Wang, J.; Depond, P.J.; Fong, A.Y.; Thampy, V.; Guss, G.M.; Kiss, A.M.; Stone, K.H.; et al. Dynamics of pore formation during laser powder bed fusion additive manufacturing. *Nat. Commun.* **2019**, *10*, 1–10. [CrossRef] [PubMed]

13. Ulbricht, A.; Mohr, G.; Altenburg, S.J.; Oster, S.; Maierhofer, C.; Bruno, G. Can potential defects in lpbf be healed from the laser exposure of subsequent layers? A quantitative study. *Metals* **2021**, *11*, 1012. [CrossRef]
14. Clijsters, S.; Craeghs, T.; Buls, S.; Kempen, K.; Kruth, J.P. In situ quality control of the selective laser melting process using a high-speed, real-time melt pool monitoring system. *Int. J. Adv. Manuf. Technol.* **2014**, *75*, 1089–1101. [CrossRef]
15. ISO. *15549:2019: Non-Destructive Testing-Eddy Current Testing-General Principles;* International Organization for Standardization (ISO): Geneva, Switzerland, 2019. [CrossRef]
16. García-Martín, J.; Gómez-Gil, J.; Vázquez-Sánchez, E. Non-destructive techniques based on eddy current testing. *Sensors* **2011**, *11*, 2525–2565. [CrossRef]
17. ASTM. *E1004-17: Standard Test Method for Determining Electrical Conductivity Using the Electromagnetic (Eddy Current) Method;* ASTM: West Conshohocken, PA, USA, 2015; pp. 3–7. [CrossRef]
18. ASTM. *E3166-20: Standard Guide for Nondestructive Examination of Metal Additively Manufactured Aerospace Parts after Build;* ASTM: West Conshohocken, PA, USA, 2020; pp. 1–63. [CrossRef]
19. Todorov, E.I.; Boulware, P.; Gaah, K. Demonstration of array eddy current technology for real-time monitoring of laser powder bed fusion additive manufacturing process. In *Characterization and Monitoring of Advanced Materials, Aerospace, Civil Infrastructure, and Transportation XI*; International Society for Optics and Photonics (SPIE): Bellingham, WA, USA, 2018; Volume 10599, pp. 190–201. [CrossRef]
20. Todorov, E.I. Non-destructive Evaluation of Additive Manufacturing Components. U.S. Patent US 2018/0266993 A1, 20 September 2018.
21. Ehlers, H.; Pelkner, M.; Thewes, R. Heterodyne Eddy Current Testing using Magnetoresistive Sensors for Additive Manufacturing Purposes. *IEEE Sens. J.* **2020**, *20*, 5793–5800. [CrossRef]
22. Carlton, H.D.; Haboub, A.; Gallegos, G.F.; Parkinson, D.Y.; MacDowell, A.A. Damage evolution and failure mechanisms in additively manufactured stainless steel. *Mater. Sci. Eng. A* **2016**, *651*, 406–414. [CrossRef]
23. Eisenbarth, D.; Stoll, P.; Klahn, C.; Heinis, T.B.; Meboldt, M.; Wegener, K. Unique coding for authentication and anti-counterfeiting by controlled and random process variation in L-PBF and L-DED. *Addit. Manuf.* **2020**, *35*, 101298. [CrossRef]
24. Obaton, A.F.; Lê, M.; Prezza, V.; Marlot, D.; Delvart, P.; Huskic, A.; Senck, S.; Mahé, E.; Cayron, C. Investigation of new volumetric non-destructive techniques to characterise additive manufacturing parts. *Weld. World* **2018**, *62*, 1049–1057. [CrossRef]
25. Hippert, D.F. High Resolution Eddy Current Inspection and Eddy Current Testing for Additive Manufacturing. Ph.D. Thesis, École Polytechnique Fédérale de Lausanne (EPFL), Lausanne, Switzerland, 2015.
26. Bowler, N. *Eddy-Current Nondestructive Evaluation;* Springer Series in Measurement Science and Technology; Springer: New York, NY, USA, 2019. [CrossRef]
27. Udpa, S.S.; Moore, P.O. (Eds.) *Electromagnetic Testing*, 3rd ed.; Number v. 5 in Nondestructive testing handbook; American Society for Nondestructive Testing: Columbus, OH, USA, 2004.
28. Stoffregen, H.; Fischer, J.; Siedelhofer, C.; Abele, E. Selective laser melting of porous structures. In Proceedings of the 22th International Solid Freeform Fabrication Symposium, Austin, TX, USA, 8–10 August 2011; Volume 19, pp. 680–695. [CrossRef]
29. Yang, K.V.; Rometsch, P.; Jarvis, T.; Rao, J.; Cao, S.; Davies, C.; Wu, X. Porosity formation mechanisms and fatigue response in Al-Si-Mg alloys made by selective laser melting. *Mater. Sci. Eng. A* **2018**, *712*, 166–174. [CrossRef]
30. Choo, H.; Sham, K.L.; Bohling, J.; Ngo, A.; Xiao, X.; Ren, Y.; Depond, P.J.; Matthews, M.J.; Garlea, E. Effect of laser power on defect, texture, and microstructure of a laser powder bed fusion processed 316L stainless steel. *Mater. Des.* **2019**, *164*, 107534. [CrossRef]
31. Domfang Ngnekou, J.N.; Nadot, Y.; Henaff, G.; Nicolai, J.; Ridosz, L. Influence of defect size on the fatigue resistance of AlSi10Mg alloy elaborated by selective laser melting (SLM). *Procedia Struct. Integr.* **2017**, *7*, 75–83. [CrossRef]
32. Bayat, M.; Mohanty, S.; Hattel, J.H. Multiphysics modelling of lack-of-fusion voids formation and evolution in IN718 made by multi-track/multi-layer L-PBF. *Int. J. Heat Mass Transf.* **2019**, *139*, 95–114. [CrossRef]
33. Mukherjee, T.; Zuback, J.S.; De, A.; DebRoy, T. Printability of alloys for additive manufacturing. *Sci. Rep.* **2016**, *6*, 1–8. [CrossRef] [PubMed]
34. Kasperovich, G.; Haubrich, J.; Gussone, J.; Requena, G. Erratum: Corrigendum to "Correlation between porosity and processing parameters in TiAl6V4 produced by selective laser melting" (Materials and Design (2016) 105 (160–170)). *Mater. Des.* **2016**, *112*, 160–161. [CrossRef]
35. Dilip, J.J.; Zhang, S.; Teng, C.; Zeng, K.; Robinson, C.; Pal, D.; Stucker, B. Influence of processing parameters on the evolution of melt pool, porosity, and microstructures in Ti-6Al-4V alloy parts fabricated by selective laser melting. *Prog. Addit. Manuf.* **2017**, *2*, 157–167. [CrossRef]
36. King, W.E.; Barth, H.D.; Castillo, V.M.; Gallegos, G.F.; Gibbs, J.W.; Hahn, D.E.; Kamath, C.; Rubenchik, A.M. Observation of keyhole-mode laser melting in laser powder-bed fusion additive manufacturing. *J. Mater. Process. Technol.* **2014**, *214*, 2915–2925. [CrossRef]
37. Chahal, V.; Taylor, R.M. A review of geometric sensitivities in laser metal 3D printing. *Virtual Phys. Prototyp.* **2020**, *15*, 227–241. [CrossRef]
38. Reijonen, J.; Revuelta, A.; Riipinen, T.; Ruusuvuori, K.; Puukko, P. On the effect of shielding gas flow on porosity and melt pool geometry in laser powder bed fusion additive manufacturing. *Addit. Manuf.* **2020**, *32*, 101030. [CrossRef]

39. Xie, Y.; Li, J.; Tao, Y.; Wang, S.; Yin, W.; Xu, L. Edge Effect Analysis and Edge Defect Detection of Titanium Alloy Based on Eddy Current Testing. *Appl. Sci.* **2020**, *10*, 8796. [CrossRef]
40. Wang, Y.; Bai, Q.; Du, W.; Zhang, B. Edge Effect on Eddy Current Detection for Subsurface Defects in Titanium Alloys. In Proceedings of the 8th International Conference on Computational Methods, Guilin, China, 25–27 July 2017.
41. Sharma, S.; Elshafiey, I.M.; Udpa, L.; Udpa, S.S. Probe design for edge-effect reduction in eddy current inspection. *Proc. SPIE-Int. Soc. Opt. Eng.* **1996**, *2945*, 14–22. [CrossRef]
42. Dodd, C.V.; Deeds, W.E. Analytical solutions to eddy-current probe-coil problems. *J. Appl. Phys.* **1968**, *39*, 2829–2838. [CrossRef]
43. Ma, X.; Peyton, A.J. Eddy current measurement of the electrical conductivity and porosity of metal foams. *IEEE Trans. Instrum. Meas.* **2006**, *55*, 570–576. [CrossRef]
44. Cuevas, F.G.; Montes, J.M.; Cintas, J.; Urban, P. Electrical conductivity and porosity relationship in metal foams. *J. Porous Mater.* **2009**, *16*, 675–681. [CrossRef]
45. Boillat, E.; Fivat, D.; Jhabvala, J.; Matthey, M.; Glardon, R. A review of different techniques to characterise the mechanical properties of SLS parts-Focus on resistivity measurements. In *Innovative Developments in Virtual and Physical Prototyping-Proceedings of the 5th International Conference on Advanced Research and Rapid Prototyping*; Taylor & Francis: Leiria, Portugal, 2011; pp. 281–286. [CrossRef]
46. GUM. Evaluation of measurement data—Guide to the expression of uncertainty in measurement. *Int. Organ. Stand. Geneva ISBN* **2008**, *50*, 134.
47. Gu, H.; Gong, H.; Dilip, J.J.; Pal, D.; Hicks, A.; Doak, H.; Stucker, B. Effects of powder variation on the microstructure and tensile strength of Ti6Al4V parts fabricated by selective laser melting. In Proceedings of the 25th Annual International Solid Freeform Fabrication Symposium; An Additive Manufacturing Conference, Austin, TX, USA, 4–6 August 2014; pp. 470–483.

metals

MDPI

Article

Process Induced Preheating in Laser Powder Bed Fusion Monitored by Thermography and Its Influence on the Microstructure of 316L Stainless Steel Parts

Gunther Mohr [1,2,*], Konstantin Sommer [1,3], Tim Knobloch [1,2], Simon J. Altenburg [1], Sebastian Recknagel [1], Dirk Bettge [1] and Kai Hilgenberg [1]

1 Bundesanstalt für Materialforschung und-prüfung (BAM), Unter den Eichen 87, 12205 Berlin, Germany; konstantin.sommer@bam.de (K.S.); tim.knobloch@bam.de (T.K.); Simon.Altenburg@bam.de (S.J.A.); sebastian.recknagel@bam.de (S.R.); Dirk.Bettge@bam.de (D.B.); kai.hilgenberg@bam.de (K.H.)

2 Institute of Machine Tools and Factory Management, Technische Universität Berlin, Straße des 17. Juni 135, 10623 Berlin, Germany

3 Institute of Materials Science and Technology, Chair of Metallic Materials, Technische Universität Berlin, Straße des 17. Juni 135, 10623 Berlin, Germany

* Correspondence: gunther.mohr@bam.de

Citation: Mohr, G.; Sommer, K.; Knobloch, T.; Altenburg, S.J.; Recknagel, S.; Bettge, D.; Hilgenberg, K. Process Induced Preheating in Laser Powder Bed Fusion Monitored by Thermography and Its Influence on the Microstructure of 316L Stainless Steel Parts. *Metals* **2021**, *11*, 1063. https://doi.org/10.3390/met11071063

Academic Editor: Sergey N. Grigoriev

Received: 7 June 2021
Accepted: 29 June 2021
Published: 1 July 2021

Abstract: Undetected and undesired microstructural variations in components produced by laser powder bed fusion are a major challenge, especially for safety-critical components. In this study, an in-depth analysis of the microstructural features of 316L specimens produced by laser powder bed fusion at different levels of volumetric energy density and different levels of inter layer time is reported. The study has been conducted on specimens with an application relevant build height (>100 mm). Furthermore, the evolution of the intrinsic preheating temperature during the build-up of specimens was monitored using a thermographic in-situ monitoring set-up. By applying recently determined emissivity values of 316L powder layers, real temperatures could be quantified. Heat accumulation led to preheating temperatures of up to about 600 °C. Significant differences in the preheating temperatures were discussed with respect to the individual process parameter combinations, including the build height. A strong effect of the inter layer time on the heat accumulation was observed. A shorter inter layer time resulted in an increase of the preheating temperature by more than a factor of 2 in the upper part of the specimens compared to longer inter layer times. This, in turn, resulted in heterogeneity of the microstructure and differences in material properties within individual specimens. The resulting differences in the microstructure were analyzed using electron back scatter diffraction and scanning electron microscopy. Results from chemical analysis as well as electron back scatter diffraction measurements indicated stable conditions in terms of chemical alloy composition and austenite phase content for the used set of parameter combinations. However, an increase of the average grain size by more than a factor of 2.5 could be revealed within individual specimens. Additionally, differences in feature size of the solidification cellular substructure were examined and a trend of increasing cell sizes was observed. This trend was attributed to differences in solidification rate and thermal gradients induced by differences in scanning velocity and preheating temperature. A change of the thermal history due to intrinsic preheating could be identified as the main cause of this heterogeneity. It was induced by critical combinations of the energy input and differences in heat transfer conditions by variations of the inter layer time. The microstructural variations were directly correlated to differences in hardness.

Keywords: additive manufacturing; laser powder bed fusion; selective laser melting; laser beam melting; in-situ process monitoring; thermography; heat accumulation; inter layer time; cellular substructure

1. Introduction

Additive manufacturing (AM) technologies provide promising advantages for the production of highly individual and complex structures, mass customization, the integration of functional designs, and the reduction of lead times [1,2]. The working principles of various metallic AM processes are described, e.g., by DebRoy et al. [3]. Although laser powder bed fusion (L-PBF) is the most prevalent AM technology for metal part production [4], the homogeneity of the material produced is still problematic. Inhomogeneity of the microstructure, defect density, and resulting mechanical properties within parts or in comparison of different parts have been alluded by several authors [3,5–7]. Microstructural variations in AM components are a major challenge, especially for safety-critical components [6,8,9].

A high degree of freedom in design in L-PBF offers the chance to produce complex shaped geometries. However, the geometry itself can influence the thermal history of a part during manufacturing as it might change the conditions of heat dissipation [10–12]. A detrimental change of the heat conduction through the part towards the base plate, as well as a significant change of the inter layer time (ILT), can lead to severe heat accumulation of the part or areas of local overheating. This, in turn, results in deviations of the thermal history and eventually affects part quality [5,10,13]. In addition to the geometry, there are many other influencing factors on the thermal history of a L-PBF component. These comprise, for instance, processing parameters, scanning strategies, support design, or ratio of area exploitation [3,5,11,12]. The thermal history of an L-PBF process is an important factor for the development of the microstructure, as it is influenced by the transient temperature fields during manufacturing. It is well known that variations of temperature gradients can significantly affect the microstructural development. Lower cooling rates are generally expected to develop a coarser microstructure than higher cooling rates [1,14].

The microstructure of 316L processed by L-PBF typical shows features that can be observed over a broad range of length scales. The features include melt pool boundaries, grains and sub-grains, cellular substructures of grains, segregations, dislocation networks at the boundaries of cell structures, and nanoscale precipitations [15–18]. The shape of melt pool boundaries is influenced by the scanning strategy and process parameters. Their shape and penetration depth strongly depend on the melting mode [19]. Patel und Vlasea [20] reported on the occurrence of deep penetration mode welding (keyhole mode welding) and transition mode welding over a broad range of process parameters in L-PBF processing of 316L. According to Krakhmalev et al. [16], the grains consist of cellular substructures due to high cooling rates, whereby the cells grow epitaxially, starting at melt pool boundaries. The cellular substructures grow competitively based on their crystallographic orientation and the local thermal gradients inside the melt pool [16]. The growth of cellular substructures in 316L processed by L-PBF typically leads to segregation of Mn, Mo, and Cr and dislocation networks at cell boundaries [15,16]. The features of these cellular substructures, namely, the cell size and the occurrence of micro-segregations, strongly depend on the local directional solidification conditions [9]. Pinomaa et al. [9] recently quantified the influence of the local thermal gradient and local melt pool solidification rate on these features as well as on the mode of growth by conducting phase field simulations.

David et al. [21] investigated the effect of rapid solidification on the weld metal microstructures in different stainless steel compositions in the late 1980s. They integrated the influence of the cooling rate into the Schaeffler diagram [21]. It can be derived from their work that 316 stainless steel solidifies as fully austenitic at cooling rates above 0.28×10^6 K·s^{-1}. Due to the lower austenite stabilizing carbon content in 316L stainless steel, these cooling rates might shift to slightly higher values for 316L. Bajaj et al. [22] conducted an intense review on steels in L-PBF and direct energy deposition. They reported fully austenitic phase for 316L when processed by L-PBF. Krakhmalev et al. [16] also mentioned a fully austenitic phase with some very exceptional cases of the occurrence of a ferritic phase.

Additionally, the occurrence of spherical nano-sized oxide inclusions has been reported for 316L produced by L-PBF [16]. Lou et al. [23] investigated the influence of these Si- and Mn-rich oxide inclusions. They concluded that these oxide inclusions can be responsible for the initiation of micro-voids with detrimental effects on impact toughness.

They compared the measured impact toughness with values from literature for specimens produced by powder metallurgy route and wrought material, revealing distinct beneficial effects when the oxygen content was below 0.02%.

In addition to these crystallographic features of the microstructure, internal defects such as delamination, cracks, and pores can occur [24]. Pores can be categorized into so-called lack-of-fusion defects and gas porosity [25]. Lack of fusion defects are irregularly-shaped pseudo pores, which essentially are cavities resulting from insufficient melting and insufficient material cohesion due to deficient melt pool dimensions or inadequate choice of processing parameters such as the hatch distance between single melt tracks [25,26]. A main source of spherical gas porosity can be found in detached and entrapped vapor bubbles of the vapor capillary in an instable keyhole welding mode [25,27].

The possibility to influence the temperature gradients occurring in the process has been shown in different studies. On the one hand, there are approaches to tailor a microstructure during the L-PBF process by adjusted sets of processing parameters [28,29] or modifications of the laser beam shape and intensity profile [30,31]. On the other hand, there are approaches to decrease residual stresses or crack susceptibility by, e.g., remelting or platform preheating adjustments [32,33]. These approaches have in common the aim to purposely influence the initial resulting microstructure of an L-PBF specimen or component either throughout the entire geometry or within particular regions. However, there are further situations where variations of microstructures in L-PBF processes should be considered. This includes unplanned microstructural variations that are induced due to heat accumulation during the process.

The accumulation of heat during the build-up of an L-PBF part essentially alters the preheating condition at the surface to be coated by the new powder layer which is subsequently exposed by laser radiation. Therefore, the initial thermal conditions prior to the exposure by laser radiation are altered. As a result, the thermal gradients during melting and solidification might change with varying initial preheating temperature. A significant change of the preheating temperature of the part can alter the melt pool dimensions and their solidification conditions [5,34]. As pointed out by Krakhmalev et al. [16], the cellular mode of solidification in 316L processed by L-PBF occurs at high solidification rates and steep thermal gradients. The resulting microstructural feature sizes such as cell spacing depend on these conditions [16]. They might change when the preheating temperature increases due to heat accumulation. Depending on the magnitude of the variation in preheating condition, the induced variations on the microstructure can be strong enough to affect the mechanical properties of a component [5]. In addition to potential changes of the melt pool shape and the melt pool dimensions, as well as potential changes of crystallographic features, an increase of the preheating temperature may also be able to shift the melting mode to an unstable region with propensity to develop detrimental keyhole porosity [5].

The authors [5] investigated process conditions where heat accumulation was provoked to occur during the L-PBF fabrication of simple cuboid specimens of 316L stainless steel at application relevant build heights, i.e., the specimens height was bigger than 100 mm. Using a mid-wavelength infrared thermography camera as an in-situ thermal monitoring device, significant differences in cooling behavior were revealed. Specimens were produced applying three distinct ILT and three distinct volumetric energy densities (VED). The build height was also identified as an affecting factor. The authors correlated the differences in cooling conditions with differences in apparent sub-grain sizes measured by light microscopy, melt pool depths, and hardness values. To refine the knowledge of process-property-relationship and to prove the relationships measured by light microscopy, a more detailed analysis of the influence of in-situ heat accumulation on the microstructure of 316L components is required. Therefore, this study pursues the examinations of the same specimens used in [5] and investigates the microstructure by means of electron back scattered diffraction (EBSD) in greater detail. In addition, scanning electron microscopy (SEM) is used to compare the feature size of the cellular substructure. The alloy composition of the produced specimens as well as a potential oxygen intake is measured by different methods of chemical analysis. By applying recently conducted temperature adjustments of the infrared (IR) monitoring set-up [35], real temperatures of the powder surfaces could be quantified. The layer-wise increase of the preheating temperatures of the specimens was measured in-situ over the entire build process of the specimens. The results are discussed with respect to recent publications. Results of hardness measurements were taken from [5].

Although there have been extensive investigations on L-PBF of 316L, a quantification of process-induced preheating of the specimens during manufacturing and its correlation to changes in crystallographic features is currently missing. Information about the magnitude of microstructural changes induced by critical but still realistic process conditions is important. It will help to improve the evaluation of a real part of the production and the comparability of test coupons that are manufactured under certain processing conditions. Effects of processing parameters such as scanning velocity or laser power have been studied extensively. However, the build height and the ILT are often not considered. Their influence may not be considered significant in the case of typical 10 mm cubic specimens or in the case of high ratios of area exploitation within the powder bed. They may become an affecting factor in the case of complex real part geometries with varying area exploitations over the build height. Additionally, the current trends in the development of new L-PBF machines (e.g., multi laser machines) are expected to decrease the ILT, which increases the need for reliable data about potential microstructural heterogeneity.

2. Materials and Methods

2.1. Material and Specimen Manufacturing by L-PBF

In this section, the key facts about the L-PBF processing conditions and specimens are mentioned. Details were published in [5] where the same set of specimens was examined. Upstanding cuboid-shaped specimens of the dimensions $(13 \times 20 \times 114.5)$ mm^3 were manufactured on a commercial L-PBF single laser system of type SLM280HL (SLM Solutions Group AG, Lübeck, Germany) using a commercial 316L stainless steel powder. Table 1 shows the chemical composition of the powder material according to supplier's information. Figure 1 depicts the geometry of the specimens and contains a schematic of the applied bidirectional scanning strategy with 90° rotation between layers. It also highlights the parts of the specimen that were taken for the deeper analysis in this study in grey color. These were basically the volumes of the lower 12.5 mm (including excess material for part removal) and of the upper 10 mm of each specimen.

Table 1. Chemical composition of the 316L raw powder material according to supplier's information and the respective min. and max. values as per the material specification by DIN EN 10088-3 [36]. The figures express mass fractions in %.

Specification	C	Si	Mn	P	S	Cr	Mo	Ni	N	Fe
Min.	-	-	-	-	-	16.5	2.0	10.0	-	bal.
Max.	0.03	1.0	2.0	0.045	0.03	18.5	2.5	13.0	0.1	bal.
Powder	0.017	0.6	0.92	0.012	0.004	17.7	2.35	12.6	0.1	bal.

The specimens were manufactured at three distinct ILT and at each ILT at three distinct levels of VED by varying the scanning velocity v_s, i.e., nine different types of specimens were built. Each specimen type was built up twice, as stated in [5]. The other manufacturing parameters were kept constant. All process parameters are summarized in Table 2.

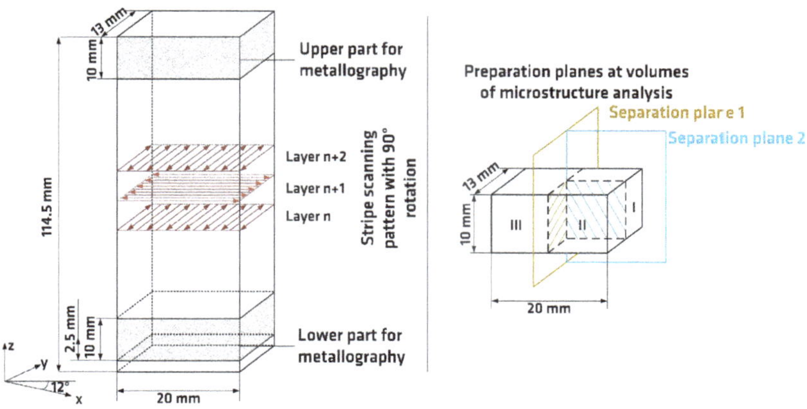

Figure 1. Specimens' geometry. Grey volumes were designated for microstructure analysis and were cut according to the preparation planes in the schematic. Adapted from ref. [5].

Table 2. Processing parameters.

Processing Parameters	Level	
Layer thickness	0.05 mm	
Laser power	275 W	
Hatch distance	0.12 mm	
Platform preheating temperature	100 °C	
Inter layer time	Short: 18 s	
	Intermediate: 65 s	
	Long: 116 s	
Volumetric energy density	Low: 49.12 J·mm^{-3}	v_s = 933 mm·s^{-1} (75% of basis VED)
	Basis: 65.48 J·mm^{-3}	v_s = 700 mm·s^{-1}
	High: 81.85 J·mm^{-3}	v_s = 560 mm·s^{-1} (125% of basis VED)

The ILT of layer number n was defined in [5] and is explained by Equation (1).

$$\text{ILT}_{\text{layer } n} = \text{time for powder recoating} + \text{time for laser exposing in layer } n \quad (1)$$

The ILT values for specimen production were chosen according to calculated values from a real part production as compared in [5]. The basis VED parameters represent parameters for the machine and material recommend by the machine's manufacturer but with a simplified scanning strategy. Low VED and high VED parameters were chosen to broaden the energy input, adjusting the VED by plus 25% and minus 25%.

Table 3 gives an overview of the combinations of variable parameters used for specimen production. It also contains the applied methods of analysis, which are described in the following subsections.

Table 3. Matrix of parameter combinations and methods of analysis.

Level of Inter Layer Time	Level of Volumetric Energy Density	Method of Analysis				
		In-Situ Thermographic Measurment	EBSD Grain Size Measurement	Cell Structure Analysis with SEM	Chemical Analysis	Analysis of Oxygen Content
Short ILT	High VED	x	x	x	x	x
	Basis VED	x	x	x	x	-
	Low VED	x	x	x	-	-
Intermediate ILT	High VED	x	x	-	-	-
	Basis VED	x	x	x	x	x
	Low VED	x	x	-	-	-
Long ILT	High VED	x	x	-	-	-
	Basis VED	x	x	x	-	-
	Low VED	x	x	-	x	-

The specimens were heat-treated under argon gas atmosphere before removal from the base plate. The heat treatment was conducted at 450 °C for 4 h after the process to relieve residual stresses without changing the as-built microstructure.

2.2. In-Situ Thermographic Monitoring and Temperature Analysis

The production of the specimens was in-situ monitored using an off-axis infrared camera of type ImageIR8300 (InfraTech GmbH, Dresden, Germany), which was installed on top of the L-PBF system as schematically shown in Figure 2. The camera was calibrated by its vendor for black body radiation. A temperature adjustment was conducted by a determination of emissivity values of 316L powder layers and 316L L-PBF surfaces for the same set-up in previous work [35]. The camera was sensitive in the spectral range from 2 µm to 5.7 µm. The cooled InSb-focal-plane-array of the camera was of size 640 pixel × 512 pixel. The frame rate of the camera was 300 Hz for full frame. The measurements were conducted using a subframe image of 160 pixel × 200 pixel. The resulting spatial resolution in the field of view corresponded to approximately 420 µm/pixel. The subframe measurements were conducted at a frame rate of 600 Hz, and a bit resolution of 14 bits was used. The layer-wise recording was triggered by the first overall infrared signal rise above a predefined threshold value. Then, a predefined number of 40 bygone time steps were taken as start of the recording by using a circular buffer. The duration of the recording was set by the definition of a certain number of frames to acquire. Further details on the thermographic set-up can be found in [5,10], which show qualitative comparisons of the same thermographically gained process information using this set-up.

During the IR measurements of the processes, various internal black body calibration ranges of the camera were used to capture the IR data, since the relevant apparent temperature range succeeded the dynamic temperature range of a single calibration range. The change of the calibration ranges had to be done manually using the camera control software. The following black body calibration ranges were used at a converter resolution of 14 bit: 60 °C–200 °C at an integration time of 89 µs, 125 °C–300 °C at an integration time of 27 µs, 200 °C–400 °C at an integration time of 193 µs, and 300 °C–600 °C at an integration time of 45 µs. They are referred to the following abbreviation scheme: IR-CB 60–200 for the calibration range of 60 °C–200 °C. The conversion of the received IR signal values (apparent temperatures) into temperatures was conducted using a MATLAB (The Mathworks Inc., Natick, MA, USA) routine considering the experimentally determined apparent emissivity values of 316L powder from previous work [35]. For simplification, a constant apparent emissivity value was used over each respective calibration range, i.e., $\varepsilon = 0.33$ for IR-CB 60–200 and for IR-CB 125–300, $\varepsilon = 0.43$ for IR-CB 200–400 and for IR-CB 300–600. Additional information about the temperature adjustment and emissivity determination using this set-up as well as some theoretical background on that matter can be found in [35].

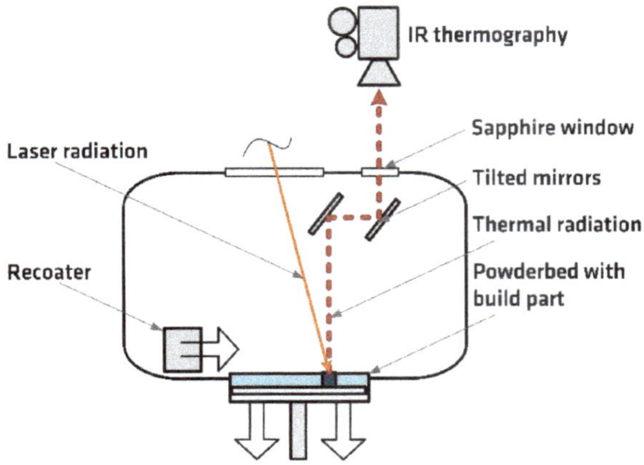

Figure 2. Schematic of the off-axis thermographic monitoring. Reprinted from ref. [10].

In this study, the surface temperatures of the specimens before laser exposure were investigated using the IR camera. Therefore, uncorrected IR signals of an area of 11 pixels × 11 pixels in the manually selected center of each specimen were averaged and processed using a GNU Octave (open source software) routine. A peak detection was implemented to get comparable sampling times and related values of the preheating IR signals of each specimen for each layer. To this end, the slope of the averaged IR signals was derived and smoothed by using a moving average method. If the slope of the smoothed IR signal is rising above a predefined threshold, it describes the start of the scan process in that area and thereby the interval of interest. The affiliated IR signal value and the timestamp are extracted from the minimum of this interval. These values describe comparable states of each specimen's surface coated with a new powder layer right before the start of the laser exposure of the respective area. Additionally, if there is a temporary drop in the IR signal values the peak detection can be heavily disturbed. This applies, for instance, when the recoater is moving through the field of view during the IR signal recording. The recoating process is accidentally recorded in some cases when, e.g., the layer-wise recording time is too long after the IR camera has been triggered. In this case, the averaged IR signals are filtered beforehand by an optional recoating filter with manually predefined parameter settings. As a last step, the extracted IR value of the comparable state of the specimen is converted to a temperature, using the emissivity values mentioned above. This value is then defined as current preheating temperature of the respective specimen. The measurement uncertainty of the emissivity determination has to be considered [35].

2.3. *Analysis of Microstructure Using Electron Back Scatter Diffraction (EBSD)*

Electron back scatter diffraction (EBSD) measurements were performed, to investigate the grain structure and phases of the produced specimens. For the measurements, the part II cross sections (see separation plane 1 in Figure 1) were ground with 180, 320, 600, and 1200 grits emery papers and polished using clothes with 3 μm and 1 μm particle suspensions, followed by MasterMet-2 (Buehler, ITW Test & Measurement GmbH, Esslingen am Neckar, Germany) amorphous 0.02 μm colloidal silica suspension. The microscopic measurements were executed on a scanning electron microscope (SEM) Tescan VEGA 3 (TESCAN ORSAY HOLDING a.s., Brno-Kohoutovice, Czech Republic) equipped with an EBSD detector Nordlys (Oxford Instruments plc, Abingdon, England). For acquisition, indexing, and post-processing, the software Aztec 4.1 (Oxford instruments plc, Abingdon, England) was used. An area of 2.25 mm × 3 mm was measured for every cross section, using an acceleration voltage of 20 keV, a beam current of approximately 10 nA, a step size of 5 μm, and a pattern size of 168 × 128 pixels. The low-angle grain boundary (LAGB) criterion was set to 5° to discriminate distinct sub-grains. In addition, for grain discrimination, the high-angle grain boundary (HAGB) criterion was set to 15°.

2.4. Analysis of Cellular Substructures by Scanning Electron Microscopy (SEM)

The investigation of cellular substructures within the grains was performed on the same cross sections after the EBSD measurements. The polished surfaces were etched with Bloech and Wedl II agent (50 mL H_2O, 50 mL HCl, and 0.6 g $K_2S_2O_5$) [37] to contrast the cellular substructure. The measurements were performed on a scanning electron microscope (SEM) Leo Gemini 1530VP (Leo Electron Microscopy Inc., New York, NY, USA) detecting back scattered electrons. Electrons of 20 keV energy were used.

SEM captures of these substructures were made at different locations of the cross-sections, to qualitatively estimate the size distribution of the cells. Cells that grew parallel or close to parallel to the preparation plane were used for the measurement. The cell walls can be imaged due to the topographic effect (bright lines in Figure 3) after etching. The measurement was focused on the number of cell walls within a defined distance in the style of the well-known metallographic grain size measurements by a line intercept method as described in, e.g., DIN EN ISO 643 [38]. To this end, five lines per SEM image with 10 μm length were placed perpendicular to the cell walls which were cut parallel or close to parallel to the preparation plane. Then, the number of intersections with the cellular walls were counted. The number of intersections can be used for relative comparison of the cell size. The lower the number of intersections, the wider are the cells. Three different regions were investigated for each section. In addition, a quantitative estimation can also be derived by dividing the length of the lines by the counted number. Figure 3 depicts an example of the measuring procedure, also presenting the number of intersections with each individual line.

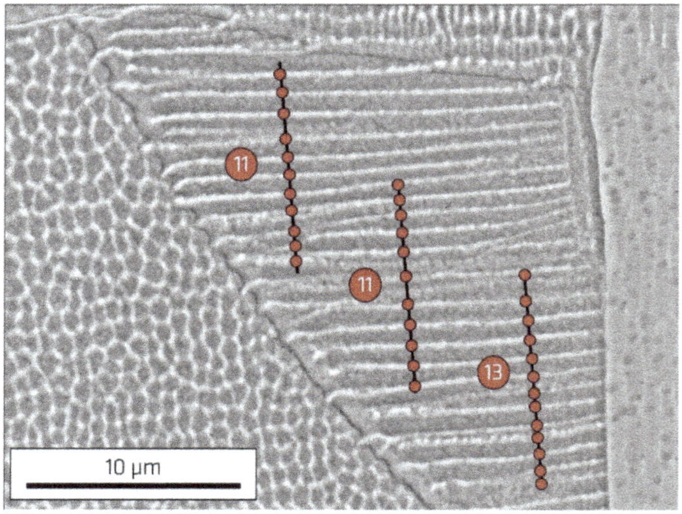

Figure 3. Example of SEM captures of cellular substructures used for estimation of their size distribution by a line intercept method. The numbers in the circles depict the number of intersections of the respective line.

Deng et al. [39] also used such a line intercept method to determine cell sizes of 316L. In contrast to the approach of this study, they applied the line intercept method to cells cut perpendicular to their growth direction. These cells, therefore, appeared in a honeycomb like structure, as visible in the left lower region of the SEM image in Figure 3.

However, the sensitivity to measurement errors seems to be higher in this case as compared to a line intercept measurement through the parallel cell walls. This is due to a minor effect of a potential tilt angle of the cross section induced by the mechanical preparation would have in the latter case. Pinomaa et al. [9] also measured cell sizes derived from simulations by line interception through parallel cell wall regions, as this was conducted in the present study.

2.5. Analysis of Chemical Composition

The mass fractions of the elements given in Table 1 and of oxygen were determined for selected specimens (see Table 3) using the measurement techniques listed in Table 4. The alloying elements Cr, Ni, Mn, and Mo were determined using X-ray fluorescence spectrometry. Two certified reference materials were used for calibration (ECRM 284-2 and ECRM 284-3, BAM, Berlin, Germany). Since this method is not sufficiently sensitive for each alloying element, traces of Si and P were determined using inductively coupled plasma optical emission spectrometry after decomposition of the material using microwave digestion. For the determination of the non-metals C, S, O, and N, element-analyzers were used, calibrated with different certified reference materials. The analysis was conducted using material of section III in the lower part and upper part of the specimen (see Figure 1). Oxygen measurements were conducted at two separate specimens manufactured under the same processing conditions also using material of section III of the respective parts.

Table 4. Measurement techniques used for precise determination of chemical composition for selected specimens.

Measurement Technique	Measuring Device	Chemical Element
Combustion/IR-detection	Elementrac CS-i (Eltra GmbH, Haan, Germany)	C
		S
Carrier gas hot extraction	G8 Galileo (Bruker Corporation, Billerica, MA, USA)	N
		O
X-ray fluorescence spectrometry	NITON XL3t (Thermo Fisher Scientific Inc., Waltham, MA, USA)	Mn
		Cr
		Mo
		Ni
Inductively coupled plasma optical emission spectrometry	Spectro Arcos (SPECTRO Analytical Instruments GmbH, Kleve, Germany)	Si
		P

3. Results

3.1. Surface Temperatures

The in-situ preheating temperature evolution over the build-up process of the specimens is depicted in three different diagrams (Figures 4–6), each showing the preheating temperature over the layer number at a fixed ILT level and different VED levels. In addition, Figure 7 displays the same plots for the basis VED at different ILT levels within one diagram for easier comparison. The captured IR signals were not exploitable for every parameter combination over the entire part due to the narrow band of the set calibration ranges of the IR camera as described in [35]. Only exploitable signals were converted into real temperatures and plotted in the diagrams. Hence, there are some missing parts within some of the curves. Especially, the lower temperature regions were often not resolved within the set calibration ranges. Therefore, considerable shares of the specimens produced with intermediate ILT and long ILT could not be resolved properly in their lower sections.

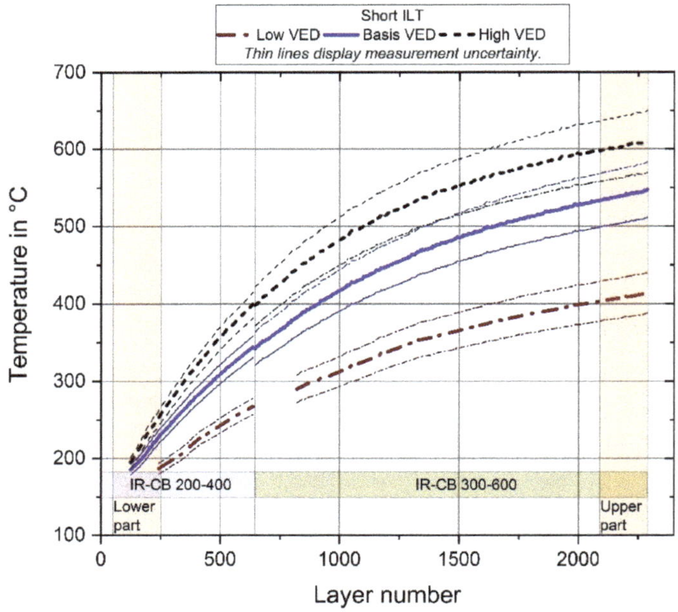

Figure 4. Temperature-layer-number-plots for short ILT. The orange sections highlight the layers of the specimen volumes used for deeper analysis as depicted in Figure 1. The colored bars show the IR calibration ranges (IR-CB) used in the respective layers. Missing parts of the curves (e.g., in the low VED specimen) are due to the narrow bands of the calibration ranges of the IR camera.

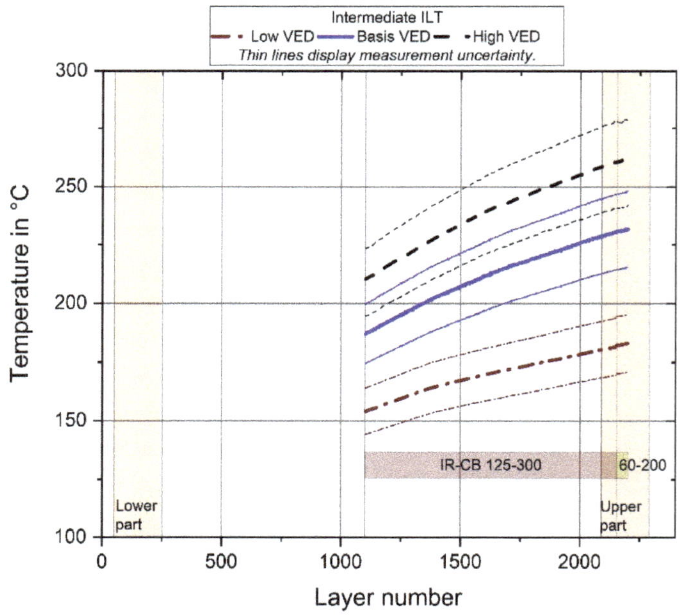

Figure 5. Temperature-layer-number-plots for intermediate ILT. The orange sections highlight the layers of the specimen volumes used for deeper analysis as depicted in Figure 1. The colored bars show the IR calibration ranges (IR-CB) used in the respective layers.

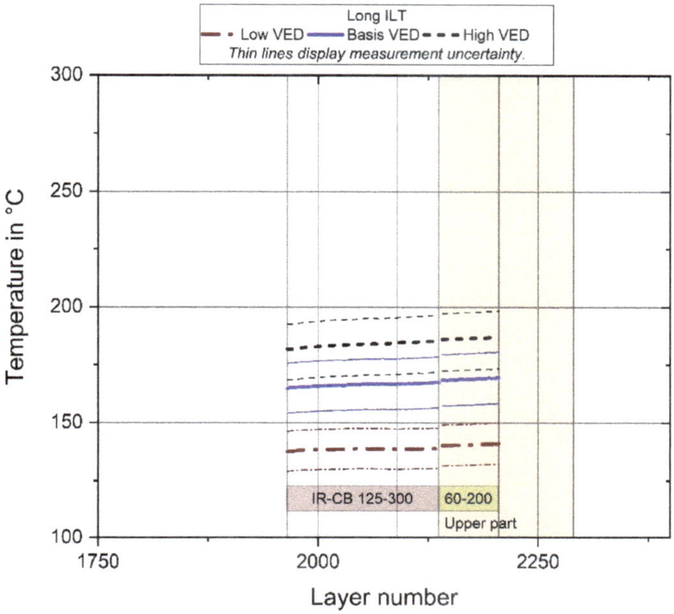

Figure 6. Temperature-layer-number-plots for long ILT. The orange sections highlight the layers of the specimen volumes used for deeper analysis as depicted in Figure 1. The colored bars show the IR calibration ranges (IR-CB) used in the respective layers.

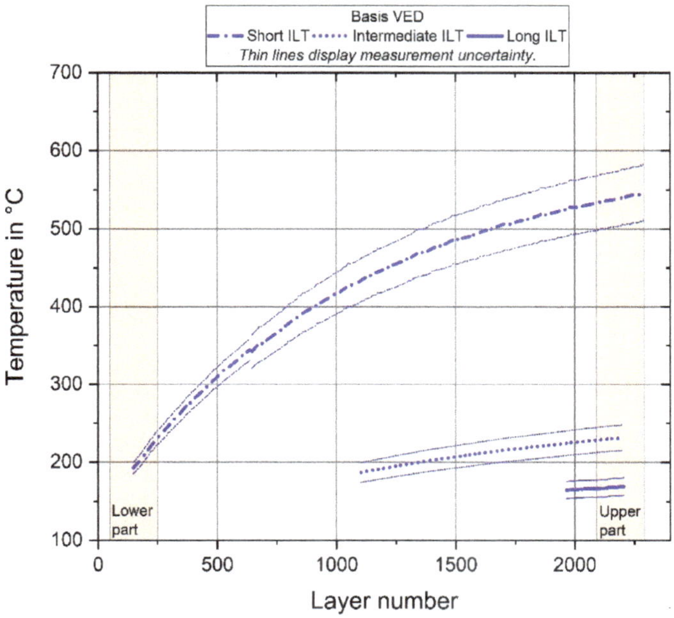

Figure 7. Temperature-layer-number-plots for basis VED. The orange sections highlight the layers of the specimen volumes used for deeper analysis as depicted in Figure 1. Missing parts of the curves are due to the narrow bands of the calibration ranges of the IR camera.

3.2. Grain Size Analysis

The EBSD measurements revealed differences in HAGB grain sizes and LAGB sub-grain sizes between the sections of the window of investigations. The mean LAGB sub-grain sizes and the mean HAGB grain sizes are depicted in Figures 8 and 9 for each combination of ILT and VED (see Table 3) and of the upper part and lower part volumes of investigation (see Figure 1). The corresponding inverse pole figure maps for the specimens' sections marked with capital letters in Figure 8 are shown in Figure 10.

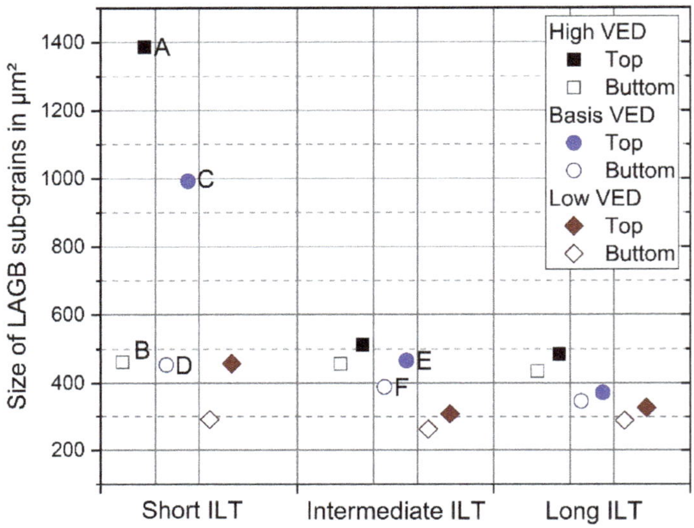

Figure 8. Comparison of sub-grain sizes (LAGB 5°) for each parameter combination. The capital letters correspond to the respective inverse pole figure maps in Figure 10.

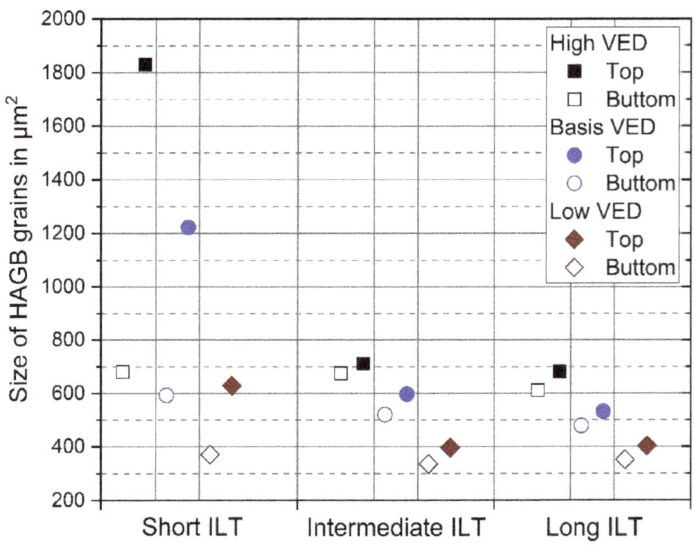

Figure 9. Comparison of grain sizes (HAGB 15°) for each parameter combination.

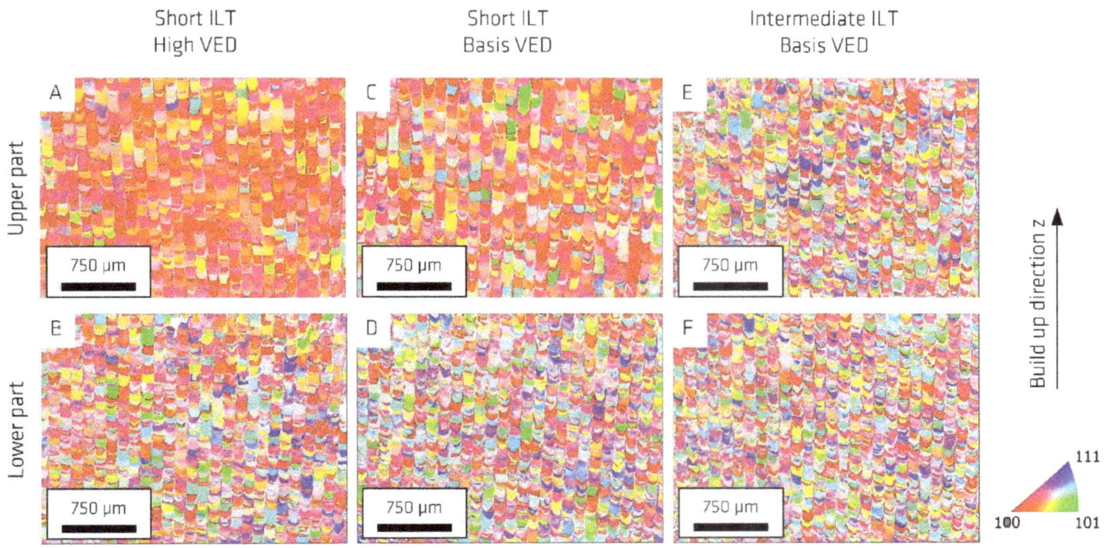

Figure 10. Inverse pole figure maps comparison with crystallographic orientation related to direction normal to the map plane. The capital letters correspond to the respective sub-grain sizes in Figure 8.

The measured grain size distribution is not Gaussian but follows a log-normal distribution. Hence, the mean and the standard deviation were calculated from logarithmic transformed measuring data. When retransformed, the standard deviation has to be considered multiplicatively, which results in asymmetric error bars with pronounced overlapping upper parts for the individual measuring points. Despite significant overlapping, the changes in the mean values are to be discussed. The mean values and standard deviations of the individual data points presented in Figures 8 and 9 are additionally given in Tables 5 and 6.

Table 5. LAGB sub-grain sizes.

ILT	VED	sub-Grain Size in µm² Lower Part		Sub-Grain Size in µm² Upper Part	
		Mean Value	Standard Deviation	Mean Value	Standard Deviation
Short	High	462.4	3.6	1386.4	5.1
	Basis	454.6	4.0	992.9	5.3
	Low	291.5	3.2	457.7	4.5
Intermediate	High	456.8	4.2	511.4	4.4
	Basis	389.4	3.7	466.5	4.0
	Low	262.5	3.1	308.8	3.3
Long	High	435.2	4.0	484.5	4.2
	Basis	347.6	3.6	532.3	4.2
	Low	289.4	3.2	403.4	3.8

Table 6. HAGB grain sizes.

ILT	VED	Grain Size in µm² Lower Part		Grain Size in µm² Upper Part	
		Mean Value	Standard Deviation	Mean Value	Standard Deviation
Short	High	689.9	4.1	1830.6	5.9
	Basis	592.6	4.3	1222.5	5.8
	Low	370.9	3.5	628.6	5.0
Intermediate	High	674.2	4.6	709.9	4.8
	Basis	520.0	4.0	596.2	4.3
	Low	335.2	3.4	396.4	3.7
Long	High	610.4	4.3	682.0	4.6
	Basis	478.6	4.0	532.3	4.2
	Low	351.2	3.4	403.4	3.8

3.3. Cellular Substructure

The SEM images revealed a cellular growth mode for all examined parameter combinations. The intersection counts as well as the calculated average cell size are depicted in Figure 11. Although there are huge overlapping areas of the deviation bars, two trends can be recognized from the results, as they appear consistently. First, the cellular size appears to be increased at the upper part of the specimens manufactured at short ILT for all VED levels. At the same time, no differences can be noticed between upper and lower part at longer ILT and standard VED. Second, the cell size increases with increasing VED (decreasing scanning velocity v_s) irrespective of the build height. This can only be reported for the short ILT level, as for the other ILT levels only the standard VED was considered in this measurement. The average cell size within this investigation is between 0.52 µm and 0.76 µm.

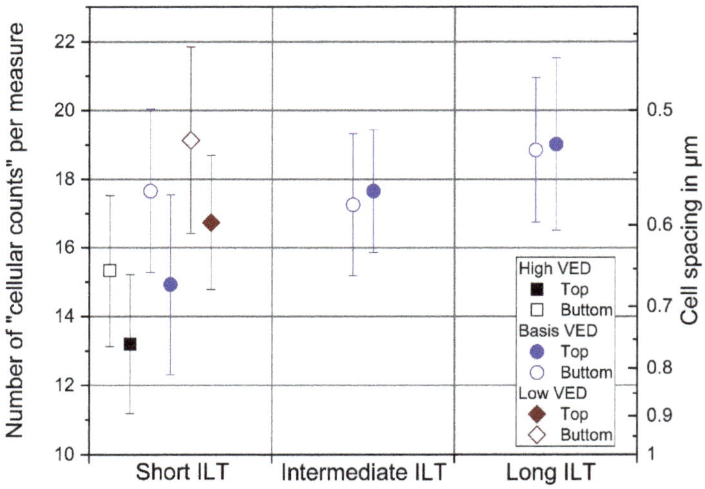

Figure 11. Comparison of cell spacing intersection counts for different parameter combinations. The right y-axis corresponds to the calculation of the cell spacing.

3.4. Chemical Composition

The EBSD measurements showed no hint for phases other than austenite. The measured chemical composition of the examined specimens is depicted in Table 7.

Table 7. Chemical composition of selected specimens in weight %.

Element	Uncertainty	Short ILT High VED		Short ILT Basis VED		Intermediate ILT Basis VED		Long ILT Low VED	
		Lower Part	Upper Part	Lower Part	Upper Part	Lower Part	Upper Part	Lower Part	Upper Part
C	0.0020	0.0149	0.0142	0.0161	0.0152	0.0156	0.0159	0.0168	0.0171
Si	0.05	0.59	0.59	0.61	0.52	0.55	0.52	0.62	0.56
Mn	0.03	0.89	0.92	0.91	0.94	0.87	0.90	0.91	0.94
Cr	0.4	17.9	17.9	18.0	17.9	17.9	18.0	17.9	17.9
Mo	0.04	2.38	2.40	2.42	2.39	2.37	2.42	2.39	2.41
Ni	0.3	12.8	12.8	12.9	12.8	12.7	12.9	12.8	12.9
N	0.005	0.075	0.077	0.078	0.077	0.079	0.078	0.082	0.083
P	0.002	0.008	0.009	0.010	0.009	0.008	0.009	0.009	0.009
S	0.0004	0.0043	0.0041	0.0042	0.0043	0.0042	0.0042	0.0042	0.0042
Fe	-	bal.	bal.	bal.	bal.	bal.	bal.	bal.	bal.

The measured chemical composition of the examined specimens was compared to the material specification listed in Table 1. In all cases, the chemical composition meets the specifications. The chemical composition of the lower part was compared to the chemical composition of the upper part for each specimen and for each chemical element. Considering the measurement uncertainty, no significant differences between the upper and lower part of each specimen can be recognized. In addition, no significant differences to the chemical composition of the powder can be recognized.

The results of the measurement of the oxygen content conducted at two separate specimens are depicted in Table 8.

Table 8. Oxygen content in weight %.

Measurement Uncertainty	Short ILT High VED		Intermediate ILT Basis VED	
	Lower Part	Upper Part	Lower Part	Upper Part
0.004	0.036	0.034	0.031	0.031

4. Discussion

4.1. Surface Temperatures

The preheating temperature plots reveal three clear trends. First, the preheating temperature increases over the entire build height of the specimens. This confirms similar results from previous work [5] and the work of Williams et al. [34]. As long as the preheating temperature rises over the build height, there is no equilibrium between the rate of heat input and the rate of heat dissipation [34]. Heat dissipates mainly by thermal conduction into the build and the base plate due to strong insulating effects of the surrounding powder [22]. Therefore, the heat dissipation is mainly governed by thermal conductivity and the geometry of the build. For a constant geometry and a given material, the heat dissipation via heat conduction can be shortened by reducing the time before the next energy input, i.e., by reducing the ILT.

This directly leads to the second trend observed in the plots: the increase of preheating temperature is significantly affected by a change of the time for heat dissipation through varying ILT. Shorter ILTs allow for a shorter time for heat dissipation resulting in increasing preheating temperatures. A massive heat accumulation is observable in the most extreme case (short ILT and high VED) of the parameter matrix, leading to temperatures of up to approximately 600 °C. In contrast, the preheating temperature in the least extreme case (long ILT and low VED) level was up to approximately 140 °C. The magnitude of this intrinsic preheating effect is material-specific due to material-specific thermal conductivity. It would be expected to be reduced for materials with higher thermal conductivity.

Third, the increase of preheating temperature also depends significantly on the energy input varied by the distinct processing parameters. This results in increased temperatures at a reduced scanning velocity. It is well known that the energy input is higher at slower scanning velocities resulting in higher specimen temperatures [40,41].

Oxidation layers may drastically change the emissivity of metallic surfaces, which is a well-known phenomenon [42]. Oxidation phenomena were argued to be responsible for drastic changes in emissivity values at temperatures above 580 °C in the previously conducted experiments for the determination of emissivities [35]. Furthermore, oxidation-driven tempering colors could be noticed for short ILT specimens, especially for those at basis VED and high VED [5]. Therefore, the interpretation of calculated temperatures above 580 °C should be considered very carefully in this study. However, there are two aspects that back the reliability of the measured temperature values despite oxidation of the specimen's bulk surface. First, oxidation thickness growth depends always on atmosphere, temperature and time [35]. The atmosphere can be assumed to be the same for all individual specimens of the different ILT levels, as they were produced within the same build process with a low oxygen content (below 0.1% [5]). The suspect temperatures are not far beyond the revealed detrimental temperature threshold of 580 °C. The time for oxidation of the recoated powder layer before the measurement signal extraction is comparably short, i.e., below 15 s in the case of the L-PBF process with short ILT. In comparison, Janssen [43] studied oxidation processes at austenitic stainless steel AISI 304 in air and noticed the start of slight yellow annealing colors by eye at 550 °C at 5 min holding time. In addition, the new recoated powder layer did not undergo the temperature cycle of the L-PBF bulk material. Therefore, it did not face very high temperatures prone to oxidation. Second, the relative comparison of the preheating temperatures of the distinct VED levels shows a constant ratio irrespective of the individual ILT. Hence, within the light of the given measurement uncertainty, the presented temperature values can be directly used for comparison. Potential oxidation is assumed to not have affected the emissivity of the powder surface significantly before the recording of the extracted IR signal.

4.2. Grain Size Analysis

Three clear trends can be derived from the measurements: First, HAGB grain sizes and LAGB sub-grain sizes show the same qualitative differences regarding VED, ILT and build height. Second, the grain sizes and, respectively, the sub-grain sizes increased with increasing VED for every ILT. Third, for every parameter combination, the upper part sections exhibit a higher grain size and sub-grain size as compared to the lower sections. However, this increase is small for long ILT and intermediate ILT compared to short ILT specimens. For the short ILT specimens, the difference in mean grain sizes, respectively, mean sub-grain sizes, can be higher than a factor of two, as can be seen for basis VED and high VED. The measured mean HAGB grain sizes as well as mean LAGB grain sizes are in the same order of magnitude as examined elsewhere.

In the previous publication [5], no significant difference in mean values of the sub-grain sizes of the lower part sections and upper part sections of the short ILT specimens could be examined by the applied manual measurement via line interception of light microscopy images. This was obviously due to a very high degree of measurement uncertainty. An in-situ heat treatment during the process was proposed as a potential explanation. However, this proposition can now be clearly disproven by the EBSD results, as clear differences in the mean grain size and the mean sub-grain size between upper and lower part can be seen whenever the preheating temperature was also increased strongly. The average preheating temperature of each specimen between layer 2140 and layer 2200, which is within the upper part sections, and the LAGB sub-grain sizes are depicted within the same diagram in Figure 12. This visualizes the high degree of correlation between the preheating temperature as a boundary condition for solidification and grain size development. The noticeable discrepancy between a comparatively high preheating temperature and still small mean grain size in the specimen of short ILT and low VED as compared to, e.g., the specimen of long ILT and high VED, should not be considered without referring to changes in melt pool depth as presented in [5]. The preheating temperature shows an effect on the development of the microstructure. The latter is also known to be strongly affected by the melt pool dimensions [3], which are comparatively small for the low VED value due to the higher scanning velocity [5]. Therefore, the temperature influence should be rather considered within the individual VED levels for direct comparison.

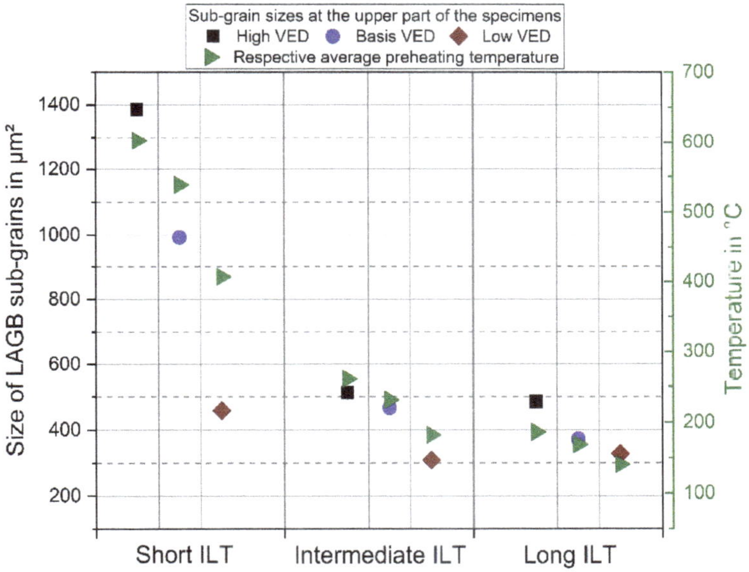

Figure 12. Comparison of LAGB sub-grain sizes and preheating temperatures for each parameter combination in the upper part sections. The preheating temperature values depict the average of the measured preheating temperature of the respective specimen between layer 2140 and 2200.

4.3. Cellular Substructure

The calculated average cell size is in the same order of magnitude as reported in other literature, where cell sizes in the range of 0.5 µm to 1 µm or 1.5 µm, depending on processing parameters, were measured [15,39,44,45]. Leicht et al. [46] presented slightly smaller cell sizes in a range of 0.36 µm to 0.58 µm.

Roehling et al. [30] described the propagation rate of the solid liquid interface being linked with the scanning velocity by its product with the cosine of the angle between the laser scanning direction and the solidification direction. Typical values are in the range of 0.012 $m \cdot s^{-1}$ to 0.12 $m \cdot s^{-1}$ and, therefore, significantly smaller than the scanning velocity v_s [9]. These values can be estimated for scanning velocities of about 700 $mm \cdot s^{-1}$ assuming an angle between the direction of the maximum heat flow and the build direction between 0° and 10°, as supposed by DebRoy et al. [3]. Thermal gradients G are reported in the range from 10^4 $K \cdot m^{-1}$ to 10^7 $K \cdot m^{-1}$ and being rather at the top of this range in the case of L-PBF as compared to direct energy deposition [9]. Pinomaa et al. [9] conducted a phase field simulation of the rapid directional solidification of 316L and measured the cell sizes for different local melt pool solidification rates R and different thermal gradients G. They examined a clear cellular growth mode of solidification over a broad range of the typically reported thermal gradient ranges and solidification rates [9].

In the case of cellular growth mode, which also appeared in the examined specimens of the present study, a general trend of a decrease of the cellular size with increasing solidification rate as well as increasing thermal gradients could be revealed from the work of Pinomaa et al. [9]. Their 2D phase field model brought up cell sizes in the range of 0.64 µm to 4.1 µm for the pure cellular growth mode. When comparing the values of Figure 11 to their simulation results, the figures are again in the same order of magnitude for their higher R and different thermal gradients G. Qualitatively, the relationship between increasing cell size with decreasing thermal gradients tends to appear for the short ILT, suggesting that the increased preheating temperature (see Figure 4) decreased the thermal gradient of cooling. Additionally, at lower scanning velocities v_s (higher VED) the cell size increased accordingly.

The good agreement to the simulation results, published in [9], seems to be surprising since the melt pool geometry is reported to have a huge impact on the heat and mass transfer within the melt pool [47,48]. Differences in melt pool depth were examined for the varying processing conditions in the previous study [5]. In addition, the solidification rate R as well as the local thermal gradient G vary over the cross section of the melt pool [30,47,48]. Yadroitsev et al. [49] have shown the sensitivity of the cell spacing to the scanning velocity and the location of measurement within single track melt pools. At a first glance, this would appear to complicate valid comparisons of cell size measurements as the selection of the measurement region within the cross sections might affect the result strongly. In fact, the measurement regions in the SEM images of the same cross section subjectively appeared to show huge variation in the cell size. This is assumed to be the reason for the comparably huge deviation bars in Figure 11, as similarly concluded by Leicht et al. [46]. However, the potential melt pool cross section areas to be investigated within the bulk of the specimens are limited to fragments of the lower part of the melt pool in this study. This is due the layer-wise remelting and overlapping of melt pools. Yadroitsev et al. [49] measured gradual increasing differences in cell spacing of up to a factor of 2 between upper and lower part of the melt pool. A comparison of the lower part and the middle section of the melt pool showed only a difference of a factor of 1.3. Therefore, it can be assumed that the variations in the measurable cell spacings are reduced due to remelting. Additionally, it is assumed that the primary solidification structures remain stable and are not affected significantly in any secondary heat cycle. This was also one of the critical model assumptions of the phase field simulation by Pinomaa et al. [9].

Deng et al. [39] conducted recently very fundamental investigations on the thermal stability of the cellular substructure which consists of dislocations. They hold 316L L-PBF specimens at elevated temperatures of 500 °C, 600 °C, and 700 °C for up to 150 h. No recrystallization was observed at these temperatures. They eliminated a lack of knowledge about the behavior of the dislocation network at elevated temperatures below the often-reported dislocation dissolution temperatures above 850 °C [16]. The cellular substructure remains stable at 600 °C for up to 100 h. At 700 °C, the decomposition of the dislocation cells was already visible after a 10 h annealing. The dislocation cells showed a uniform growth along all directions at this heating condition. This growth was related to a rearrangement and coarsening of dislocation structures. It did not occur homogeneously over the entire cross section areas that were investigated. The growth proceeded very slightly when increasing the annealing time up to 150 h. The findings of Deng et al. [39] exclude a potential in-situ annealing effect as a reason for the measured differences in the cell spacing in this study. This was suggested as a potential reason for differences in hardness values in the previous study [5]. However, this does not apply since the measured preheating temperatures are well below the threshold of 700 °C. Therefore, the differences in the feature size of the cellular substructure of this study are assumed to completely develop during solidification. This also supports the consideration of differences of G and R for being the main cause of the differences in feature size as discussed above.

4.4. Chemical Composition

No significant differences between the upper and lower part of each specimen or to the chemical composition of the powder can be recognized. However, small differences in the oxygen content of the "short ILT, high VED" specimen and the "intermediate ILT, basis VED" specimen are detected. Although these differences (0.003 weight % to 0.005 weight %) barely exceed the measurement uncertainty, a closer look into these differences seems to be reasonable.

At the first glance, this would even correlate well with the different preheating temperatures (see Figures 4 and 5) of the parts examined here and with the different annealing colors at the outer surfaces of the specimens observed in [5]. However, looking in more detail, such a perceived view cannot explain the difference between the lower part of the short ILT specimen and the upper part of the intermediate ILT specimen. Both face similar preheating conditions as shown in Section 3.1. It cannot explain a perceived slightly higher oxygen content in the lower part of the short ILT specimen compared to its upper part, although oxidation of the L-PBF surface in its upper part was clearly visible. In addition, the preheating temperatures were in a temperature range prone to surface oxidation. Surface oxides were not measured in the oxygen analysis since hydrochloric acid etching was conducted before the measurement. Hence, only the oxygen intake in the bulk material was measured.

However, short ILT and intermediate ILT specimens were manufactured in two different processes. Therefore, the logging data of the internal oxygen lambda probes of the L-PBF system have to be examined. As the experienced L-PBF user of this specific machine knows, the process starts when the residual oxygen concentration in the process chamber is below 0.1%. For the working principle of the gas flushing regarding the maintenance of a low oxygen content, one is referred to the detailed explanation given by Pauzon et al. [50] for another L-PBF machine with a similar principle. After reaching 0.1%, the concentration measured at the lambda probes usually levels down in the beginning of the process to approximately 0.02%. Figures 13 and 14 depict the oxygen concentration in the atmosphere of the process chamber during the first 300 min of the short ILT process and intermediate ILT process respectively. The upper x-axis of these plots shows the respective build height, which must be substantially higher within the same process time for short ILT. It can be derived from the two diagrams that the residual oxygen content during manufacturing of the lower ex-situ volumes of analysis (build height between 2.5 mm and 12.5 mm, see Figure 1) was higher for the short ILT process compared to the intermediate ILT process. Therefore, this can be presumed to cause the higher weight % of oxygen in the lower part of the short ILT specimen measured by chemical analysis, as listed in Table 8. This can be supported by findings of similar magnitude by Dietrich et al. [51]. They reported an increase in oxygen content by 0.0188 weight % in L-PBF Ti6Al4V bulk material manufactured in a process chamber atmosphere of 0.0977 weight % oxygen concentration compared to a process with 0.0002 weight % oxygen concentration.

The reason for the difference in the time span until the oxygen content in the process chamber leveled down cannot be clarified within the frame of the experimental set-up of this study. Potentially, the residual oxygen consumption in the process chamber is affected by the number of parts or (more precisely) the ratio of area exploitation. While only the three specimens, which were in the field of view of the IR camera, were manufactured in the short ILT process, additional 15 specimens of the same size were manufactured in the intermediate ILT process. Hence, the faster decrease of residual oxygen concentration during the build-up could be related to a bigger surface area available for oxygen intake in the intermediate ILT process.

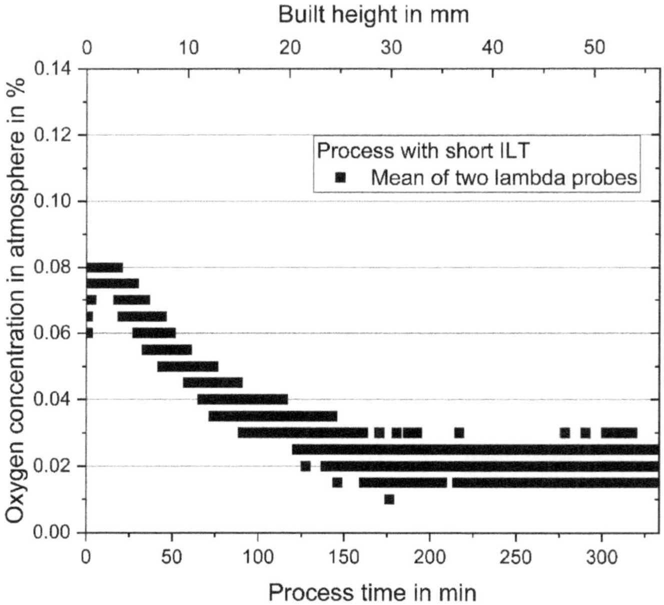

Figure 13. Oxygen concentration in the process chamber during the first 300 min of the short ILT process.

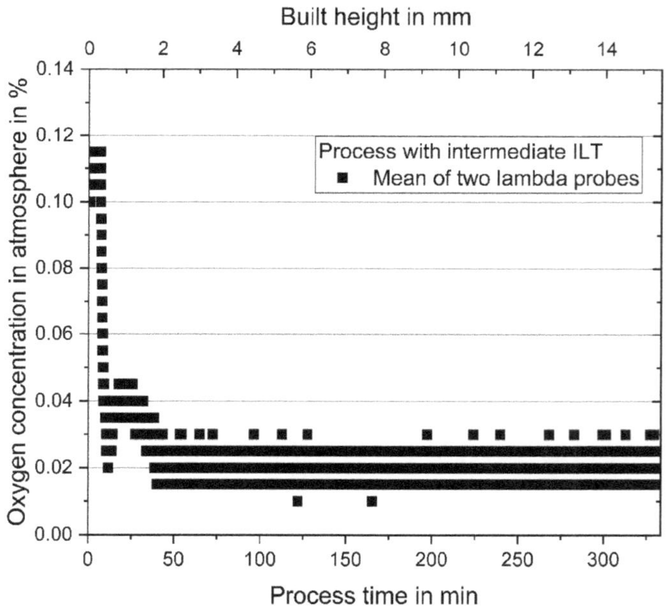

Figure 14. Oxygen concentration in the process chamber during the first 300 min of the intermediate ILT process.

After leveling down of the oxygen concentration in the process chamber to approximately 0.02%, these values kept constant over the rest of each build process. Hence, the small differences in the upper part of the two specimens, which are smaller than the measurement uncertainty, could be a result of higher oxygen intake into the L-PBF bulk due to different preheating temperatures. The specimens with shorter ILT showed slightly higher oxygen content than specimens with longer ILT. This suggestion can be supported by recent findings of Pauzon et al. [52]. They reported that a reduced ILT can lead to increased oxygen pick-up during L-PBF manufacturing of Ti6Al4V specimens. They measured differences in oxygen pick-up of approximately 0.05 weight % for specimens at 70 mm build height manufactured with different ILT, when the oxygen concentration in the process atmosphere was below 0.1 weight %. However, as the measurement uncertainty is high in comparison to the discussed difference, further examinations should be done in future work.

The measured quantities of oxygen content are well within the range of published values in the literature for the same material and process. A recent study from Pauzon et al. [50], dedicated to process gas influence during L-PBF processing of 316L, revealed oxygen concentrations of approximately 0.0424 weight % in the bulk of their specimens. This is not only in good agreement in terms of magnitude but also quite close to the values examined here. Lou et al. [23] measured an oxygen content of about 0.0384 weight % at 316L L-PBF specimens. Pauzon [53] also highlighted the solubility limit of oxygen in austenite being estimated by Kitchener et al. [54] at about 0.003 weight % +/− 0.003 weight %. Oxygen content above this value is expected to be connected to secondary phase oxide inclusions, mainly with elements such as Cr, Mn, or Si [53]. Hints about the existence of such nanosized precipitations in 316L can be found, e.g., in the work of Liverani et al. [17], Saedi et al. [55], and Sun et al. [18]. Krakhmalev et al. [16] described the size of these particles in the range of 15 nm to 100 nm. Detrimental influences on impact toughness were discussed by Lou et al. [23].

4.5. Subsumption of the Results with Regard to Hardness

In this section, the variations of the analyzed features are discussed with regard to hardness values obtained in the previous study [5]. A significant hardness drop in the upper part sections of the short ILT specimens as compared to the lower part sections was revealed [5], despite relatively low defect densities. At intermediate ILT, only a slight decrease in hardness over the build height was recognized. The range of hardness values was at a similar level in the upper and the lower part sections at long ILT. The hardness values in the upper part sections are transcribed from [5] into a diagram that again shows the sub-grain sizes of the respective upper parts, see Figure 15. The hardness values are discussed in the following as representative for material strength. The strength of metallic materials is the sum of the following contributions: contributions from dislocations, from grain boundaries, from solid solutions, and from precipitations [56]. Hence, the identified differences in grain sizes as well as in cell spacing should be discussed with respect to hardness.

An inverse relationship between hardness values and LAGB sub-grain sizes can be seen in the direct comparison diagram in Figure 15. The same relationship would be visible also for the HAGB grain sizes, as a comparison between the results in Section 3.2 indicates (see Figures 8 and 9). Both microstructural feature sizes seem to obey a Hall-Petch type relationship between grain sizes and material strength. However, as reported at many places the high dislocation density of the cellular substructure within the sub-grains plays an additional and very important role in strengthening of 316L material produced by L-PBF [9,15,39,57]. Besides the measured grain sizes, the measured cell spacing also shows an inverse relationship regarding hardness. For the most extreme parameter combination (short ILT, high VED, and upper part), a contribution of keyhole porosity to the hardness drop cannot be excluded entirely [5]. However, for all other specimens the defect density was low enough for not being assumed to affect hardness.

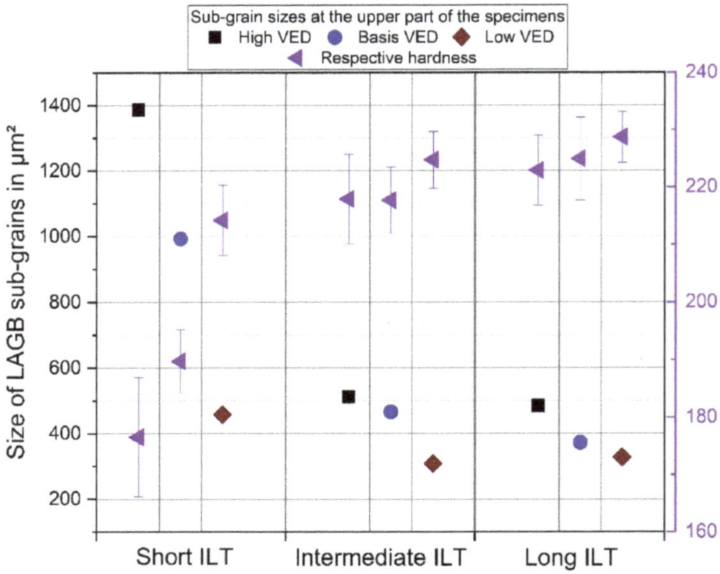

Figure 15. Comparison of LAGB sub-grain sizes of the upper part sections and the respective hardness values (Adapted from ref. [5]) for each parameter combination.

Deng et al. [39] conducted very fundamental annealing experiments. They showed that a decrease in hardness already appears before the onset of any recrystallization as a matter of changes in dislocation cell size. Under the assumption of changes in dislocation cell size only, the cell size is allowed to be used for a direct correlation to strength based on a Hall-Petch relationship as proofed valid by Deng et al. [39]. However, they also showed that the correlation factor k, which is normally assumed to be a constant factor in a Hall-Petch type relationship, can change non-linearly due to different dislocation annihilation behavior depending, e.g., on the annealing time. This is already relevant below the starting temperature of a complete annihilation of dislocations, below 800 °C. In simplified terms, not only the cell size of the dislocation structure but also the dislocation density within the cell boundaries contribute to the material strength. However, this can only be studied by intense use of transmission electron microscopy.

A direct transfer from the annealing experiments of Deng et al. [39] to the results of the present study does not work since it can be assumed that the differences in microstructural features of the specimens of this study are not caused by annealing but by differences in the initial conditions of solidification. This assumption is supported by differences in grain sizes between upper part and lower part sections of the same specimens as discussed in Section 4.2. This cannot be the result of any recrystallization or annealing grain growth, as otherwise the cellular substructure would have been dissolved. Therefore, the primary cause of the differences in the feature size of the cellular structure as well as in the LAGB and HAGB grain sizes is assumed to be related to the differences in solidification rate and thermal gradients. Both are affected by the interplay of VED and ILT variations as well as by the build height. The evolution of the preheating conditions over the build height for the different parameter combinations clearly indicates a change of the initial thermal conditions of solidification. An increase of grain size with decreasing cooling rates was already reported by Zitelli et al. [1] and Keshavarzkermani et al. [14].

Interestingly, Bang et al. [58] recently investigated changes in microstructure and hardness i.a. of small 10 mm cubic L-PBF specimens of 316L over a broad range of parameter combinations of laser power (80 W to 480 W) and scanning velocity (493 $mm \cdot s^{-1}$ to 2958 $mm \cdot s^{-1}$). For all parameter combinations with resulting high density, their microstructural characterization revealed an increase of cell size and grain size in proportion to the applied VED except for one outlier at the highest VED condition. A decrease in hardness from approximately 220 HV0.5 to approximately 180 HV0.5 was exhibited with increasing VED, which they related to the inverse relationship between hardness and grain sizes. As mentioned elsewhere, a direct comparison of VED values is not acceptable in its entirety and has to be done very carefully, see, e.g., [59,60]. However, when considering a comparison of the hardness values of Bang et al. [58] with the values of the present study—its pre-study [5]—it becomes clear that Bang et al. [58] used a much broader range of processing parameters (laser power and scanning velocity) than in the present study. Thereby, they provoked similar conditions in terms of resulting hardness and trends of increasing microstructural feature size at their 10 mm cubic specimens. These conditions were induced by a relatively narrow variation of the VED in combination with an application relevant build height of over 100 mm and decreasing ILTs in the present study.

This in total emphasizes the importance of ILT in combination with build height on the development of the microstructure and the resulting mechanical properties in L-PBF, as together they seem to be able to drastically shift the processing window, gained from typical process parameter studies.

5. Conclusions

An in-depth analysis of the microstructural feature sizes of 316L specimens produced at different VED levels and different ILT levels has been conducted. Furthermore, the evolution of the intrinsic preheating temperature over the complete build-up of specimens was monitored by use of a thermographic in-situ monitoring set-up.

Several influencing factors and their implications have been identified.

1. Preheating temperature: An increase of VED can rise the preheating temperature as the heat input into the material is increased. The rise of preheating temperature by a decrease of ILT is much more significant. This is related to the reduced time for heat dissipation. Temperature measurements have revealed a change in preheating temperature by more than a factor of 2 in the upper part of the specimens for short ILT (18 s) compared to longer ILT (65 s and 116 s). Intrinsic preheating temperatures of up to about 600 °C were revealed. In turn, this resulted in heterogeneity of the microstructure and differences in material properties within the same specimen, as specified below.
2. Grain sizes: A significant increase of grain sizes and sub-grain sizes was identified in sections of specimens with increased preheating temperature. Differences in grain size of more than a factor of 2.5 were found within the same specimen, which was attributed to the variations in build height and parameter combination.
3. Spacing of cellular substructure: The measurement of cell spacing is handicapped by significant measurement uncertainty due to the high degree of local changes within very small areas, i.e., within the size of individual melt pools. Despite this scatter, a trend to increasing cell sizes was observed and was related to differences in solidification rate and thermal gradients induced by differences in scanning velocity and preheating temperature. The average cell size within this investigation was between 0.52 µm and 0.76 µm, depending on the parameter combination.
4. Hardness: Eventually, the examined and discussed differences in grain sizes and cell sizes were related to differences in hardness examined in a previous study [5]. A general trend of decreasing hardness (from 221 HV1 to 176 HV1) with increasing microstructural feature size was revealed.

Furthermore, there was a slight tendency of increasing oxygen intake in regions of high preheating temperature. However, the basis for these oxygen measurements is quite limited, and further investigations in this field are required. This will also include better control of the boundary conditions of oxygen content within the L-PBF process.

The findings of this study are strongly linked to intrinsic changes in preheating temperature during the L-PBF process. The causes of these significant changes were identified to be related to processing parameters such as scanning velocity (affecting VED) but also to build height and ILT, which are overlooked in many other cases in the literature. When considering real part geometries and current trends in the development of new L-PBF machines (e.g., multi laser machines), a decrease in ILT can be expected during manufacturing. This will make the issue of differences in microstructure and mechanical properties due to intrinsic preheating temperature changes more severe. Therefore, the authors want to close with the recommendation to always include the ILT into the process documentation to enhance the comparability of measurement results of L-PBF products.

Author Contributions: Conceptualization, G.M.; methodology, G.M., K.S. and T.K.; software, G.M. and T.K.; validation, G.M., K.S. and T.K.; formal analysis, G.M., K.S., T.K. and S.R.; investigation, G.M., K.S., T.K. and S.R.; resources, S.J.A., K.H.; data curation, G.M., K.S. and T.K.; writing—original draft preparation, G.M., T.K.; writing—review and editing, G.M., K.S., T.K., S.J.A., S.R., D.B. and K.H,.; visualization, G.M.; supervision, K.H.; project administration, S.J.A. All authors have read and agreed to the published version of the manuscript.

Funding: This research was funded by BAM within the focus area materials.

Institutional Review Board Statement: Not applicable.

Informed Consent Statement: Not applicable.

Data Availability Statement: Not applicable.

Acknowledgments: The authors would like to thank M. Ostermann for XRF measurements, A. Meckelburg for determination of non-metals, J. Roik for ICP-OES measurements, R. Saliwan Neumann for SEM images, and A. Charmi for support in generating IPF maps.

Conflicts of Interest: The authors declare no conflict of interest. The funders had no role in the design of the study; in the collection, analyzes, or interpretation of data; in the writing of the manuscript; or in the decision to publish the results.

References

1. Zitelli, C.; Folgarait, P.; di Schino, A. Laser powder bed fusion of stainless steel grades: A review. *Metals* **2019**, *9*, 731. [CrossRef]
2. Gralow, M.; Weigand, F.; Herzog, D.; Wischeropp, T.; Emmelmann, C. Biomimetic design and laser additive manufacturing—A perfect symbiosis? *J. Laser Appl.* **2020**, *32*, 021201. [CrossRef]
3. DebRoy, T.; Wei, H.L.; Zuback, J.S.; Mukherjee, T.; Elmer, J.W.; Milewski, J.O.; Beese, A.M.; Wilson-Heid, A.; De, A.; Zhang, W. Additive manufacturing of metallic components—Process, structure and properties. *Prog. Mater. Sci.* **2018**, *92*, 112–224. [CrossRef]
4. Wycisk, E.; Munsch, M.; Schmidt-Lehr, M. *AMPOWER Report 2021: Additive Manufacturing Market*; Report; AMPOWER GmbH & Co. KG: Hamburg, Germany, 2021.
5. Mohr, G.; Altenburg, S.J.; Hilgenberg, K. Effects of inter layer time and build height on resulting properties of 316L stainless steel processed by laser powder bed fusion. *Addit. Manuf.* **2020**, *32*, 101080. [CrossRef]
6. Sanaei, N.; Fatemi, A.; Phan, N. Defect characteristics and analysis of their variability in metal L-PBF additive manufacturing. *Mater. Des.* **2019**, *182*, 108091. [CrossRef]
7. Gorelik, M. Additive manufacturing in the context of structural integrity. *Int. J. Fatigue* **2017**, *94*, 168–177. [CrossRef]
8. Zerbst, U.; Bruno, G.; Buffiere, J.-Y.; Wegener, T.; Niendorf, T.; Wu, T.; Zhang, X.; Kashaev, N.; Meneghetti, G.; Hrabe, N.; et al. Damage tolerant design of additively manufactured metallic components subjected to cyclic loading: State of the art and challenges. *Prog. Mater. Sci.* **2021**, 100786. [CrossRef]
9. Pinomaa, T.; Lindroos, M.; Walbrühl, M.; Provatas, N.; Laukkanen, A. The significance of spatial length scales and solute segregation in strengthening rapid solidification microstructures of 316L stainless steel. *Acta Mater.* **2020**, *184*, 1–16. [CrossRef]
10. Mohr, G.; Scheuschner, N.; Hilgenberg, K. In situ heat accumulation by geometrical features obstructing heat flux and by reduced inter layer times in laser powder bed fusion of AISI 316L stainless steel. *CIRP Proceedia* **2020**, *94*, 155–160. [CrossRef]
11. Seifi, M.; Gorelik, M.; Waller, J.; Hrabe, N.; Shamsaei, N.; Daniewicz, S.; Lewandowski, J.J. Progress towards metal additive manufacturing standardization to support qualification and certification. *Jom* **2017**, *69*, 439–455. [CrossRef]
12. Ranjan, R.; Ayas, C.; Langelaar, M.; van Keulen, F. Towards design for precision additive manufacturing: A simplified approach for detecting heat accumulation. In Proceedings of the Advancing Precision in Additive Manufacturing: 2018 ASPE and EUSPEN Summer Topical Meeting, Berkely, CA, USA, 22–25 July 2018.
13. Ranjan, R.; Ayas, C.; Langelaar, M.; van Keulen, F. Fast detection of heat accumulation in powder bed fusion using computationally efficient thermal models. *Materials* **2020**, *13*, 4576. [CrossRef]

14. Keshavarzkermani, A.; Marzbanrad, E.; Esmaeilizadeh, R.; Mahmoodkhani, Y.; Ali, U.; Enrique, P.D.; Zhou, N.Y.; Bonakdar, A.; Toyserkani, E. An investigation into the effect of process parameters on melt pool geometry, cell spacing, and grain refinement during laser powder bed fusion. *Opt. Laser Technol.* **2019**, *116*, 83–91. [CrossRef]
15. Wang, Y.M.; Voisin, T.; McKeown, J.T.; Ye, J.; Calta, N.P.; Li, Z.; Zeng, Z.; Zhang, Y.; Chen, W.; Roehling, T.T.; et al. Additively manufactured hierarchical stainless steels with high strength and ductility. *Nat. Mater.* **2018**, *17*, 63–71. [CrossRef] [PubMed]
16. Krakhmalev, P.; Fredriksson, G.; Svensson, K.; Yadroitsev, I.; Yadroitsava, I.; Thuvander, M.; Peng, R. Microstructure, solidification texture, and thermal stability of 316 L stainless steel manufactured by laser powder bed fusion. *Metals* **2018**, *8*, 643. [CrossRef]
17. Liverani, E.; Toschi, S.; Ceschini, L.; Fortunato, A. Effect of selective laser melting (SLM) process parameters on microstructure and mechanical properties of 316L austenitic stainless steel. *J. Mater. Process. Technol.* **2017**, *249*, 255–263. [CrossRef]
18. Sun, Z.; Tan, X.; Tor, S.B.; Yeong, W.Y. Selective laser melting of stainless steel 316L with low porosity and high build rates. *Mater. Des.* **2016**, *104*, 197–204. [CrossRef]
19. Aggarwal, A.; Patel, S.; Kumar, A. Selective laser melting of 316L stainless steel: Physics of melting mode transition and its influence on microstructural and mechanical behavior. *Jom* **2018**, *71*, 1105–1116. [CrossRef]
20. Patel, S.; Vlasea, M. Melting modes in laser powder bed fusion. *Materialia* **2020**, *9*, 100591. [CrossRef]
21. David, S.; Vitek, J.; Reed, R.; Hebble, T. *Effect of Rapid Solidification on Stainless Steel Weld Metal Microstructures and Its Implications on the Schaeffler Diagram*; Technical Report; Oak Ridge National Lab.: Oak Ridge, TN, USA, 1987.
22. Bajaj, P.; Hariharan, A.; Kini, A.; Kürnsteiner, P.; Raabe, D.; Jägle, E.A. Steels in additive manufacturing: A review of their microstructure and properties. *Mater. Sci. Eng. A* **2020**, *772*, 138633. [CrossRef]
23. Lou, X.; Andresen, P.L.; Rebak, R.B. Oxide inclusions in laser additive manufactured stainless steel and their effects on impact toughness and stress corrosion cracking behavior. *J. Nucl. Mater.* **2018**, *499*, 182–190. [CrossRef]
24. Grasso, M.; Colosimo, B.M. Process defects and in situ monitoring methods in metal powder bed fusion: A review. *Meas. Sci. Technol.* **2017**, *28*, 0440051. [CrossRef]
25. Gordon, J.V.; Narra, S.P.; Cunningham, R.W.; Liu, H.; Chen, H.; Suter, R.M.; Beuth, J.L.; Rollett, A.D. Defect structure process maps for laser powder bed fusion additive manufacturing. *Addit. Manuf.* **2020**, *36*, 101552. [CrossRef]
26. Mukherjee, T.; DebRoy, T. Mitigation of lack of fusion defects in powder bed fusion additive manufacturing. *J. Manuf. Process.* **2018**, *36*, 442–449. [CrossRef]
27. Bayat, M.; Thanki, A.; Mohanty, S.; Witvrouw, A.; Yang, S.; Thorborg, J.; Tiedje, N.S.; Hattel, J.H. Keyhole-induced porosities in Laser-based Powder Bed Fusion (L-PBF) of Ti6Al4V: High-fidelity modelling and experimental validation. *Addit. Manuf.* **2019**, *30*, 100835. [CrossRef]
28. Mohr, G.; Johannsen, J.; Knoop, D.; Gärtner, E.; Hummert, K.; Emmelmann, C. Processing of a high-strength Al-Fe-Ni alloy using laser beam melting and its potential for in-situ graded mechanical properties. In Proceedings of the Lasers in Manufacturing Conference, Munich, Germany, 26–29 June 2017.
29. Xu, W.; Lui, E.W.; Pateras, A.; Qian, M.; Brandt, M. In situ tailoring microstructure in additively manufactured Ti-6Al-4V for superior mechanical performance. *Acta Mater.* **2017**, *125*, 390–400. [CrossRef]
30. Roehling, T.T.; Wu, S.S.Q.; Khairallah, S.A.; Roehling, J.D.; Soezeri, S.S.; Crumb, M.F.; Matthews, M.J. Modulating laser intensity profile ellipticity for microstructural control during metal additive manufacturing. *Acta Mater.* **2017**, *128*, 197–206. [CrossRef]
31. Shi, R.; Khairallah, S.A.; Roehling, T.T.; Heo, T.W.; McKeown, J.T.; Matthews, M.J. Microstructural control in metal laser powder bed fusion additive manufacturing using laser beam shaping strategy. *Acta Mater.* **2020**, *184*, 284–305. [CrossRef]
32. Kempen, K.; Vrancken, B.; Thijs, L.; Buls, S.; van Humbeeck, J.; Kruth, J.-P. Lowering thermal gradients in selective laser melting by pre-heating the baseplate. In Proceedings of the Solid Freeform Fabrication Symposium Proceedings, Austin, TX, USA, 12–14 August 2013.
33. Shiomi, M.; Osakadaa, K.; Nakamuraa, K.; Yamashitaa, T.; Abeb, F. Residual stress within metallic model made by selective laser melting process. *CIRP Ann.* **2004**, *53*, 195–198. [CrossRef]
34. Williams, R.J.; Piglione, A.; Rønneberg, T.; Jones, C.; Pham, M.-S.; Davies, C.M.; Hooper, P.A. In situ thermography for laser powder bed fusion: Effects of layer temperature on porosity, microstructure and mechanical properties. *Addit. Manuf.* **2019**, *30*, 100880. [CrossRef]
35. Mohr, G.; Nowakowski, S.; Altenburg, S.J.; Maierhofer, C.; Hilgenberg, K. Experimental determination of the emissivity of powder layers and bulk material in laser powder bed fusion using infrared thermography and thermocouples. *Metals* **2020**, *10*, 1546. [CrossRef]
36. Günter Briefs. *Nichtrostende Stähle—Teil 3: Technische Lieferbedingungen für Halbzeug, Stäbe, Walzdraht, Gezogenen Draht, Profile und Blankstahlerzeugnisse aus korrosionsbeständigen Stählen für allgemeine Verwendung*; DIN: Berlin, Germany, 2014.
37. Elisabeth Leistner. *Metallographic Instructions for Colour Etching by Immersion—Part III: Non-Ferrous Metals, Cemented Carbides and Ferrous Metals, Nickel-Base and Cobalt-Base Alloys*; D.V.S.-Verlag: Düsseldorf, Germany, 1998.
38. DIN EN ISO 643. *Mikrophotographische Bestimmung der Erkennbaren Korngröße*; DIN: Berlin, Germany, 2019.
39. Deng, P.; Yin, H.; Song, M.; Li, D.; Zheng, Y.; Prorok, B.C.; Lou, X. On the Thermal stability of dislocation cellular structures in additively manufactured austenitic stainless steels: Roles of heavy element segregation and stacking fault energy. *Jom* **2020**, *72*, 4232–4243. [CrossRef]
40. Heigel, J.C.; Lane, B.M. Measurement of the melt pool length during single scan tracks in a commercial laser powder bed fusion process. *J. Manuf. Sci. Eng.* **2018**, *140*, 051012–051017. [CrossRef]

41. Mohr, G.; Altenburg, S.J.; Ulbricht, A.; Heinrich, P.; Baum, D.; Maierhofer, C.; Hilgenberg, K. In-situ defect detection in laser powder bed fusion by using thermography and optical tomography—Comparison to computed tomography. *Metals* **2020**, *10*, 103. [CrossRef]
42. Vollmer, M.; Möllmann, K.-P. *Infrared Thermal Imaging: Fundamentals, Research and Applications*; John Wiley & Sons: Berlin, Germany, 2017.
43. Joachim, J. Untersuchung von Oxidationsprozessen an Oberflächen von FeCr-Legierungen und Austenitstahl Mittels Röntgenabsorptionsspektroskopie unter Streifendem Einfall. Ph.D. Thesis, Rheinische Friedrich-Wilhelms-Universität Bonn, Bonn, Germany, 2003.
44. Zhong, Y.; Liu, L.; Wikman, S.; Cui, D.; Shen, Z. Intragranular cellular segregation network structure strengthening 316L stainless steel prepared by selective laser melting. *J. Nucl. Mater.* **2016**, *470*, 170–178. [CrossRef]
45. Choo, H.; Sham, K.-L.; Bohling, J.; Ngo, A.; Xiao, X.; Ren, Y.; Depond, P.J.; Matthews, M.J.; Garlea, E. Effect of laser power on defect, texture, and microstructure of a laser powder bed fusion processed 316L stainless steel. *Mater. Des.* **2019**, *164*, 107534. [CrossRef]
46. Leicht, A.; Rashidi, M.; Klement, U.; Hryha, E. Effect of process parameters on the microstructure, tensile strength and productivity of 316L parts produced by laser powder bed fusion. *Mater. Charact.* **2020**, *159*, 110016. [CrossRef]
47. David, S.; Babu, S.; Vitek, J. Welding: Solidification and microstructure. *Jom* **2003**, *55*, 14–20. [CrossRef]
48. Stoudt, M.R.; Williams, M.E.; Levine, L.E.; Creuziger, A.; Young, S.A.; Heigel, J.C.; Lane, B.M.; Phan, T.Q. Location-specific microstructure characterization within IN625 additive manufacturing benchmark test artifacts. *Integr. Mater. Manuf. Innov.* **2020**, *9*, 54–69. [CrossRef]
49. Yadroitsev, I.; Krakhmalev, P.; Yadroitsava, I.; Johansson, S.; Smurov, I. Energy input effect on morphology and microstructure of selective laser melting single track from metallic powder. *J. Mater. Process. Technol.* **2013**, *213*, 606–613. [CrossRef]
50. Pauzon, C.; Hryha, E.; Forêt, P.; Nyborg, L. Effect of argon and nitrogen atmospheres on the properties of stainless steel 316 L parts produced by laser-powder bed fusion. *Mater. Des.* **2019**, *179*, 107873. [CrossRef]
51. Dietrich, K.; Diller, J.; Dubiez-Le Goff, S.; Bauer, D.; Forêt, P.; Witt, G. The influence of oxygen on the chemical composition and mechanical properties of Ti-6Al-4V during laser powder bed fusion (L-PBF). *Addit. Manuf.* **2020**, *32*, 100980. [CrossRef]
52. Pauzon, C.; Dietrich, K.; Forêt, P.; Hryha, E.; Witt, G. Mitigating oxygen pick-up during laser powder bed fusion of Ti-6Al-4V by limiting heat accumulation. *Mater. Lett.* **2021**, 129365. [CrossRef]
53. Pauzon, C. The Process Atmosphere as a Parameter in the Laser-Powder Bed Fusion Process. Licentiate Thesis, Chalmers University of Technology, Göteborg, Sweden, 2019.
54. Kitchener, J.; Bockris, J.; Gleiser, M.; Evans, J. The solubility of oxygen in gamma iron. *Acta Metall.* **1953**, *1*, 93–101. [CrossRef]
55. Saeidi, K.; Gao, X.; Zhong, Y.; Shen, Z.J. Hardened austenite steel with columnar sub-grain structure formed by laser melting. *Mater. Sci. Eng. A* **2015**, *625*, 221–229. [CrossRef]
56. Bergmann, W. *Werkstofftechnik 1: Struktureller Aufbau von Werkstoffen-Metallische Werkstoffe-Polymerwerkstoffe-Nichtmetallisch-Anorganische Werkstoffe*; Carl Hanser Verlag GmbH Co. KG: Munich, Germany, 2013.
57. Krakhmalev, P.; Yadroitsava, I.; Fredriksson, G.; Yadroitsev, I. Microstructural and thermal stability of selective laser melted 316L stainless steel single tracks. *S. Afr. J. Ind. Eng.* **2017**, *28*. [CrossRef]
58. Bang, G.B.; Kim, W.R.; Kim, H.K.; Park, H.-K.; Kim, G.H.; Hyun, S.-K.; Kwon, O.; Kim, H.G. Effect of process parameters for selective laser melting with SUS316L on mechanical and microstructural properties with variation in chemical composition. *Mater. Des.* **2021**, *197*, 109221. [CrossRef]
59. Scipioni Bertoli, U.; Wolfer, A.J.; Matthews, M.J.; Delplanque, J.-P.R.; Schoenung, J.M. On the limitations of volumetric energy density as a design parameter for selective laser melting. *Mater. Des.* **2017**, *113*, 331–340. [CrossRef]
60. Prashanth, K.G.; Scudino, S.; Maity, T.; Das, J.; Eckert, J. Is the energy density a reliable parameter for materials synthesis by selective laser melting? *Mater. Res. Lett.* **2017**, *5*, 386–390. [CrossRef]

metals
MDPI
Article

Can Potential Defects in LPBF Be Healed from the Laser Exposure of Subsequent Layers? A Quantitative Study

Alexander Ulbricht [1,*], Gunther Mohr [1,2], Simon J. Altenburg [1], Simon Oster [1], Christiane Maierhofer [1] and Giovanni Bruno [1,3]

1 Bundesanstalt für Materialforschung und—Prüfung (BAM, Federal Institute for Materials Research and Testing), Unter den Eichen 87, 12205 Berlin, Germany; gunther.mohr@bam.de (G.M.); Simon.Altenburg@bam.de (S.J.A.); simon.oster@bam.de (S.O.); Christiane.Maierhofer@bam.de (C.M.); Giovanni.bruno@bam.de (G.B.)
2 Institute of Machine Tools and Factory Management, Technische Universität Berlin, Straße des 17. Juni 135, 10623 Berlin, Germany
3 Institute of Physics and Astronomy, University of Potsdam, Karl-Liebknecht-Straße 24/25, 14476 Potsdam, Germany
* Correspondence: alexander.ulbricht@bam.de

Citation: Ulbricht, A.; Mohr, G.; Altenburg, S.J.; Oster, S.; Maierhofer, C.; Bruno, G. Can Potential Defects in LPBF Be Healed from the Laser Exposure of Subsequent Layers? A Quantitative Study. *Metals* **2021**, *11*, 1012. https://doi.org/10.3390/met11071012

Academic Editors: Tomasz Czujko and Matteo Benedetti

Received: 11 May 2021
Accepted: 17 June 2021
Published: 24 June 2021

Abstract: Additive manufacturing (AM) of metals and in particular laser powder bed fusion (LPBF) enables a degree of freedom in design unparalleled by conventional subtractive methods. To ensure that the designed precision is matched by the produced LPBF parts, a full understanding of the interaction between the laser and the feedstock powder is needed. It has been shown that the laser also melts subjacent layers of material underneath. This effect plays a key role when designing small cavities or overhanging structures, because, in these cases, the material underneath is feed-stock powder. In this study, we quantify the extension of the melt pool during laser illumination of powder layers and the defect spatial distribution in a cylindrical specimen. During the LPBF process, several layers were intentionally not exposed to the laser beam at various locations, while the build process was monitored by thermography and optical tomography. The cylinder was finally scanned by X-ray computed tomography (XCT). To correlate the positions of the unmolten layers in the part, a staircase was manufactured around the cylinder for easier registration. The results show that healing among layers occurs if a scan strategy is applied, where the orientation of the hatches is changed for each subsequent layer. They also show that small pores and surface roughness of solidified material below a thick layer of unmolten material (>200 μm) serve as seeding points for larger voids. The orientation of the first two layers fully exposed after a thick layer of unmolten powder shapes the orientation of these voids, created by a lack of fusion.

Keywords: selective laser melting (SLM); additive manufacturing (AM); process monitoring; infrared thermography; optical tomography; X-ray computed tomography (XCT); healing; in situ monitoring

1. Introduction

Additive manufacturing (AM) of metals has evolved from a method for rapid prototyping to a production process applied in many industries including automotive, aerospace, and railway [1]. Among the AM methods established for metals, laser powder bed fusion (LPBF) is the most widely used, as it can produce net-shaped parts, which do not necessarily need additional surface treatments. To design LPBF parts, it is necessary to understand how the laser interacts with the layers of molten and unmolten powder. Previous studies have shown that the laser also melts, in addition to the current layer of powder, subjacent layers [2,3]. This effect is actually needed for a strong bonding between the layers and to prevent lack-of-fusion (LoF) defects [4,5]. If the volumetric energy density is high enough, the melt pool may not only re-melt solidified material, but also entrap keyhole pores into the bulk material up to several hundred of μm below the currently illuminated surface [6].

Therefore, the melting process of subjacent layers needs to be kept in mind when designing parts with cavities or overhanging structures [7]. To address this issue, LPBF machine manufacturers provide specialized downskin process parameters [8,9]. Many studies have been focussed on the determination of the part quality from in-situ monitored process signals of the build process [6,10–13]. In situ monitoring techniques based on thermographic cameras usually aim at the quantification of thermal inhomogeneities (within the current built surface or a stack of surfaces) to predict the formation of voids in the final part. The main drawback of thermographic techniques is the inability to observe the effect of the melt pool in the underlying, previously exposed layers.

To predict the existence of defects in the final part (i.e., voids whose forms, positions, and sizes might be a threat to the service life of the component) from in situ surface temperature data, it is necessary to understand the influence of the melt pool penetration depth on unmolten powder. Thus, several studies reported results from simulations [14,15] and experiments [16,17] on the interaction between melt pool and powder.

Based on these results, this study aimed to evaluate how the laser exposure of superjacent layers affects void formation in previously non-exposed areas of different heights in an LPBF specimen. A particular question is, if and how the laser exposure of superjacent layers can heal or close voids by successive melting of material in these areas.

In an LPBF specimen made of AISI 316L austenitic stainless-steel, a few consecutive layers were not directly exposed to the laser at nine different locations along the height of a cylindrical specimen. A scan strategy that rotates the orientation of a stripe scan pattern for each layer was applied to correlate the shape of detected voids to the scan pattern of distinct layers.

The build process was in situ monitored using thermographic cameras and optical tomography (the same LPBF machine and a similar optical setup as in [6]). The results from optical tomography (OT) enabled the extraction of the deployed orientations of hatches in each layer. OT was also used to study the integral intensity of the emitted thermal radiation of each built layer.

To analyse the final component, X-ray computed tomography (XCT) was used. This widely accepted non-destructive 3D measuring method enables quantitative characterisation of internal structures and volumetric irregularities; it also allows the evaluation of the geometrical precision of the built part compared with its planned design [6,18].

Additionally, we also used XCT data to capture the influence of the shape of internal surfaces and roughness on the formation of voids.

2. Materials and Methods

2.1. Material and Machine

Gas atomized powder of AISI 316L austenitic stainless steel was used as feedstock material on a commercial LPBF system SLM280 HL (SLM Solutions Group AG, Lübeck, Germany). The machine was equipped with a single 400 W continuous wave ytterbium fiber laser. In the focal position, a spot size of 80 µm at an emitted wavelength of 1070 nm was used [6]. The standard process parameters for 316L provided by the machine's manufacturer were applied for the build process: laser power P = 275 W, scan velocity vs. = 700 mm/s, layer thickness t = 50 µm, and hatching distance h = 120 µm. Sky writing was used during the build job to prevent changes in the volumetric energy density at the end of scan tracks, where the laser guiding system needed to be aligned to the orientation of the next scan track. The powder specifications given by its supplier are listed in Table 1. Argon was used as shielding gas during the build job with a resulting oxygen content of less than 0.1%.

Table 1. Powder characteristics (AISI 316L).

Apparent Density	Mean Diameter	D_{10}	D_{50}	D_{90}
4.58 g/cm^3	35 µm	18 µm	31 µm	56 µm

2.2. Specimen Design

A cylindrical specimen with a height of 12 mm and a diameter of 7 mm was designed. This resulted in a total of 240 layers based on a layer thickness of 50 µm. At nine specific heights, one quarter of the cylinder's circular surface was not exposed by the laser (see Table 2). To prevent heat accumulation during the build job, owing to possible voids in the non-exposed quarters, the quarters were distributed clockwise over the four quadrants of the cylinder and the build height. A 67° rotation of a meandering stripe scan strategy for each layer in the bulk was applied to minimize porosity [19]. A border scan and a fill contour scan were additionally applied to ensure smooth surfaces and precise part dimensions. In the layers of the non-exposed quarters, the contour scans followed the actual three-quarter shape of the exposed section. Hence, potentially unmolten powder particles could be removed through these gaps in the surface. An ultrasonic bath of the specimen was used to remove as much unmolten powder as possible from the non-exposed quarters of the manufactured specimen.

Table 2. Overview of the position and characteristics of the non-exposed quarters. The height to be exposed included the height of the non-exposed quarter plus the height of the first fully exposed layer above (50 µm).

Quarter's Number	Start Height from Cylindrical Base	Number of Non-Exposed Layers	Height of the Non-Exposed Quarters	Layer Numbers	Height to be Exposed
Q1	3 mm	1	50 µm	61	100 µm
Q2	4 mm	2	100 µm	81–82	150 µm
Q3	5 mm	3	150 µm	101–103	200 µm
Q4	6 mm	4	200 µm	121–124	250 µm
Q5	7 mm	5	250 µm	141–145	300 µm
Q6	8 mm	6	300 µm	161–166	350 µm
Q7	9 mm	7	350 µm	181–187	400 µm
Q8	10 mm	8	400 µm	201–208	450 µm
Q9	11 mm	9	450 µm	221–229	500 µm

To analyse the ability of the laser exposure of subsequent layers to fully melt subjacent layers of powder, a precise registration of the XCT results as well as of the in situ monitoring results was needed. Hence, a (helical) staircase was designed to be manufactured around the cylinder. This design was derived from Gobert et al. [20], who introduced additional staircases to enable a better registration of observable features to a distinct layer. Here, every step of the staircase represented the starting height of a non-exposed quarter of the inner cylinder. A small gap was introduced between the staircase and cylinder to enable an easy removal of the staircase and enable high resolution XCT scans. In order to improve the registration of a rotational symmetric specimen, an embossed T-shaped mark was added to the top surface (see Figure 1).

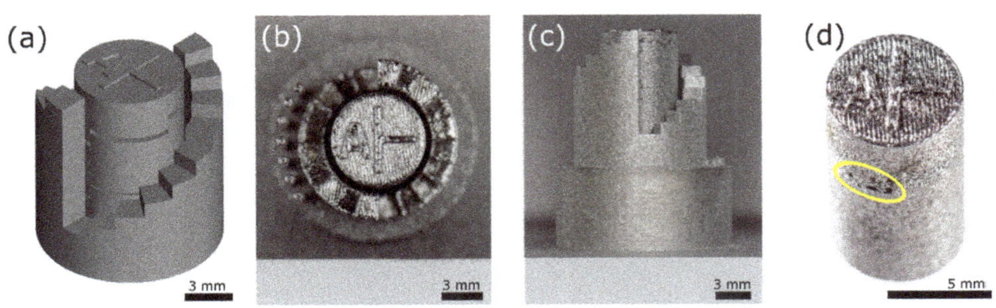

Figure 1. (**a**) CAD file of the specimen including the staircase; (**b**) Photograph of the produced specimen (top view, including the trigger pins) (**c**) Photograph of the produced specimen (side view) (**d**) Photograph of the specimen after the removal of the staircase. The yellow ellipse marks an open porosity of a non-exposed quarter.

2.3. Optical Tomography and Thermography

The build process was monitored in situ by optical tomography (OT, bulb exposure of each layer exposition [21]) and thermography. In comparison with the optical setup within the LPBF machine's build chamber previously reported in [6], the external optical setup was modified. Beam splitters divided the radiation into an OT CMOS camera (M4020, Teledyne Digital Imaging Inc., Billerica, MA, USA) and a short wave infrared (SWIR) camera (Goldeye CL-033 TEC1, Allied Vision Technology GmbH, Stadtroda, Germany). The OT camera system was sensitive to light emission in the near infrared spectral range (880 nm ± 25 nm). A spatial resolution of 40 μm per pixel was achieved, enabling visualisation of the hatching pattern with a hatching distance of 120 μm. The image data acquired by in situ OT show a map of intensity values that are proportional to the spectral radiosity of the part surface integrated in the spectral range of the used filter and over the whole layer illumination time. As the radiosity in a fixed spectral window strongly increases in the near infrared at the involved temperatures of molten steel, these intensity values are well suited as a measure for the maximum temperature reached at each point. However, as the time integral over the spectral radiosity is recorded, reduced cooling rates can also cause high OT intensities. Thus, data interpretation must be performed carefully. A comparison with the time resolved thermography data may help to clarify the origin of possible deviations in the OT intensity, despite the lower spatial resolution (factor of three).

2.4. Micro Computed Tomography

Micro computed tomography (XCT) was performed in two steps. An overview scan of the whole part was performed using the commercial CT-scanner GE v | tome | x 180/300 (GE Sensing & Inspection Technologies GmbH, Wunstorf, Germany) at a voltage of 222 kV and a current of 45 μA applied with an aluminium pre-filter of 1.0 mm thickness. A reconstructed voxel size of $(10\ \mu m)^3$ was achieved. After the removal of the staircase, the inner cylinder was scanned again using a custom-made industrial 3D micro CT scanner, which was equipped based on a 225 kV micro focus X-Ray source (X-Ray WorX GmbH, Garbsen, Germany) [22]. A voltage of 210 kV and a current of 60 μA in combination with a metal pre-filter of 1 mm copper and 0.25 mm aluminium were used. A voxel size of $(5\ \mu m)^3$ was achieved by combining the results from two measurements taken at two heights. The combined $(5\ \mu m)^3$ data were filtered using the plugin "non-local means denoise" [23,24] in the open-source imaging software Fiji [25]. For further analysis, the higher resolution—$(5\ \mu m)^3$ voxel size—data were registered onto the lower resolution—$(10\ \mu m)^3$ voxel size—data using the commercial software VG Studio MAX version 3.3.3 (Volume Graphics GmbH, Heidelberg, Germany). The same software was used for all XCT data analyses. The $(5\ \mu m)^3$ voxel size data enabled the quantitative analysis of voids with size above $(10\ \mu m)^3$. The analysis was performed in a virtual cylindrical cut-out (Ø = 6.87 mm, L = 11 mm) of the inner cylinder (Ø = 7 mm, L = 11.86 mm) to prevent surface roughness from influencing the size determination of open voids. The volumetric porosity was analysed using VG Studio MAX's built-in porosity analysis modules. A lower threshold for void detection was set to 8 voxels. To correlate the main orientation of the voids with the orientation of the stripe scan pattern, virtual lines were fitted onto the voids of virtual cuts in the VG Studio MAX software. The built-in dimensioning tool was used to calculate the angles of the LoF voids from a set 0° reference line.

3. Results

3.1. In Situ Monitoring

The OT results from layer 200 (last fully exposed layer before the beginning of Q8, see Table 1) and layer 209 (first fully exposed layer after Q8) are exemplarily depicted in Figure 2. From the OT data, the rotation angle α of the meandering stripe pattern was extracted. Exemplarily, the rotation angles of these two layers are indicated, $\alpha_{(\text{layer } 200)} = 75°$ and $\alpha_{(\text{layer } 209)} = 13°$. The comparison of the OT data revealed lower intensities in the quarter of layer 209, which covered the unexposed quadrant I. A comparison with time resolved thermography results shows that the spectral radiosity in the wavelength window of the SWIR camera was reduced in this quadrant as well. Thus, the maximum temperature reached in this quadrant was indeed lower than in the other quadrants. Ghost images appeared in the OT data, caused by the optical set-up, adding some blur to the images. The resulting offset intensity error in the superposed areas is in the range of approximately +8% related to peak intensity.

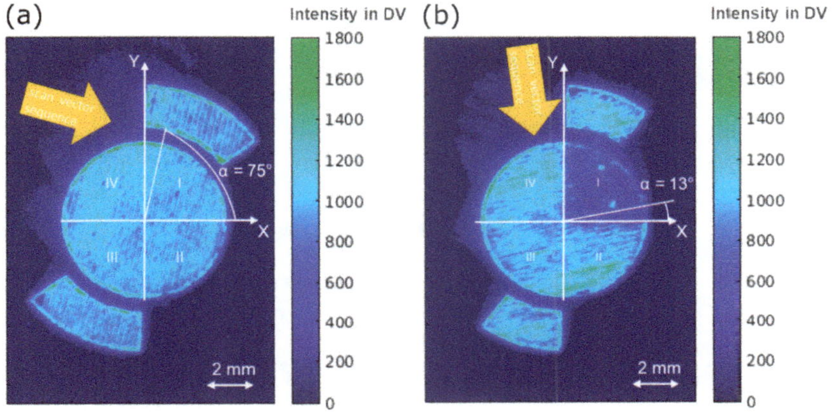

Figure 2. Optical tomography (OT) images showing the orientation of the hatches for (**a**) layer 200 (last complete layer below Q8) and (**b**) layer 209 (first complete layer above Q8).

3.2. XCT

The XCT results showed that, in all of the nine quarters, most of the powder was molten by the laser exposure (see Figure 3 and Table 3). Figure 3 shows the 3D rendering of the segmented voids in the whole cylindrical specimen. Figures 4 and 5 show the segmented voids in a virtual cut of the XCT data as well as the 3D rendered void segmentation. As shown in Figure 5e,f and Table 3, at a height of 450 µm (+50 µm of the superjacent fully exposed layer) of unmolten powder, a porosity of 5.56% was observed. This was surprisingly low as the unexposed volume was expected to contain many unmolten powder particles. Instead, solid material with LoF voids was observed.

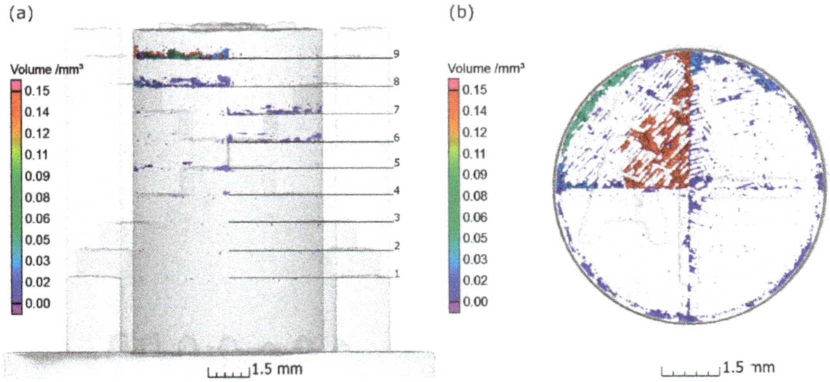

Figure 3. (**a**) Sideview of the combined rendering of the $(10\ \mu m)^3$ data and the $(5\ \mu m)^3$ data showing segmented voids in the high-resolution data; (**b**) top view of the rendering of the $(5\ \mu m)^3$ data of the cylinder showing a projection of all segmented voids.

Table 3. Porosity of the non-exposed quarters.

Quarter's Number	Q1	Q2	Q3	Q4	Q5	Q6	Q7	Q8	Q9
porosity	none	0.05%	0.04%	0.08%	0.38%	0.78%	0.80%	2.23%	5.56%

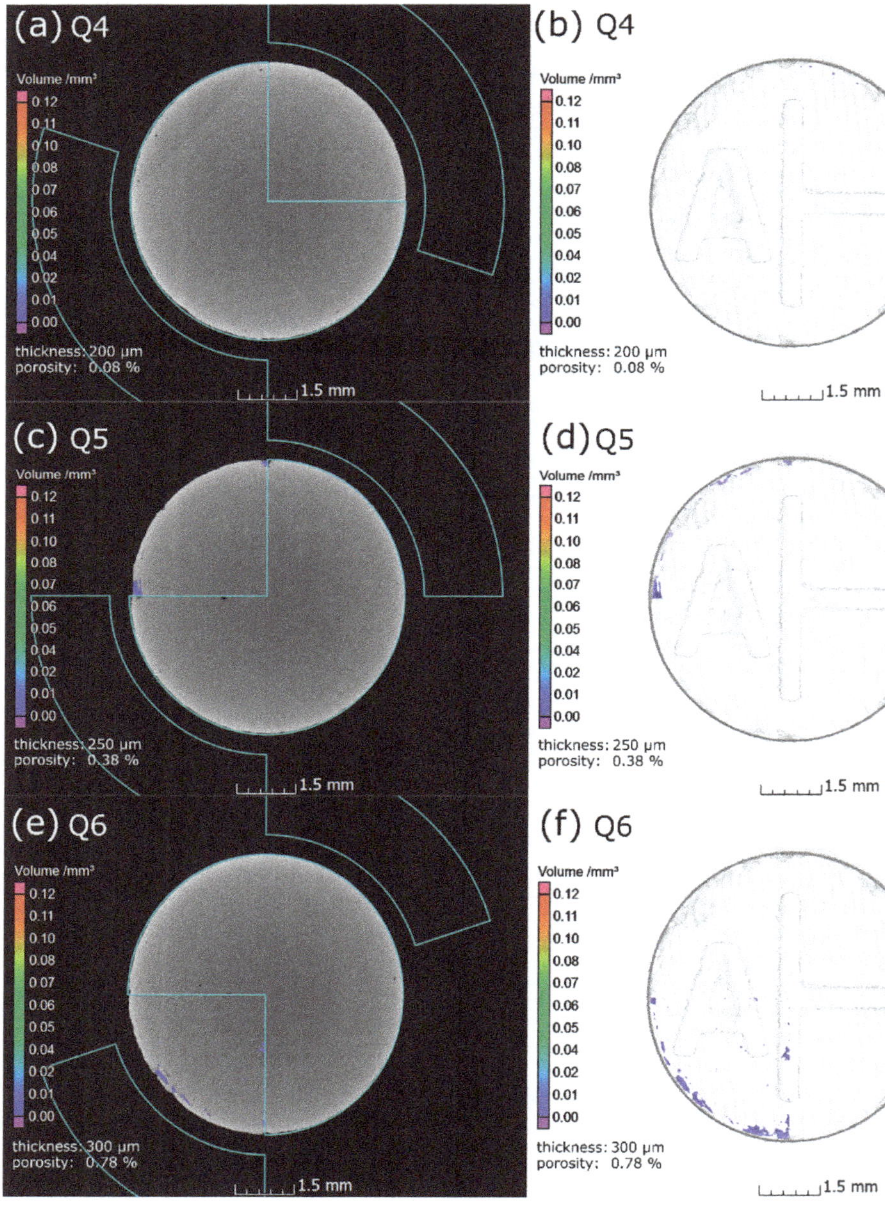

Figure 4. Void distribution at the positions of the unexposed quarters. Slice images taken at 10% of the quarter's height. (**a**) Q4: four layers unexposed; (**b**) projection of the void distribution in the unexposed layers containing Q4; (**c**) Q5: five layers unexposed; (**d**) projection of the void distribution in the unexposed layers containing Q5; (**e**) Q6: six layers unexposed; (**f**) projection of the void distribution in the unexposed layers containing Q6.

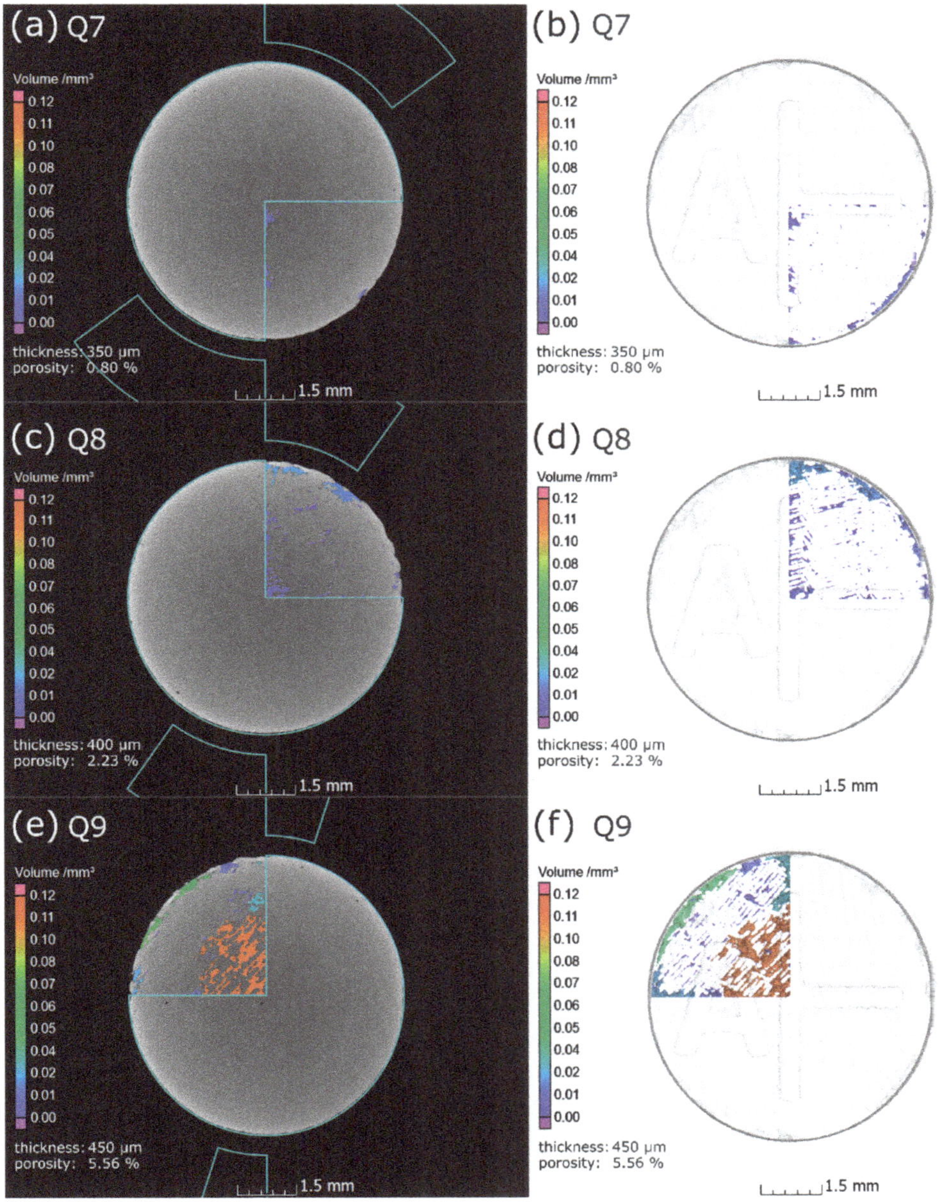

Figure 5. Void distribution at the positions of the unexposed quarters. Slice images taken at 10% of the quarter's height. (**a**) Q7: seven layers (350 µm) unexposed; (**b**) projection of the void distribution in the unexposed layers containing Q7; (**c**) Q8: eight layers unexposed; (**d**) projection of the void distribution in the unexposed layers containing Q8; (**e**) Q9: nine layers unexposed; (**f**) projection of the void distribution in the unexposed layers containing Q9.

At up to a thickness of 200 µm of unmolten powder, almost no additional porosity due to the underexposure by the laser was observed (see Figure 4a,b for 200 µm of unmolten powder in Q4). In these quarters, only pores at the interface between bulk and contour scan could be detected. This observation correlates well with the results from the literature [26]. The porosity value is directly linked to the number of pores occurring inside the quarters, not to the explicitly unmolten layer thickness. At a non-exposed thickness of 250 µm of unmolten powder (Q5), voids were detected at the positions where the outlines of the quarter intersect the perimeter of the cylinder (see Figure 4c,d). Voids at the quarters' inner perimeter were detected only above a non-exposed thickness of 300 µm (i.e., six layers, Q6, see Figure 4e,f and Figure 5a,b). At a thickness of 400 and 450 µm of unexposed powder, areal voids were detected (see Figure 5c–f, Q8 and Q9). Figure 5f shows the rendered defects; they seem to correspond to hatches of the scan pattern. As the orientation of the stripe scan pattern rotated by 67° with each layer, it was possible for Q6–Q9 to assign the orientation of LoF voids to distinct layers. As listed in Table 4, the main orientation of the LoF voids corresponds to the orientation of the first fully exposed layer above the quarter. For Q8 and Q9, a secondary orientation of LoF voids was also detected. Figure 7 combines the 3D rendering of the segmented voids with OT images of the first and second exposed layer above the quarter. The figure emphasises how the hatch orientation of these layers determined the shape of the segmented LoF voids.

Table 4. Comparison between the observed lack-of-fusion (LoF) pattern and the scan hatch orientation. OT, optical tomography.

Quarter's Number	Non-Exposed Height	Primary LoF Orientation (XCT)	Secondary LoF Orientation (XCT)	Orientation of the First Exposed Layer (OT)	Orientation of the Second Exposed Layer (OT)
Q6	350 µm	127°	-	127°	59°
Q7	400 µm	162°	-	160°	93°
Q8	450 µm	14°	121°	13°	127°
Q9	500 µm	46°	151°	44°	157°

Figure 6a shows a magnified virtual cut of a large defect at the centre of Q9 obtained from the high-resolution XCT data. Figure 6b revealed that voids were mainly found at the bottom of Q9 and large voids could be observed towards the perimeter of Q9. In contrast to the remaining quarters in Q9, a larger void could be observed in the centre of the quarter (see yellow highlighting in Figure 6a,b).

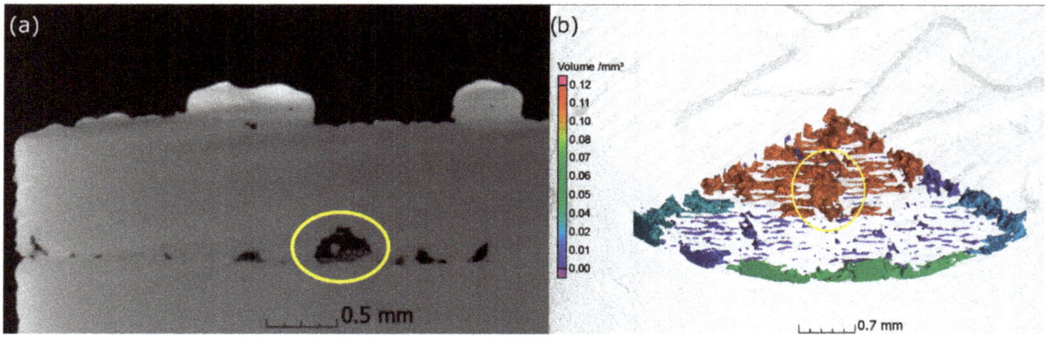

Figure 6. (**a**) Magnified sideview of the large defect at the centre of Q9 obtained from the high-resolution X-ray computed tomography (XCT) data. The largest void in Q9 is highlighted; (**b**) 3D rendering of the voids in Q9. The yellow mark highlights the position of the largest void from (**a**).

4. Discussion

The aim of this study was to analyse the effect of healing of successive laser passes during LPBF manufacturing. We thus investigate how (and how many) unmolten layers of powder were molten to form a solid with dense connection to the underlying bulk material. Although standard machine parameters of the LPBF machine based on 50 µm thick layers of powder were applied, XCT data revealed that the melt pool was deep enough to partially melt 10 layers of powder (500 µm), thereby creating lower than expected porosity values (5.6%). The lack of unmolten powder particles found in the non-exposed quarters illustrates the ability of the laser to transfer heat to powder layers below the surface. This result corresponds to the results from the in situ monitoring.

Figure 2b reveals that the thermal radiosity in quadrant I of the first fully exposed layer above Q8 seems to be lower than for the regions that have solid material below. This might be surprising at a first glance, as powder can be regarded as a heat insulator [27]. Therefore, one would expect to observe a higher radiosity owing to heat accumulation above the insulating powder. To explain the observed radiosity, the ability of the laser to transfer a sufficient amount of heat to layers below the current surface should be invoked. Foroozmehr et al. [15] had simulated the optical penetration of the laser radiation into a powder bed of AISI 316L. In agreement with previous studies [14,17], they assumed, in addition to absorption, multiple reflections of the laser radiation at the surfaces of the powder particles. Therefore, the laser–powder interaction seems to be a complex combination of transmission, scattering, and absorption of the laser energy by the powder particles. The laser radiation is not only absorbed at the surface of the powder layer, but also *within* the layer. At Q9, the missing layer thickness was 10 times higher than the single layer thickness of 50 µm. This led to a higher optical penetration depth, and thus to a lower volume energy density as in the areas with the single layer thickness. Consequently, the maximum temperature and thus the radiosity detected by the OT camera of the first solidified layer is lower.

Figure 6 and Figure 9 show that the voids created by the missing illumination of the layers were mainly found at the bottom of the quarters where the melt pool could not melt powder particles [28]. As shown in Figure 7, the melt pools of the first (and, above seven unmolten layers, also the second) fully exposed layer above the non-exposed quarter shaped the detected LoF voids. At these deeper regions, the melt pool is expected to be narrower. Therefore, in such regions, melt pools from neighbouring scan tracks did not seem to overlap each other. The combination of solidification shrinkage and surface tension effects of the melt pool (e.g., Marangoni effect [29–31]), which could draw neighbouring powder particles into the melt pool, appears to have created voids between scan tracks. Further in-depth analysis of the energy limits that form LoF voids in LPBF parts has been discussed by Biffo et al. [32–34]. In Q4 to Q7, these LoF voids between scan tracks were only observed close to the perimeter of the quarter. In Q8 and Q9, these LoF voids were also observed in the bulk of the quarters. This indicates that, up to a depth of 350 µm (below the current layer surface), neighbouring scan tracks overlap in such a way to cover the gap between them (120 µm).

Figure 7a–d shows that the orientation of the LoF voids observed in the quarters Q6–Q9 mainly corresponded to the orientation of the scan pattern of the first fully exposed layer above. This indicated that, at a depth of 350 µm and below, the melt pool widths of neighbouring scan tracks do not overlap sufficiently to prevent LoF defects. Because, in Q6 and Q7, this observation was made only at positions close to the edge of the quarters, the melt pool width seemed to vary during the laser exposure. In Q8 and Q9, a secondary orientation of LoF voids was also observed. This showed that the melt pool depth of the first solidified layer was lower than the powder layer thickness, leaving some unmolten power at the bottom of Q8 and Q9. The penetration depth of the melt pool did not seem to be constant, because, up to Q8 (powder height of 400 µm), only small voids (height < 100 µm) at the bottom of the quarter were detected. In Q9 (quarter height of 450 µm), a network of large voids with a height of up to 300 µm was observed (see highlighted void in Figure 6b). The lack of large voids in quarters Q1–Q5 indicates that the laser energy density was sufficient to melt several layers of powder in the non-exposed quarters.

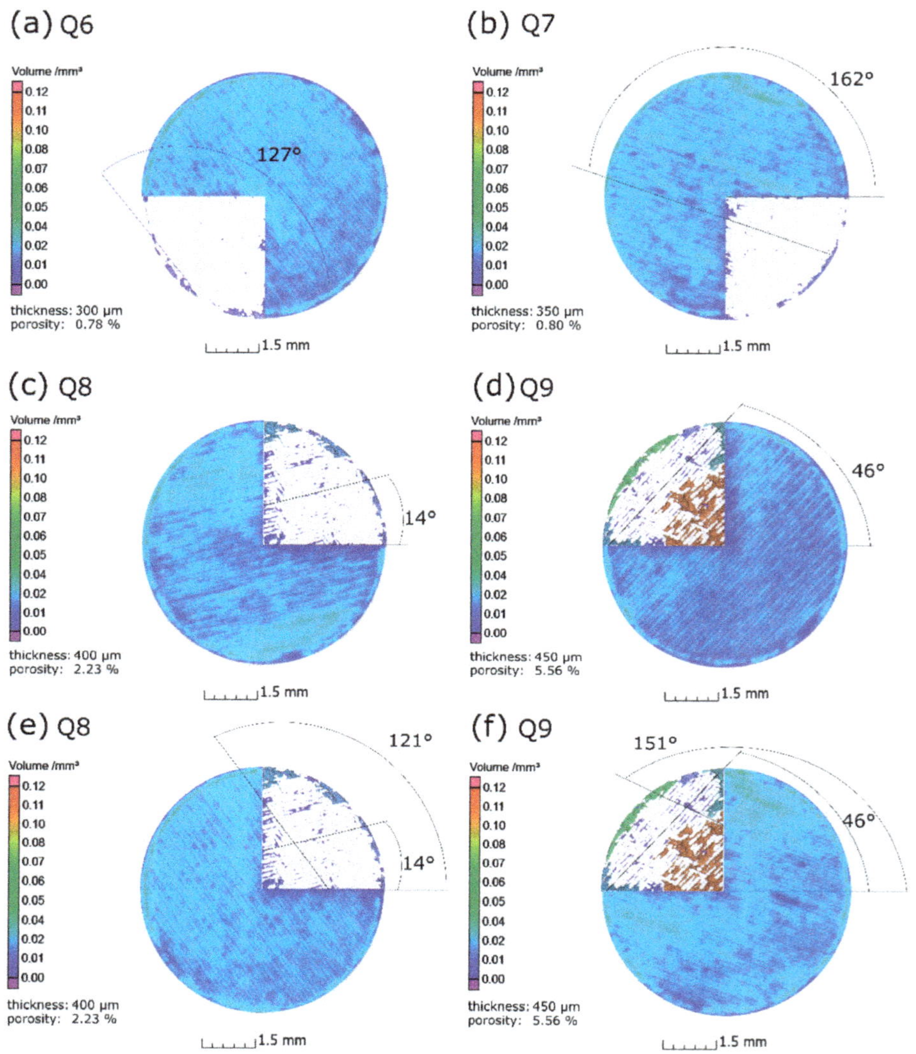

Figure 7. Combination of optical tomography images of the first fully exposed layers above the quarters and the orientation of LoF voids, as detected by XCT. (**a**) Void distribution of Q6 combined with the OT of the first layer above; (**b**) void distribution of Q7 combined with the OT of the first layer above; (**c**) void distribution of Q8 combined with the OT of the first layer above; (**d**) void distribution of Q9 combined with the OT of the first layer above; (**e**) void segmentation of Q8 combined with the OT of the second layer above; (**f**) void distribution of Q9 combined with the OT of the second layer above.

Figure 8 shows a third orientation of LoF voids close to the bottom surface of Q8 and Q9. The combination of XCT and OT data revealed that their orientation corresponds to the orientation of layer numbers 200 (in the case of Q8) and 220 (in the case of Q9). These layers were the last fully exposed layers below the non-exposed quarters. The virtual cuts presented in Figure 8 were taken 60 µm above the bottom of Q8 and 10 µm below the bottom of Q9. Still, at that height, the orientation of the underlying solidified surface seemed to influence the orientation of LoF voids. It can be assumed that the surface roughness of these layers served as seeding points for the larger LoF voids in the quarters.

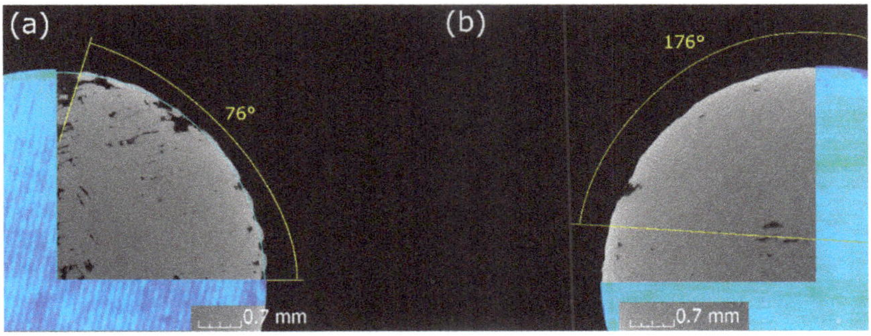

Figure 8. (**a**) Virtual cut taken at 60 µm above the bottom of Q8 combined with an optical tomography image showing the hatch orientation of layer number 200 (last completely exposed layer before Q8). The latter seems to dictate the orientation of the voids more than the layers above. (**b**) Virtual cut taken at 10 µm below the bottom of Q9 combined with an optical tomography image showing the hatch orientation of layer number 220 (last completely exposed layer before Q9). The latter seems to have formed the first voids in the bulk of Q9.

In Figure 9, the top yellow line was drawn from the shape of the cylinder's top surface. This profile was also applied to the internal surfaces of Q9 and Q8. The correspondence between this template and inner surfaces could indicate that this uneven surface was present throughout the whole build process. This indicates that the height difference between the centre and perimeter of the cylinder also existed at lower built heights. The slope explains why the imperfections of layer number 220 were visible in both images of Figure 8. During the build job of the specimen, sky writing was used to ensure a constant scan speed for the whole hatch length. To achieve a constant speed, the laser was turned off at the end of the scan track before repositioning the guiding mirrors. Hence, deceleration and acceleration effects could be avoided. Suddenly turning off the laser resulted in the reported elevated end-sections of scan tracks, followed by a dent as reported by Yeung et al. [35], and corresponds to simulations of the melt flow by Khairallah et al. [31]. According to them, these dents could form pores at the end of scan tracks and might explain the observation of pores at the intersection of the bulk and contour scan in this study. A ring of elevated end-sections of scan tracks has formed throughout the build height of the specimen owing to the 67° rotation of the applied stripe pattern in the bulk (see Figure 9).

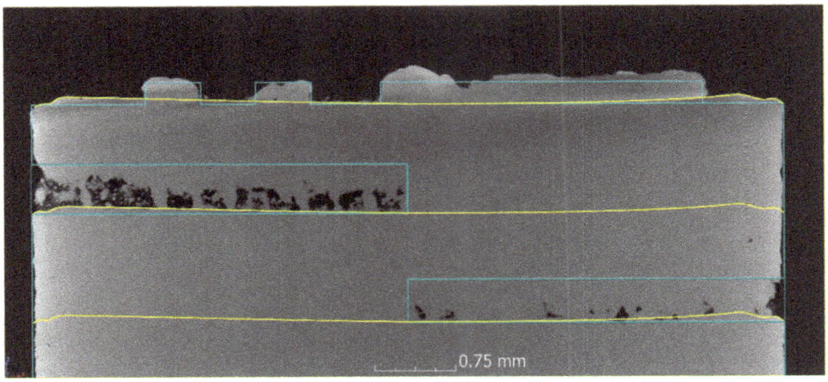

Figure 9. Cross section taken in the middle of the core cylinder showing voids at the bottom edges of Q8 (blue marked area at the bottom of the right side) and Q9 (blue marked area at the left side). The blue lines represent the nominal size taken from the CAD file. Yellow lines emphasize the structure of the internal surface.

5. Conclusions

In this study, we investigated the effect of unmolten powder layers on the defect formation in LPBF AISI 316L. In particular, we determined how many laser-unexposed layers of AISI 316L powder could be molten using a set of basis parameters provided by the LPBF machine's manufacturer. We found that, up to a thickness of 200 μm of unexposed powder (i.e., four powder layers), no additional porosity was observed, aside from pores between the bulk and contour scan. Therefore, we conclude that the heat input from the melt pool could sufficiently melt an amount of powder of four layers. We presume that this occurs because of a sufficient melt pool penetration depth, which allows fusion to the underlying solid material, and by being able to allow gas bubbles entrapped between powder particles to escape from the melt.

Further process parameter optimization (not within the scope of this study) might be able to make the process more efficient by, e.g., a slight increase of the scanning velocity. However, for large components, a slightly excessive melting depth can render the build process less prone to process irregularities (e.g., heterogeneities of the powder recoating process).

The present results will enable the interpretation of the signals acquired by in situ monitoring systems, allowing the LPBF users to decide whether or not an irregularly observed signal during the build process will be locked as a defect in the final part.

Our results also suggest that healing among layers occurs only if a scan strategy is applied, where the orientation of the hatches is changed for each subsequent layer. In fact, the porosity observed between the bulk and border scan indicates that healing does not occur if the subsequent layers are applied using the same hatching orientation.

Finally, the present study shows that small pores and surface roughness of solidified material below a thick layer of unmolten material (>200 μm) could serve as seeding points for larger voids. The orientation of the first two layers that are fully exposed after a thick layer of unmolten powder shapes the orientation of these voids, created by a lack of fusion.

Author Contributions: Conceptualization, A.U., G.M. and S.J.A.; methodology, A.U., S.J.A. and S.O.; formal analysis, A.U., S.J.A. and S.O.; investigation, A.U., S.J.A. and S.O.; writing—original draft preparation, A.U., G.M., S.J.A., S.O. and C.M.; writing—review and editing, A.U., G.M., S.J.A., S.O., C.M. and G.B.; visualization, A.U. and S.O.; supervision, S.J.A. and G.B.; project administration, S.J.A. and C.M. All authors have read and agreed to the published version of the manuscript.

Funding: This research received no external funding.

Institutional Review Board Statement: Not applicable.

Informed Consent Statement: Not applicable.

Acknowledgments: This work has been funded by the BAM Focus Area Materials project ProMoAM "Process monitoring of Additive Manufacturing". We are thankful for the financial support and the fruitful cooperation with all partners.

Conflicts of Interest: The authors declare no conflict of interest.

References

1. Frazier, W.E. Metal Additive Manufacturing: A Review. *J. Mater. Eng. Perform.* **2014**, *23*, 1917–1928. [CrossRef]
2. Roehling, T.T.; Wu, S.S.Q.; Khairallah, S.A.; Roehling, J.D.; Soezeri, S.S.; Crumb, M.F.; Matthews, M.J. Modulating laser intensity profile ellipticity for microstructural control during metal additive manufacturing. *Acta Mater.* **2017**, *128*, 197–206. [CrossRef]
3. Mohr, G.; Altenburg, S.J.; Hilgenberg, K. Effects of inter layer time and build height on resulting properties of 316L stainless steel processed by laser powder bed fusion. *Addit. Manuf.* **2020**, *32*, 101080. [CrossRef]
4. Tang, M.; Pistorius, P.C.; Beuth, J.L. Prediction of lack-of-fusion porosity for powder bed fusion. *Addit. Manuf.* **2017**, *14*, 39–48. [CrossRef]
5. Metelkova, J.; Kinds, Y.; Kempen, K.; de Formanoir, C.; Witvrouw, A.; Van Hooreweder, B. On the influence of laser defocusing in Selective Laser Melting of 316L. *Addit. Manuf.* **2018**, *23*, 161–169. [CrossRef]
6. Mohr, G.; Altenburg, S.J.; Ulbricht, A.; Heinrich, P.; Baum, D.; Maierhofer, C.; Hilgenberg, K. In-Situ Defect Detection in Laser Powder Bed Fusion by Using Thermography and Optical Tomography-Comparison to Computed Tomography. *Metals* **2020**, *10*, 103. [CrossRef]

7. Pradeep, P.I.; Kumar, V.A.; Sriranganath, A.; Singh, S.K.; Sahu, A.; Kumar, T.S.; Narayanan, P.R.; Arumugam, M.; Mohan, M. Characterization and Qualification of LPBF Additively Manufactured AISI-316L Stainless Steel Brackets for Aerospace Application. *Trans. Indian Natl. Acad. Eng.* **2020**, *5*, 603–616. [CrossRef]
8. Shange, M.; Yadroitsava, I.; Yadroitsev, S.P.I.; du Plessis, A. Determining the effect of surface roughness and porosity at different inclinations of LPBF parts. In Proceedings of the 20th Annual International RAPDASA Conference, Bloemfontein, South Africa, 6–8 November 2019.
9. Andreau, O.; Pessard, E.; Koutiri, I.; Peyre, P.; Saintier, N. Influence of the position and size of various deterministic defects on the high cycle fatigue resistance of a 316L steel manufactured by laser powder bed fusion. *Int. J. Fatigue* **2021**, *143*. [CrossRef]
10. Druzgalski, C.L.; Ashby, A.; Guss, G.; King, W.E.; Roehling, T.T.; Matthews, M.J. Process optimization of complex geometries using feed forward control for laser powder bed fusion additive manufacturing. *Addit. Manuf.* **2020**, *34*, 101169. [CrossRef]
11. Repossini, G.; Laguzza, V.; Grasso, M.; Colosimo, B.M. On the use of spatter signature for in-situ monitoring of Laser Powder Bed Fusion. *Addit. Manuf.* **2017**, *16*, 35–48. [CrossRef]
12. Forien, J.-B.; Calta, N.P.; DePond, P.J.; Guss, G.M.; Roehling, T.T.; Matthews, M.J. Detecting keyhole pore defects and monitoring process signatures during laser powder bed fusion: A correlation between in situ pyrometry and ex situ X-ray radiography. *Addit. Manuf.* **2020**, *35*, 101336. [CrossRef]
13. Scheuschner, N.; Altenburg, S.J.; Gumenyuk, A.; Maierhofer, C. In-situ thermographic monitoring of the laser metal deposition process. In Proceedings of the Sim-AM 2019: II International Conference on Simulation for Additive Manufacturing, Pavia, Italy, 11–13 September 2019; pp. 246–255.
14. Gusarov, A.; Smurov, I. Modeling the interaction of laser radiation with powder bed at selective laser melting. *Phys. Procedia* **2010**, *5*, 381–394. [CrossRef]
15. Foroozmehr, A.; Badrossamay, M.; Foroozmehr, E.; Golabi, S.i. Finite Element Simulation of Selective Laser Melting process considering Optical Penetration Depth of laser in powder bed. *Mater. Design* **2016**, *89*, 255–263. [CrossRef]
16. Ma, M.; Wang, Z.; Gao, M.; Zeng, X. Layer thickness dependence of performance in high-power selective laser melting of 1Cr18Ni9Ti stainless steel. *J. Mater. Process. Technol.* **2015**, *215*, 142–150. [CrossRef]
17. Streek, A.; Regenfuss, P.; Exner, H. Fundamentals of energy conversion and dissipation in powder layers during laser micro sintering. *Phys. Procedia* **2013**, *41*, 858–869. [CrossRef]
18. du Plessis, A.; Yadroitsev, I.; Yadroitsava, I.; Le Roux, S.G. X-ray Microcomputed Tomography in Additive Manufacturing: A Review of the Current Technology and Applications. *3D Print. Addit. Manuf.* **2018**, *5*, 227–247. [CrossRef]
19. Marattukalam, J.J.; Karlsson, D.; Pacheco, V.; Beran, P.; Wiklund, U.; Jansson, U.; Hjörvarsson, B.; Sahlberg, M. The effect of laser scanning strategies on texture, mechanical properties, and site-specific grain orientation in selective laser melted 316L SS. *Mater. Design* **2020**, *193*, 108852. [CrossRef]
20. Gobert, C.; Reutzel, E.W.; Petrich, J.; Nassar, A.R.; Phoha, S. Application of supervised machine learning for defect detection during metallic powder bed fusion additive manufacturing using high resolution imaging. *Addit. Manuf.* **2018**, *21*, 517–528. [CrossRef]
21. Bamberg, J.; Zenzinger, G.; Ladewig, A. In-process control of selective laser melting by quantitative optical tomography. In Proceedings of the 19th World Conference on Non-Destructive Testing, Munich, Germany, 3–17 June 2016.
22. Oesch, T.; Weise, F.; Meinel, D.; Gollwitzer, C. Quantitative In-situ Analysis of Water Transport in Concrete Completed Using X-ray Computed Tomography. *Transp. Porous Media* **2019**, *127*, 371–389. [CrossRef]
23. Buades, A.; Coll, B.; Morel, J.-M. Non-Local Means Denoising. *Image Process. Line* **2011**, *1*. [CrossRef]
24. Darbon, J.; Cunha, A.; Chan, T.F.; Osher, S.; Jensen, G.J. Fast nonlocal filtering applied to electron cryomicroscopy. In Proceedings of the 2008 5th IEEE International Symposium on Biomedical Imaging: From Nano to Macro, Paris, France, 14–17 May 2008; pp. 1331–1334.
25. Schindelin, J.; Arganda-Carreras, I.; Frise, E.; Kaynig, V.; Longair, M.; Pietzsch, T.; Preibisch, S.; Rueden, C.; Saalfeld, S.; Schmid, B.; et al. Fiji: An open-source platform for biological-image analysis. *Nat. Methods* **2012**, *9*, 676–682. [CrossRef] [PubMed]
26. Ertay, D.S.; Ma, H.; Vlasea, M. Correlative Beam Path and Pore Defect Space Analysis for Modulated LPBF Process. In Proceedings of the 2018 Annual International Solid Freeform Fabrication (SFF) Symposium—An Additive Manufacturing (AM) Conference, Austin, TX, USA, 13–15 August 2018.
27. Denlinger, E.R.; Jagdale, V.; Srinivasan, G.V.; El-Wardany, T.; Michaleris, P. Thermal modeling of Inconel 718 processed with powder bed fusion and experimental validation using in situ measurements. *Addit. Manuf.* **2016**, *11*, 7–15. [CrossRef]
28. Li, E.L.; Wang, L.; Yu, A.B.; Zhou, Z.Y. A three-phase model for simulation of heat transfer and melt pool behaviour in laser powder bed fusion process. *Powder Technol.* **2021**, *381*, 298–312. [CrossRef]
29. Qiu, C.; Panwisawas, C.; Ward, M.; Basoalto, H.C.; Brooks, J.W.; Attallah, M.M. On the role of melt flow into the surface structure and porosity development during selective laser melting. *Acta Mater.* **2015**, *96*, 72–79. [CrossRef]
30. Clark, S.J.; Leung, C.L.A.; Chen, Y.; Sinclair, L.; Marussi, S.; Lee, P.D. Capturing Marangoni flow via synchrotron imaging of selective laser melting. *IOP Conf. Ser. Mater. Sci. Eng.* **2020**, *861*. [CrossRef]
31. Khairallah, S.A.; Anderson, A.T.; Rubenchik, A.; King, W.E. Laser powder-bed fusion additive manufacturing: Physics of complex melt flow and formation mechanisms of pores, spatter, and denudation zones. *Acta Mater.* **2016**, *108*, 36–45. [CrossRef]
32. Biffi, C.A.; Fiocchi, J.; Valenza, F.; Bassani, P.; Tuissi, A. Selective Laser Melting of NiTi Shape Memory Alloy: Processability, Microstructure, and Superelasticity. *Shape Mem. Superelasticity* **2020**, *6*, 342–353. [CrossRef]

33. Fiocchi, J.; Biffi, C.A.; Tuissi, A. Selective laser melting of high-strength primary AlSi9Cu3 alloy: Processability, microstructure, and mechanical properties. *Mater. Design* **2020**, *191*, 108581. [CrossRef]
34. Biffi, C.A.; Bassani, P.; Fiocchi, J.; Albu, M.; Tuissi, A. Selective laser melting of AlCu-TiB2 alloy using pulsed wave laser emission mode: Processability, microstructure and mechanical properties. *Mater. Design* **2021**, *204*. [CrossRef]
35. Yeung, H.; Lane, B.M.; Donmez, M.A.; Fox, J.C.; Neira, J. Implementation of Advanced Laser Control Strategies for Powder Bed Fusion Systems. *Procedia Manuf.* **2018**, *26*, 871–879. [CrossRef]

metals

MDPI

Article

Radiographic Visibility Limit of Pores in Metal Powder for Additive Manufacturing

Gerd-Rüdiger Jaenisch [1], Uwe Ewert [2], Anja Waske [1] and Alexander Funk [1,*]

[1] Bundesanstalt für Materialforschung und-prüfung (BAM), Unter den Eichen 87, D-12205 Berlin, Germany; gerd-ruediger.jaenisch@bam.de (G.-R.J.); anja.waske@bam.de (A.W.)

[2] KOWOTEST GmbH, Solinger Straße 186, D-40764 Langenfeld, Germany; uwe@ewert-net.de

* Correspondence: alexander.funk@bam.de; Tel.: +49-30-8104-3251

Received: 11 November 2020; Accepted: 30 November 2020; Published: 4 December 2020

Abstract: The quality of additively manufactured (AM) parts is determined by the applied process parameters used and the properties of the feedstock powder. The influence of inner gas pores in feedstock particles on the final AM product is a phenomenon which is difficult to investigate since very few non-destructive measurement techniques are accurate enough to resolve the micropores. 3D X-ray computed tomography (XCT) is increasingly applied during the process chain of AM parts as a non-destructive monitoring and quality control tool and it is able to detect most of the pores. However, XCT is time-consuming and limited to small amounts of feedstock powder, typically a few milligrams. The aim of the presented approach is to investigate digital radiography of AM feedstock particles as a simple and fast quality check with high throughput. 2D digital radiographs were simulated in order to predict the visibility of pores inside metallic particles for different pore and particle diameters. An experimental validation was performed. It was demonstrated numerically and experimentally that typical gas pores above a certain size (here: 3 to 4.4 μm for the selected X-ray setup), which could be found in metallic microparticles, were reliably detected by digital radiography.

Keywords: additive manufacturing; feedstock powder; porosity; radiography; digital detector array; numerical simulation; detectability

1. Introduction

The reliability of additively manufactured (AM) parts depends to a large degree on the defects and irregularities they contain [1]. Typical flaws in AM parts include delaminations, cracks, inclusions, and pores. For the laser powder bed fusion (LPBF) process, most defects are formed during the dynamic melting process, when the laser interacts with the solid and the melt pool [2,3]. The AM community distinguishes between lack of fusion pores, which are flat and irregular pores (voids) between non-fused powder layers, keyhole pores, which are irregular pores (voids) created by the laser and a high energy input, and gas pores, which are mostly spherical pores formed by adding gas to the melt pool [2,4–8]. A detailed review of pores and voids created during the AM build process was recently published by Sola et al. [2].

Another possible source for gas pores in AM parts is pores already present in the feedstock powder [6,9–12]. Due to the fast heating and cooling rate of the melt pool up to 10^6 K/s [13] during the LPBF process, gas pores in the powder become bubbles in the melt pool, which under certain circumstances do not have sufficient time to reach the surface of the melt by convection [2,14]. Hence, it is important to check the quality of the feedstock powders, for instance, by determining its overall porosity and pore size distribution [6,15,16]. However, determining the porosity of powders experimentally, e.g., by pycnometry or metallography, is a challenging task [16]. For standard AM materials like steels, aluminum, magnesium, titanium, and nickel alloys, typical porosities of the feedstock powder are below 1 vol.% [10,16]. In particular, high feedstock powder porosities and single particles with large pores influence the melting behavior of the powder and are likely to merge and solidify together with the melt pool and may become a defect in the final part [2,6]. Several testing methods for the characterization of AM feedstock powders are standardized [17,18]. However, since currently no common quality requirements or certification for AM powders exist, there is the uncertainty that the quality of the powder might vary between different manufacturers and batches. This is particularly true for non-standard AM powder materials, for instance, functional materials like shape-memory alloys [19], permanent magnets [20], or magnetic refrigerants [21], that are now being adapted to AM. For non-standard powders, the production of the feedstock powder is particularly challenging, as the atomization parameters need to be optimized. Large overall porosities up to 24 vol.%, large pore sizes, and shape irregularities leading to a poor powder flowability are also reported for standard AM materials [6,9,12], which require quality control of every powder batch before using them for AM [1].

Feedstock powders for AM are usually checked for their size and shape distribution using sieve-based, light scattering, or direct imaging methods [15,22]. Since these rely on the outer contour of the geometry or optical imaging, internal defects of metal powders, like gas-filled pores, are not accessible. Very accurate determination of internal defects, as well as size and shape distributions of a powder, can be carried out using 3D X-ray computed tomography (XCT) [6,9,16,23]. However, the scan time for a typical powder sample with a commercial XCT device is rarely below one hour [16]. Depending on the size of the powder, the system magnification, and the detector size, only about 1000 to 1,000,000 particles can be imaged during one scan when the highest resolution is required, since the field of view of XCT is limited. This number of particles is equivalent to a few milligrams of metal powder. When considering that several hundred grams and up to several kilograms of powder are needed to manufacture AM parts, milligram samples are not representative for the evaluation of feedstock powder batches and hence for meaningful quality assurance. A technology capable of screening larger amounts of particles in a shorter time is needed for this task.

In-line digital radiography of products and goods is already used, e.g., in fine casting inspection in industry, weld inspections, the food industry, and for luggage inspection at airports [24]. Correspondingly, 2D X-ray inspection was developed for speed and high throughput. In this paper, the potential and requirements of digital radiography for the quality assurance of metal powder by detecting spherical gas-filled pores in metal particles are evaluated. Nowadays, commercially available laboratory XCT scanners are used for measurements, which are available in most AM laboratories worldwide to inspect the final parts produced. However, in the present article, instead of collecting projections from different angles of the powder sample for a reconstruction of a 3D tomographic dataset, 2D radiographs of single layers of the powder are used for inspection. In practice, this would later require that, e.g., a conveyor belt-like or particle stream apparatus is used for moving the particles into the inspection region, similar to the ones used in light scattering devices or direct imaging devices for powder, like a commercial Camsizer® (Microtrac Retsch GmbH, Haan, Germany) [22,25,26]. For the study presented here, a metal alloy is used, but the pore imaging principle applies to other materials, as well. In this paper, a criterion based on a minimal contrast to noise ratio and a basic spatial image resolution is applied, which must be fulfilled in order to detect pores of certain sizes in a metallic particle by digital radiography.

2. Materials and Methods

2.1. Experimental Sample Preparation

The tested spherical gas-atomized powder material is a MnFePSi-based alloy [27], with a nominal particle size range of diameters between 100 μm and 150 μm. Note, the particle size range for numerical simulations is broader (see Section 2.5). The experimentally investigated powder represents the geometrical constitution of powders used in AM processes [1]. Both a multi-particle and single-particle sample were radiographed. For sample preparation, one side of a double-sided adhesive tape was placed on a polymer substrate to ensure mechanical stiffness, whereas the other side was covered with particles. The multi-particle sample was fixed on a sample rod.

After testing the multi-particle sample, a single particle containing several pores was identified and isolated from the multi-particle sample using a micro-manipulation tool, radiography, tomography, and light microscopy. The isolated particle was fixed again on a sample rod.

2.2. Experimental Acquisition of Radiographic Images and Tomograms

Radiographic images were acquired employing a commercial X-ray computed tomography (XCT) device, GE nanotom m 180, and using the image acquisition software datos|x 2.2 acquisition (Waygate Technologies (Baker Hughes Digital Solutions GmbH), Wunstorf, Germany) [28]. A set of radiographic projections, taken at discrete angular positions of the observed sample, served as input data for a subsequent volume reconstruction using the image reconstruction software datos|x 2.2 reconstruction (Waygate Technologies, Wunstorf, Germany) [28]. For the experiments, a phoenix|X-ray micro-focus tube (Waygate Technologies, Wunstorf, Germany) with a tungsten transmission target on a diamond window and a digital detector array (DDA) of type DXR500L (Waygate Technologies, Wunstorf, Germany) was used [28]. The X-ray radiation was not filtered at the source side in order to achieve the highest possible intensity. The X-ray tube can be operated in different operation modes, varying the size of the focal spot. For the selected tube current and acceleration voltage range, nominal spot size of $f = 5$ µm in mode 0 and $f = 1.3$ µm in mode 2 are provided. In this work, mode 2 was used.

Both samples, the multi-particle sample (a) and the single-particle sample (b), were imaged using similar X-ray system parameters, as summarized in Table 1. For sample size reasons, the source to object distance (SOD) and the source to detector distance (SDD) were chosen to be 3 mm/600 mm (a) and 2 mm/400 mm (b), respectively. As a result, the effective pixel size of the radiographs and the voxel size of the volume reconstruction are 0.5 µm. For the tomographic reconstruction, 1700 projections (a) and 800 projections (b) were recorded while rotating the sample. An acquisition time of about 45 min is required for medium image quality for sample (a). To improve the image quality of the XCT projections (noise reduction), a skip of two projections and a frame averaging of five projections on each angular position were selected, which prolonged the total acquisition time by a factor of seven to about 5 h.

Table 1. Summary of experiments and their image acquisition parameters.

Scan Identifier	Tube Mode	Voltage (kV)	Current (µA)	SOD (mm)	SDD (mm)	M	Pixel/Voxel Size (µm)	Frame Averaging	Skip	No. of Projections	Frame Time (s)
2D (a)	2	130	100	3	600	200	0.5	10	none	1	1.5
3D (a)	2	130	100	3	600	200	0.5	5	2	1700	1.5
2D (b)	2	130	100	2	400	200	0.5	10	none	1	10
3D (b)	2	130	100	2	400	200	0.5	5	2	800	10

For the analysis of 3D volumes and the particle and pore size distributions, as well as for visualization, Thermo Fisher Scientific Avizo 9.2 software (Waltham, MA, USA) [29] was applied. The used analysis approaches to detect particles and pores in volumetric datasets can be found in [16].

2.3. Prediction of Pore Detectability by Human Observers

The basics for detecting small holes in objects, examined by human observers using radiography, were published in [30]. Here, a minimal contrast to noise ratio $\mathrm{CNR_{min}}$ is derived as a criterion to determine the detection limit of hole plate image quality indicators at a magnification of one, depending on the basic spatial resolution of the detector $\mathrm{SR_b^{detector}}$ and the hole diameter d_{hole} given by the equation:

$$\mathrm{CNR_{min}} = \frac{10 \cdot \mathrm{SR_b^{detector}}}{d_{\mathrm{hole}}} \quad (1)$$

This concept was extended considering the magnification M, spot size f, and the pore geometry [31]

$$\mathrm{CNR_{min}} = \frac{10}{M} \cdot \frac{2}{\pi d_{\mathrm{pore}}} \sqrt{((M-1)f)^2 + \left(2\mathrm{SR_b^{detector}}\right)^2} \tag{2}$$

with d_{pore}—pore diameter. Note, the equation is independent of the material tested.

2.4. Image Analysis of Radiographs Based on the Contrast to Noise Ratio

The digital image analysis was carried out using the software ISee! version 1.11.1 (BAM, Berlin, Germany) [32] for the measurement of the contrast to noise ratio (CNR) in simulated and measured radiographs. The profile was drawn over the center of the powder particle pore to measure the contrast in the radiograph (see Figure 1). The noise was measured in a region of interest in the free beam area. The procedure for the noise measurement implemented in ISee! is described in [33]. The resulting CNR was compared to the theoretical minimal required $\mathrm{CNR_{min}}$ for pore detection (see Section 2.3). The pore is visible to human observers if the determined CNR exceeds $\mathrm{CNR_{min}}$, as calculated by Equation (2).

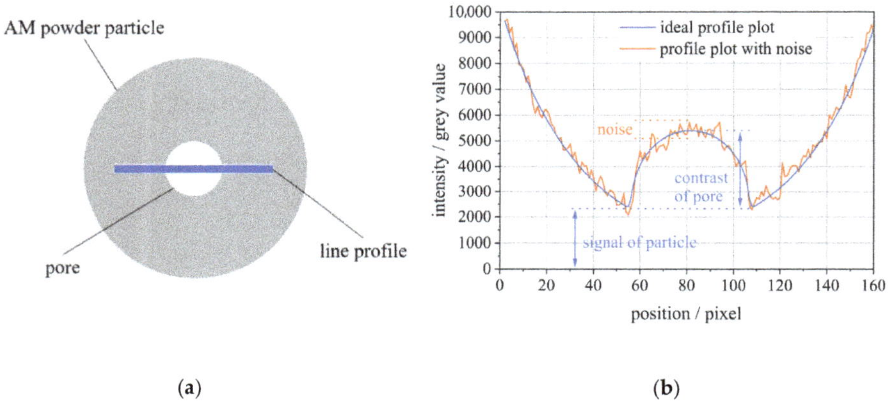

Figure 1. Scheme for measuring the contrast to noise ratio (CNR): the position of the line profile for the CNR measurement (**a**) and representation of the corresponding profile with and without noise to measure the CNR (**b**).

2.5. Numerical Simulation of Radiographic Images

The radiographic simulation software *aRTist* (BAM, Berlin, Germany) [34–36] was used to model different inspection scenes of the feedstock powders, using physical models for the production, interaction, and detection of X-rays. The graphical user interface of *aRTist* allows for setting up experimental scenes with different components like the radiation source, different detectors, and geometries of the object under investigation, as shown in Figure 2a. The software combines analytical and Monte Carlo methods to efficiently model the interaction of X-rays with matter. Hence, the way in which X-rays are absorbed and scattered in an object can be simulated and the resulting radiographic image at the detector can be calculated.

For simulation, mode 2 of the XCT scanner (see Section 2.2) with a nominal focal spot size of f = 1.3 µm was applied, together with an acceleration voltage of 130 kV and maximum gray values of $\mathrm{GV_{max}}$ = 4200, in accordance with the available experiments, and $\mathrm{GV_{max}}$ = 10,000 to examine the influence of the exposure time on the detectability of the pores.

The following parameters were used for the simulation with *aRTist*:

1. Source:
    - Tungsten transmission target with a thickness of 5 µm;
    - Diamond window with a thickness of 200 µm;
    - Focal spot size of $f = 1.3$ µm;
    - Tube voltage of 130 kV and spectrum resolution of 1 keV, no external filtering on tube side (for resulting spectrum, see Figure 2b).
2. Objects:
    - Powder particles: MnFePSi alloy, density of 6.4 g/cm^3;
    - Pores: air: N_2 0.755, O_2 0.231, Ar 0.013 (in mass percent), density of 0.001293 g/cm^3.
3. Detector:
    - Pixel size of 100 µm;
    - Basic spatial resolution of the detector: $SR_b^{detector} = 130$ µm;
    - Covered by a 750 µm thick carbon fiber-reinforced plastic plate;
    - Maximum gray value of (a) GV_{max} = 4200 and (b) GV_{max} = 10,000;
    - Internal scatter ratio of 15%, internal scatter ratio correlation length of 10 mm, gray response of 1000 GV/mGy, maximal signal-to-noise ratio (SNR) of 1400, efficiency of 500 $SNR_N/mGy^{1/2}$.
4. Geometry:
    - SDD = 400 mm;
    - SOD = 2 mm;
    - M = 200.

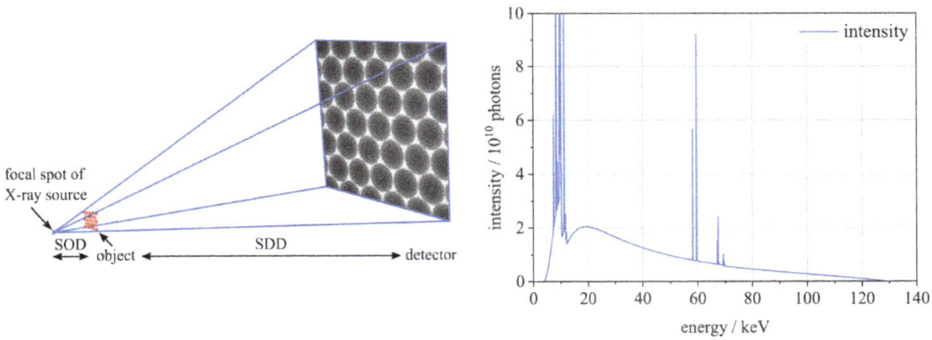

(**a**) (**b**)

Figure 2. Schematic representation of the radiographic setup with source, object, and detector in (**a**), and the 130 kV spectrum plot used in the simulation in (**b**).

For the numerical simulation, an idealized geometrical description was used to represent the feedstock particles in a size range between 10 µm and 150 µm, to cover the full range of commonly used AM feedstock powders [1]. The particles were assumed to be spherical and arranged in a single layer. The pores are described as spheres filled with air and positioned in the center of the powder particle, which yields the minimum achievable contrast in the radiographic images. Figure 3 shows an example of the geometric arrangement and the radiographic image of 441 particles with internal pores used for simulation. The high number of spheres was selected to permit a more accurate evaluation of pore visibility. To determine the minimum detectable pore sizes, the actual pore diameter was reduced stepwise by 0.1 μm. Simulated radiographic images were analyzed after each step, as described in Section 2.4.

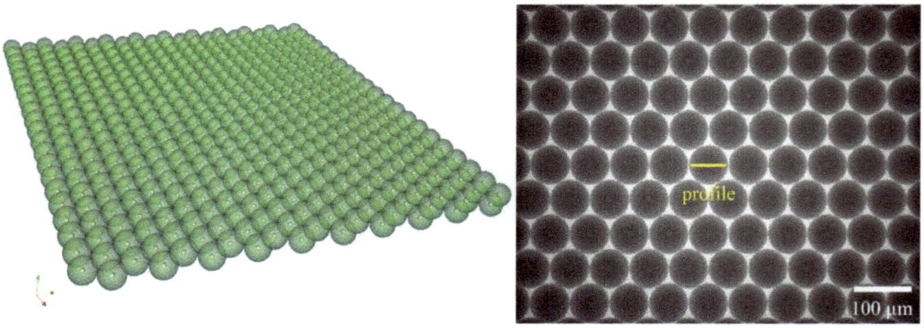

(**a**) (**b**)

Figure 3. Representation of a single layer of a 21 × 21 arrangement of metallic particles of 50 μm in diameter and with 5 μm air-filled pores inside, taken from the *aRTist* simulation scene (**a**). Radiographic image of the identical geometry with the profile (yellow inset) to investigate the visibility as function of CNR (**b**). The pores are the light gray areas in the center of the spheres.

3. Results and Discussion

3.1. Minimal Visible Pore Size

Figure 4 shows the CNR_{min} according to Equation (2) for two different focal spot sizes and $SR_b^{detector} = 130$ μm. For $f = 5$ μm (Figure 4a), the minimum required CNR_{min} is almost independent of the magnification in the selected range, while for $f = 1.3$ μm (Figure 4b) for small pores and micro- or nano-focus tubes, the CNR_{min} depends significantly on the magnification. Additionally, the CNR_{min} for $f = 5$ μm is generally higher than the smaller focal spot size of $f = 1.3$ μm, meaning that for a real experimental setup with known $SR_b^{detector}$ and proper exposure, small pores are more difficult to detect with a big focal spot size (Figure 4).

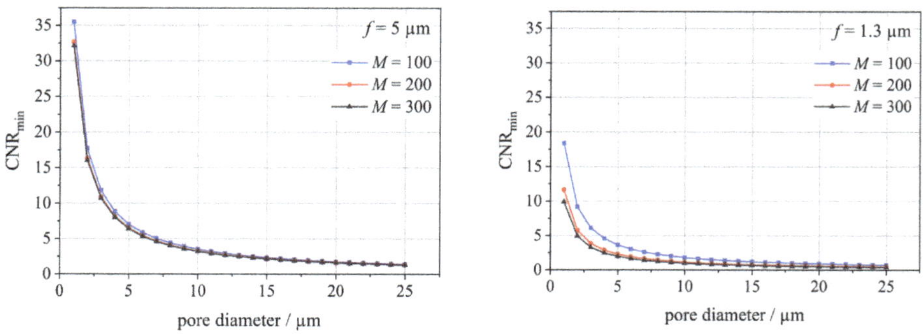

(**a**) (**b**)

Figure 4. Minimal contrast to noise ratio CNR_{min} for pore visibility according to Equation (2) with $SR_b^{detector} = 130$ μm, different magnifications M and the focal spot sizes: $f = 5$ μm (**a**) and $f = 1.3$ μm (**b**).

The diameter of a just visible pore, d_{pore_min}, in a radiograph of spherical particles can be predicted as function of the radiographic setup and the exposure parameters as follows:

$$d_{pore\_min} = \frac{10}{M} \cdot \frac{2}{\pi \cdot \mathrm{CNR}_{min}} \sqrt{((M-1)f)^2 + \left(2\mathrm{SR}_b^{detector}\right)^2} \quad (3)$$

The magnification M, the focal spot size f, and the basic spatial detector resolution $\mathrm{SR}_b^{detector}$ depend on the radiographic setup. The detectable pore size decreases if the focal spot size and the basic spatial resolution of the detector are selected to be as small as possible.

Furthermore, the detectable pore size also decreases if the CNR can be increased by the exposure conditions and M is optimized. The operator controls the pore contrast and noise by the selection of the X-ray energy and exposure parameters. The contrast is increased when the selected tube voltage is decreased. The CNR is increased if the noise is reduced. This requires increasing the tube current and the exposure time (for DDAs, increase frame time and frame averaging number). Generally, it is concluded that the CNR increases with the square root of exposure time and tube current. This requires an effective calibration of the digital detector array (DDA) to avoid an additional noise contribution by fixed pattern noise, which appears typically due to the sensitivity differences of the detector elements (pixels) and bad pixels. An increase in the tube voltage reduces the contrast and the noise, which influences the CNR depending on the attenuation process, typically the photo absorption or the Compton attenuation. At a higher tube voltage, where the Compton effect dominates, the contrast shows a low dependence on the selected tube voltage, but the exposure time is reduced significantly since the SNR increases proportionally to the tube voltage.

The ratio of particle diameter $d_{particle}$ to pore diameter d_{pore}, referred to as DPP, influences the measured pore contrast slightly due to the measurement procedure, as shown in Figure 1. Figure 5 shows the surface curvature changes, as seen in the profile, if the DPP is changed from 2 to 10. The pore contrast is slightly reduced with decreasing DPP depending on the particle surface curvature and corresponding line profile change. For large particle diameters, compared to the pore diameters, the pore contrast does not change any more with increasing particle diameter.

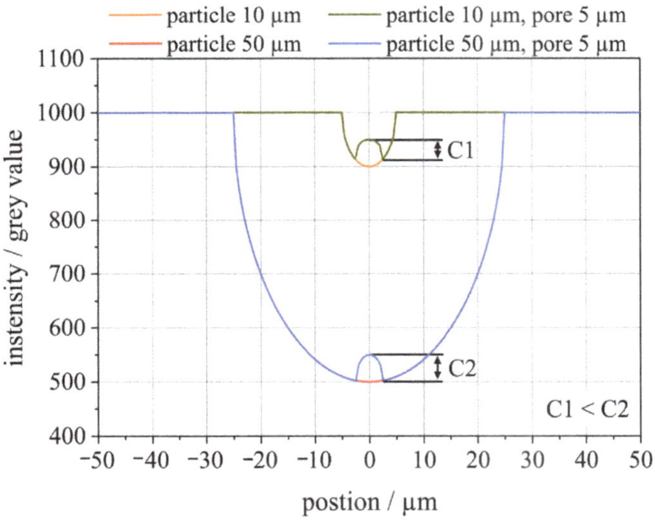

Figure 5. Contrast change as a function of the overlaid profiles of the pore and particle for different particle diameters. C1 is smaller than C2 for the same pore diameter, since the line profile of the larger particle (50 µm) is almost a flat plateau in contrast to the smaller particle.

A significant hardening of the spectrum is observed with increasing particle diameter for transmission microfocus tubes with diamond window and no prefiltering on the source side. Furthermore, the noise in the pore image increases due to the attenuation of larger particles. Consequently, the CNR is slightly reduced.

3.2. Simulation Results

By varying the size of the particles and the pores and determining whether the CNR of the pore signal in the radiographic image exceeds CNR_{min}, it can be determined whether a pore can be detected or not. The CNR_{min} value of Equation (2) as the visibility limit was verified by three operators for the simulated pores in the particles and accepted as a valid approach within a precision of about ±15% of the determined CNR, which was measured as shown in Figure 1 of the modeled images. In total, about 100 images were created and analyzed. As a result, the CNR_{min} obtained from Equation (2) was considered as valid limit by the three human observers. For the given simulation setup with $f = 1.3$ µm, a CNR_{min} for just visible pores of about 3, corresponding to pore diameters of about 4 µm, was determined. Bigger pores can be detected reliably, as the CNR will always exceed the CNR_{min}.

Based on this, Figure 6 shows the minimum pore size required in order to detect pores in metallic powder particles between 10 µm and 150 µm in diameter by means of digital radiography for the implemented X-ray setup. For the smallest particles with a diameter of 10 µm, the detectable pore diameter is equivalent to 35% of the particle diameter, while for the largest particles of 150 µm in diameter, the detectable pore has a diameter of 3% of the particle diameter (see Figure 6). The detectable minimum pore size increases slightly for very large and very small particles, with a local minimum at a particle diameter of about 70 µm. This is explained due to effects discussed in Section 3.1, e.g., the change of the X-ray spectrum (hardening) with increasing particle diameter on the right side of the graph in Figure 6 and the line profile shape (particle surface curvature), which yields a bias in the contrast measurement, for the left side of the graph in Figure 6.

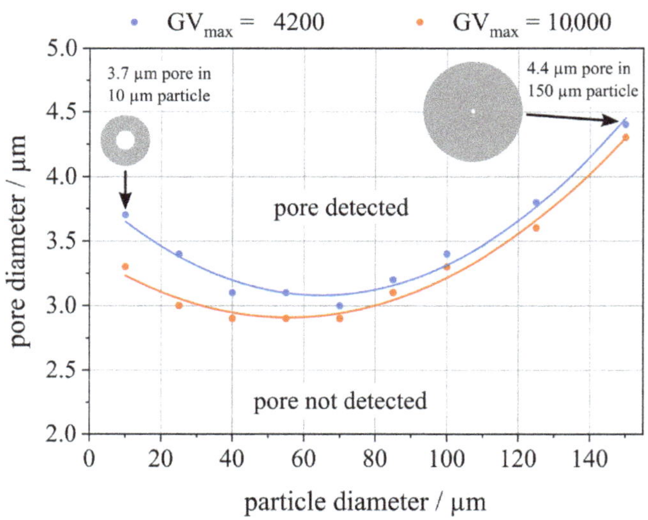

Figure 6. Diameter of the minimum pore sizes detectable by digital radiography in metal powder particles with diameters between 10 µm and 150 µm. The schematic insets are not to scale.

The minimum detectable pore sizes differ only slightly dependent on the selected exposure time of the detector with gray values of GV_{max} = 4200 and GV_{max} = 10,000 due to the limitation by the detector calibration (remaining detector fixed pattern noise) and internal backscatter in the detector (see Figure 6). The more highly illuminated image with GV_{max} = 10,000 demonstrates slightly smaller detected pores. This indicates that experimental exposure parameters leading to $GV_{max} = 4200$ are sufficient to detect the smallest pores according to Equation (2) for the present X-ray setup ($SR_b^{detector}$, detector efficiency, magnification, focal spot size, and exposure parameters).

For $GV_{max} = 4200$, the smallest detectable pore sizes for three different particle diameters are summarized in Table 2, together with an additional example for a large particle with a large pore. The corresponding simulated images are shown in Figure 7. The additional example with a large pore is in accordance with a particle and pore size which are examined in the experimental results section (see Section 3.3).

Table 2. Detectable pore size diameters for different particle sizes (at a maximum gray value of $GV_{max} = 4200$).

Particle Diameter	Pore Diameter	CNR_{min}	Measured CNR
10 µm [1]	3.7 µm [1]	3.5	3.9
70 µm [1]	3.0 µm [1]	3.9	3.9
150 µm [1]	4.4 µm [1]	2.7	2.8
135 µm [2]	40 µm [2]	0.3	30.7

[1] Smallest detectable pore for the corresponding particle diameter. [2] Large pore and particle size, which was found in the experiment (see Section 3.3).

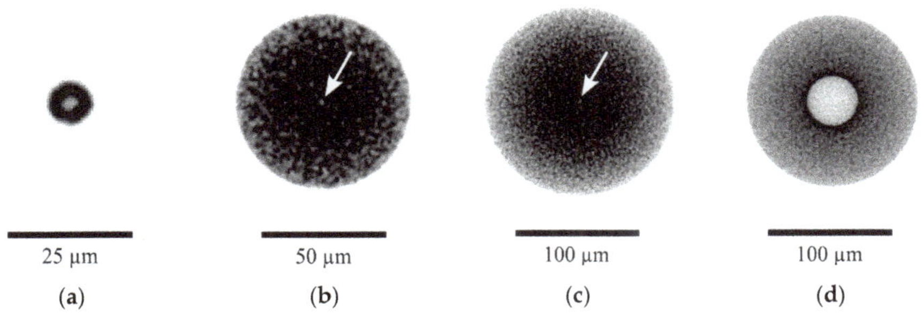

Figure 7. Simulated radiographs ($GV_{max} = 4200$) of different particle and pore diameters. The images were filtered, and the contrast was optimized for printed paper. (**a–c**) The smallest detectable pores for a particle diameter of (**a**) 10 µm, (**b**) 70 µm, and (**c**) 150 µm. Image (**d**) covers a particle diameter of 135 µm and a pore diameter of 40 µm, which is in the range of the particle and pore sizes found in the experiment (see Section 3.3). White arrows in (**b**,**c**) point to the centered pore. All images are in correspondence with the values listed in Table 2.

The smallest detectable pore sizes were found to be very similar in the range between 3.0 µm to 4.4 µm for the variety of particle sizes and $GV_{max} = 4200$. This can be explained by Equation (3) using a small focal spot size of $f = 1.3$ µm and a high magnification of $M = 200$ as well as the constant unsharpness of the system, which does not require the particle diameter.

Consequently, an almost constant pore diameter as a visibility limit, independent of the particle diameter, is expected. Due to the magnification of 200, the object's scattered radiation does not contribute to the image formation process. Therefore, only the attenuation differences in the materials (pores inside the powder particles) are essential for the contrast in the radiographic image. Basically, as it is measured here (Figure 1), the contrast can be approximated as proportional to the pore diameter, which applies to small pores. The noise is independent of the pore diameter, when measured in the free beam area. This approximation was applied since the noise could not be measured accurately behind the spheres and the noise changes only slightly for low contrast indications. The simulation did confirm that the visibility limit is just between 3.0 µm and 4.4 µm pore diameter for $GV_{max} = 4200$, for a sphere diameter range of 10 µm to 150 µm and for the selected setup.

It is shown in Figure 6 that the detectable pore size does vary by about a factor of 1.5 if the particle size varies by a factor of 15. The volume of a sphere increases with the third power of its diameter. Because of this geometrical relationship, the minimal detectable pore volume percentage of a particle of a certain size decreases significantly with the particle size. Figure 8 shows the previous results (Figure 6) on detectable pores expressed as the volume percent of the particles. For GV_{max} = 4200, it can be concluded that for the smallest powder particles of 10 µm, the smallest visible pore volume amounts to about 5.1 vol.%. For powder particles of 25 µm, the pore volume fraction decreases to 0.25 vol.% and, for the largest powder particles studied with a diameter of 150 µm, the detectable pore volume is equivalent to only 0.0025 vol.% of the particle volume. With increased exposure time and higher magnification, even smaller percentages of pore volume in relation to the particle volume can be detected.

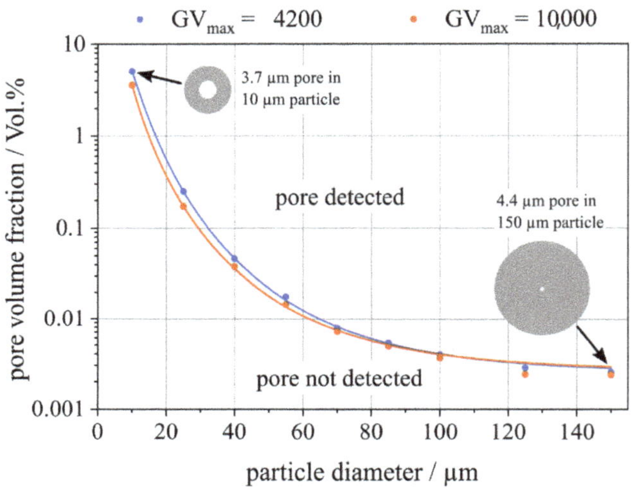

Figure 8. Minimum volume of pores detectable by digital radiography in metallic powder particles with diameters between 10 µm and 150 µm expressed as the pore volume fraction of the powder particles. The fraction of the detectable pore volume decreases rapidly with particle size. The schematic insets are not to scale.

This indicates that the contribution of the smallest detectable pore size in large powder particles to the overall porosity of the AM part is low and most likely does not affect the mechanical properties of the AM part. This is true if the volume fraction of the feedstock porosity is negligible and pore healing strategies in the AM fabrication process are applied [6,37]. A powder porosity of 5.1 vol.% is most likely critical for small particles with a diameter of 10 µm. The criticality of pore sizes in AM parts is discussed carefully in the literature, as many authors use very different materials, powder fractions, melting devices, and process parameters [1,2]. There is no strict rule for which amounts, sizes, and distributions of porosity in AM parts need to be avoided. In general, the porosity of an AM part should be as low as possible. However, it makes a difference if a single big pore or multiple very small pores create the total porosity of the AM part, as from mechanical testing it is known that the parts commonly crack at the biggest pores present [38]. Filigree and delicate AM parts, e.g., lattice structures, are especially endangered [26]. Whether a certain pore size or defect distribution is critical or not may also depend on the material, heat treatment, number of defects, arrangement of defects, location of the pores (close to surface or not), part geometry, and mechanical load. Most likely, very small pores and low total porosities are assumed to not be harmful. However, most potential harmful powder pores and large total feedstock porosities, e.g., hollow spheres or multiple medium-sized pores, can be detected by radiography for the examined particle diameters. Hence, it can be concluded that digital radiography can reliably detect the important pores (in particles with diameters ranging from 10 μm to 150 μm), which potentially have an impact on the AM product.

3.3. Experimental Results and Comparison to Simulation

Figure 9a shows a radiograph and Figure 9b a tomogram of the multi-particle monolayer-like sample. The tomogram is used here as a basic information standard. From the images, it is obvious that the radiograph already contains most information of the powder particles, e.g., the diameters, positions, and shapes of the particles (gray regions), as well as the diameters, positions, and shapes of the pores (light gray regions). This is due to the prepared monolayer-like sample, which is optimized for 2D radiography. Most of the particles are located individually. However, in the experiment, it is difficult to prepare an exact monolayer of powder particles, and hence overlapping of particles may be observed, leading to dark gray regions. The particles are not ideally arranged and do not cover the complete detector area. Therefore, the maximum possible number of particles per image is not reached. Small satellites attached to the surfaces of the particles are visible, which originate from the atomization process. The shape of several particles deviates from a sphere and they form ellipsoidal-shaped particles. The imaged sample region contains about 70 particles in the size range of about 100 μm to 150 μm known from the 3D volume analysis. Pores of very different sizes are frequently observed in this powder sample specimen, which enhances the chance to find very small pores in this sample. Even multiple pores located in one particle are visible if they do not overlap significantly. The smallest visible pore diameter found in the radiographic image (see Figure 9a) is about 10 μm. The biggest visible pore diameters are about 80 μm. Pores with a diameter above 10 μm are directly visible without any image contrast optimization, as they show a large CNR, which is discussed in detail later. Smaller pores are more difficult to detect in big particles by human observers, as the total contrast and CNR of a small pore is low, which was described earlier (see Section 3.1, Figure 5). The detection of very small pores is difficult, as there is only one sample orientation imaged. The overlap of particles and pores present here disturbs the analysis. This applies even if the exact position of the pores is known from the tomogram (see Figure 9b). Therefore, a single particle was investigated in more detail (see Figure 10 and red arrows in Figure 9). In particular, here, a higher exposure time of the detector, leading to $GV_{max} = 4200$, was used.

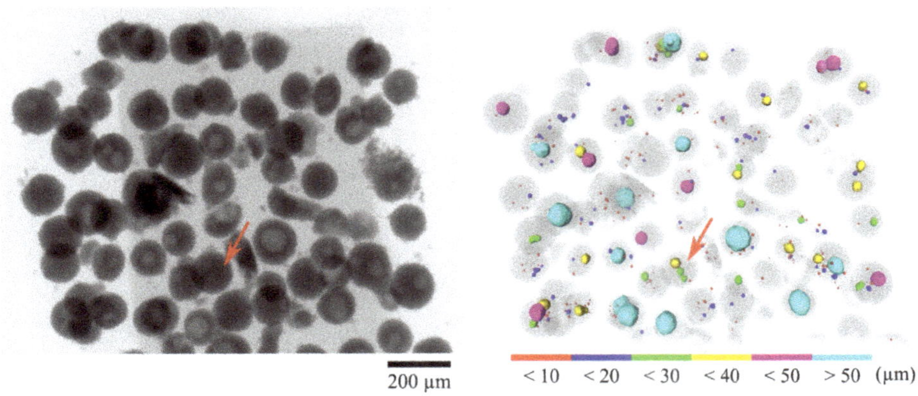

(a) (b)

Figure 9. Experimental verification of the simulation results. A monolayer-like multi-particle sample with diameters in between 100 μm and 150 μm and severe porosity inside was prepared and analyzed: (**a**) radiography and (**b**) rendering of the tomogram. A single particle from this monolayer-like sample was picked (red arrows) and was investigated in more detail (see Figure 10).

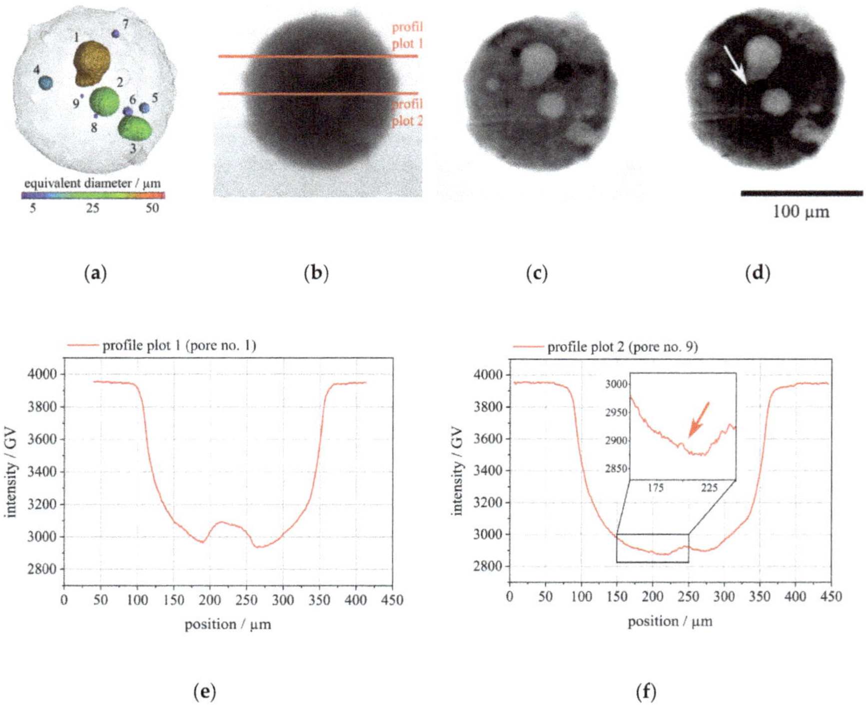

Figure 10. A rendered tomographic image and one radiographic projection of the single-particle sample with nine pores are shown in (**a**) and (**b**), respectively. In (**a**), the pores are numbered in correspondence to the index of the pores in Table 3. The contrast and brightness of (**b**) were enhanced based on filtering and a histogram correction, as presented in (**c**,**d**). In image (**b**), two red lines mark the profiles which belong to big pore no. 1 (**e**) and small pore no. 9 (**f**), respectively. Small pore no. 9 is highlighted by an arrow in (**d**,**f**).

Table 3. Measured CNR of pores in the single-particle sample.

Pore ID	Pore Equivalent Diameter XCT (µm)	Pore Diameter Radiography (µm)	CNR from Radiography
1	44	37	30
2	29	26	20
3	29	23	13.2
4	14	14	10
5	11	12	8
6	8	8	6.2
7	8	7	5.5
8	5	Not detected [1]	Not detected [1]
9	4	4	2.5

[1] Pore no. 8 was detected in the tomogram, but not in the radiograph, since it was overlapped by pore no. 2 at the selected projection angle (see Figure 10a).

The single particle covers an equivalent diameter of 137 µm. The equivalent diameter of an irregular shaped volume object (pore, particle) is the diameter of a perfect sphere of equivalent volume [39,40]. From the tomogram of the single-particle sample (Figure 10a), it is known that nine individual pores of different sizes are present (see Table 3). However, to detect all pores in the radiographic image, one must turn the particle to its optimal projection direction to ensure the projected pores are visible separately. From different projection angles, the CNR was measured for the nine pores (Table 3) when the contrast sensitivity was optimal for the observer. Two-line profile plots of a big (profile plot 1) and a small (profile plot 2) pore are shown exemplarily together with radiographic images in Figure 10b–f. The contrast of the radiographs was optimized and filtered for the observed regions in Figure 10b,d to enhance the visibility on printed paper. Larger pores above 10 µm in diameter show a CNR above 8, whereas small pores down to 4 µm in diameter are close to the detectability limit of CNR_{min} according to Equation (2) (see Section 2.3). In some cases, the determined pore diameters from the radiography were found to be slightly smaller than the calculated volume equivalent pore diameters [39,40] from the corresponding tomogram. This can be explained by the orientation of irregularly shaped pores in the 2D projection and the usage of an equivalent pore diameter from the 3D analysis.

The presented numerical simulations show that it is feasible to detect pores down to a diameter of 3.0 µm to 4.4 µm in metallic microparticles of 10 µm to 150 µm in diameter by radiographic imaging, depending on the exposure parameters and the available setup (see Section 2.5). When comparing the experimental and simulated results, it is obvious that the numerical model is an idealized approximation of centered pores but gives valuable estimates of the detectable pore sizes. In practice, an optimal projection direction is important to identify individual pores if multiple pores per particle occur. In addition, pores are rarely placed exactly in the center of a particle. Pores located out of the center and close to the surface tend to be more difficult to detect, as they are placed within a strong signal gradient originating from the particle shape. Real feedstock powders may show satellites, irregular shapes, rough surfaces, or irregularly shaped pores, which have an orientation that is not known in any case when doing radiography. This potentially modifies the detectability. The low CNR of small pores occurring in the experiment may also originate from a variation of the device parameters. According to Equation (2), this would influence the required CNR_{min} to detect a pore of a certain size and may explain the difficulties to find pores close to the detectability limit in the experimental radiographs. The smallest detectable pores have a diameter of 4 µm and a CNR of about 2.5 (see Table 3), and therefore the simulated results are confirmed by the experiment within the measurement precision. As mentioned before, small pores in large particles have a negligible contribution to the total porosity of the final AM part. In contrast, large pores with a significant impact on the AM process are detected reliably.

In a final step, the pore size distribution of the metallic particles found experimentally by means of XCT (see Figure 9b) and 3D image analysis was compared with the visibility limit calculated by the simulation. Figure 11 shows the histogram of all equivalent pore diameters found in the multi-particle sample. The theoretical visibility limit from the simulation was found to be 3.0 µm to 4.4 µm for the used device and exposure conditions and is indicated as a red line. It is concluded from Figure 11 that more than 99 vol.% of the pore volume present in this powder sample can be detected using digital radiography within the 95% confidence bound.

3.4. The Potential of Digital Radiography in Comparison to Other Methods for Powder Analysis

AM powders are usually characterized by their shape and size distribution using methods like sieving, laser diffraction, or light microscopy (Camsizer® [22,25]) methods. However, as highlighted in this article, the investigation of internal powder defects, e.g., pores, is not necessarily in the scope of standard investigations. However, the density of a powder is an independent quality characteristic. To access the density or the pore size distribution of a powder, typically pycnometry, metallography, or XCT are used.

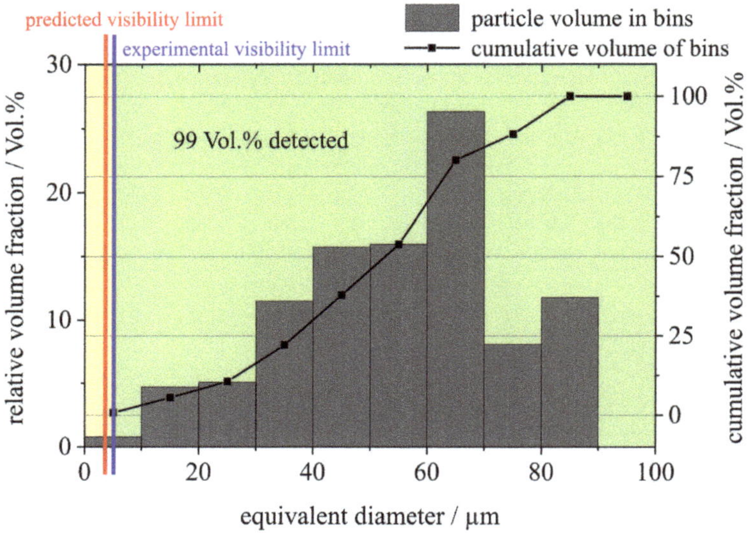

Figure 11. Histogram of the pore size distribution found in the metal powder by means of 3D X-ray computed tomography (XCT). Digital radiography was able to detect pores down to about 5 µm considering the 95% confidence bound at the selected setup (blue line). The predicted value for centered pores is 3 µm to 4.4 µm (red line). More than 99 vol.% of the pores volume present in this sample were detected experimentally.

All of the named methods have their benefits and drawbacks. An AM powder customized radiography technique comprises the potential to combine the best benefits of the discussed methods while minimizing the drawbacks. Sieving, laser diffraction, and light microscopy methods can screen large amounts of powder in a short time, but internal defects are not accessible. Pycnometry can also characterize large quantities of powder for its density, but not for its pore size distribution. Metallography or XCT can investigate the density and pore size distribution simultaneously. Still, both methods screen only small amounts of powder and need special attention, e.g., sample preparation or image analysis. For metallography, the sample preparation including polymer embedding, grinding, and polishing, is time-consuming. In addition, the imaging and image analysis takes time. When metallography is compared to high resolution XCT, it potentially investigates more particles, since XCT is a volume-based method and its quality of information can be treated as a "gold standard" for AM powders. However, XCT is a time-consuming method as well, including an optimal sample preparation, scanning, reconstruction, and image analysis. High throughputs of large powder batches are difficult to achieve by means of XCT.

As XCT takes projections from various angles of a sample, the data basis for the reconstructed volume and correspondingly the information content of internal sample features are much larger compared to digital radiography. However, considering the important information for AM powders, that pores are present or not, and that the actual position of the pores is not relevant per se, a radiographic imaging setup should be sufficient. The particle throughput of a radiographic setup compared to XCT is larger, even when radiography and XCT employ the same X-ray device setup. XCT needs a set of hundreds of projections from the sample for one scan. At the same time, radiography can image more particles. This is due to a geometrical relationship between the powder particle sizes and the X-ray device detector. One XCT scan can cover a cylindrical sample full of particles fitting the detector size. Imaging monolayer-like arranged powder particles fitting the complete detector by radiography leads to a larger throughput, when the XCT scan time or the amount of XCT projections is used as a time equivalent. The radiographic setup needs to be optimized for throughput, sensitivity, and resolution.

Therefore, digital radiography potentially combines the benefits of the methods, e.g., a high throughput and detailed internal powder information (pores). Drawbacks exist, as a complex sample preparation and image analysis need to be replaced by an automatic powder supply apparatus and dynamic and automated image analysis software, respectively, similar to the already employed techniques in light imaging and laser diffraction methods.

4. Summary and Outlook

The potential of digital radiography for the quality control of metal feedstock powders for AM in comparison to XCT was investigated in this work. It was shown numerically that for AM feedstock powder particles in the size range of 10 μm to 150 μm, pores down to 3.0 μm to 4.4 μm can be detected by X-ray radiography for the used commercial device in a typical setup and relevant exposure conditions. In most cases, the detectable pore size is sufficient for a quality control of the AM powder before it is processed. The numerical results were confirmed by the experiments. However, the selected X-ray setup is essential to detect smallest pores in particles. It can be improved by using a smaller focal spot size, higher magnification, longer exposure time, and a detector with smaller inherent detector unsharpness (lower $SR_b^{detector}$ value). Experimentally, it turned out that the orientation of the particles and pores relative to the radiographic projection direction influences the detectability, e.g., when pores or particles overlap in the radiograph. The minimal detectable pore volume amounts to around 5.1 vol.% of the particle volume for the smallest particle size studied in this work (10 μm). For the largest particle size (150 μm), this value decreases down to 0.0025 vol.%. Relevant particle pores, e.g., pores with a large pore to particle diameter ratio or large total feedstock powder porosities (multiple pores), with a potential negative impact on the AM process, are detected with high reliability by digital radiography.

As an outlook, an advanced digital radiography device can be designed to detect pores in AM powders. The powder would need to be moved on a conveyor belt or in a free-falling setup in order to achieve a high throughput. In addition, to speed up the imaging process, short exposure times are required, which can be achieved by faster and more efficient detectors and X-ray sources with high power and small focal spots, e.g., flash tubes. Such an advanced radiographic imaging setup would outrange competitive techniques and enable the opportunity to become a standard route for the quality assurance of AM powders.

Author Contributions: Conceptualization, G.-R.J. and A.W.; methodology, G.-R.J., A.F., and U.E.; software, G.-R.J., and A.F.; validation, G.-R.J., A.F., and U.E.; formal analysis, G.-R.J., A.F., and U.E.; investigation, G.-R.J., A.F., and U.E.; resources, A.W.; data curation, G.-R.J., A.F., and U.E.; writing—original draft preparation, G.-R.J., A.F., U.E., and A.W.; writing—review and editing, G.-R.J., A.F., U.E., and A.W.; visualization, G.-R.J., A.F., and U.E.; supervision, A.W.; project administration, A.W.; funding acquisition, none. All authors have read and agreed to the published version of the manuscript.

Funding: This research received no external funding.

Acknowledgments: The authors gratefully thank Uwe Zscherpel, Jörg Beckmann, David Schumacher, Alexander Ulbricht, Lina Pavasaryte, and Tobias Gustmann for fruitful discussions and proofreading of the manuscript. BASF is acknowledged for providing testing material.

Conflicts of Interest: The authors declare no conflict of interest.

References

1. Vock, S.; Klöden, B.; Kirchner, A.; Weißgärber, T.; Kieback, B. Powders for powder bed fusion: A review. *Prog. Addit. Manuf.* **2019**, *4*, 383–397. [CrossRef]
2. Sola, A.; Nouri, A. Microstructural porosity in additive manufacturing: The formation and detection of pores in metal parts fabricated by powder bed fusion. *J. Adv. Manuf. Process.* **2019**, *1*, 1. [CrossRef]
3. Martin, A.A.; Calta, N.P.; Khairallah, S.A.; Wang, J.; Depond, P.J.; Fong, A.Y.; Thampy, V.; Guss, G.M.; Kiss, A.M.; Stone, K.H.; et al. Dynamics of pore formation during laser powder bed fusion additive manufacturing. *Nat. Commun.* **2019**, *10*, 1987. [CrossRef] [PubMed]

4. Sinclair, L.; Leung, C.L.A.; Marussi, S.; Clark, S.J.; Chen, Y.; Olbinado, M.P.; Rack, A.; Gardy, J.; Baxter, G.J.; Lee, P.D. In situ radiographic and ex situ tomographic analysis of pore interactions during multilayer builds in laser powder bed fusion. *Addit. Manuf.* **2020**, *36*, 101512. [CrossRef]
5. Gould, B.; Wolff, S.; Parab, N.; Zhao, C.; Lorenzo-Martin, M.C.; Fezzaa, K.; Greco, A.; Sun, T. In Situ Analysis of Laser Powder Bed Fusion Using Simultaneous High-Speed Infrared and X-ray Imaging. *JOM* **2020**, 1–11. [CrossRef]
6. Cunningham, R.; Nicolas, A.; Madsen, J.; Fodran, E.; Anagnostou, E.; Sangid, M.D.; Rollett, A.D. Analyzing the effects of powder and post-processing on porosity and properties of electron beam melted Ti-6Al-4V. *Mater. Res. Lett.* **2017**, *5*, 516–525. [CrossRef]
7. Cunningham, R.; Zhao, C.; Parab, N.; Kantzos, C.; Pauza, J.; Fezzaa, K.; Sun, T.; Rollett, A.D. Keyhole threshold and morphology in laser melting revealed by ultrahigh-speed X-ray imaging. *Science* **2019**, *363*, 849–852. [CrossRef]
8. Hojjatzadeh, S.M.H.; Parab, N.; Guo, Q.; Qu, M.; Mohammad, S.; Hojjatzadeh, H.; Parab, N.D.; Xiong, L.; Zhao, C.; Escano, L.I.; et al. Direct observation of pore formation mechanisms during LPBF additive manufacturing process and high energy density laser welding. *Int. J. Mach. Tools Manuf.* **2020**, *153*, 103555. [CrossRef]
9. Ahsan, M.N.; Bradley, R.; Pinkerton, A.J. Microcomputed tomography analysis of intralayer porosity generation in laser direct metal deposition and its causes. *J. Laser Appl.* **2011**, *23*, 022009. [CrossRef]
10. Choo, H.; Sham, K.L.; Bohling, J.; Ngo, A.; Xiao, X.; Ren, Y.; Depond, P.J.; Matthews, M.J.; Garlea, E. Effect of laser power on defect, texture, and microstructure of a laser powder bed fusion processed 316L stainless steel. *Mater. Des.* **2019**, *164*, 107534. [CrossRef]
11. Morrow, B.M.; Lienert, T.J.; Knapp, C.M.; Sutton, J.O.; Brand, M.J.; Pacheco, R.M.; Livescu, V.; Carpenter, J.S.; Gray, G.T., III. Impact of Defects in Powder Feedstock Materials on Microstructure of 304L and 316L Stainless Steel Produced by Additive Manufacturing. *Metall. Mater. Trans. A* **2018**, *49*, 3637–3650. [CrossRef]
12. Sun, P.; Fang, Z.Z.; Zhang, Y.; Xia, Y. Review of the Methods for Production of Spherical Ti and Ti Alloy Powder. *JOM* **2017**, *69*, 1853–1860. [CrossRef]
13. Hooper, P.A. Melt pool temperature and cooling rates in laser powder bed fusion. *Addit. Manuf.* **2018**, *22*, 548–559. [CrossRef]
14. Leung, C.L.A.; Marussi, S.; Atwood, R.C.; Towrie, M.; Withers, P.J.; Lee, P.D. In situ X-ray imaging of defect and molten pool dynamics in laser additive manufacturing. *Nat. Commun.* **2018**, *9*, 1355. [CrossRef] [PubMed]
15. Slotwinski, J.A.; Garboczi, E.J.; Stutzman, P.E.; Ferraris, C.F.; Watson, S.S.; Peltz, M.A. Characterization of Metal Powders Used for Additive Manufacturing. *J. Res. Natl. Inst. Stand. Technol.* **2014**, *119*, 460–493. [CrossRef] [PubMed]
16. Heim, K.; Bernier, F.; Pelletier, R.; Lefebvre, L.P. High resolution pore size analysis in metallic powders by X-ray tomography. *Case Stud. Nondestruct. Test. Eval.* **2016**, *6*, 45–52. [CrossRef]
17. ISO. *ISO/ASTM 52907:2019-11—Additive Manufacturing—Feedstock Materials—Methods to Characterize Metal Powders*; ISO: Geneva, Switzerland, 2019.
18. ASTM. *ASTM F3049-14—Standard Guide for Characterizing Properties of Metal Powders Used for Additive Manufacturing Processes*; ASTM: West Conshohocken, PA, USA, 2014.
19. Gustmann, T.; dos Santos, J.M.; Gargarella, P.; Kühn, U.; Van Humbeeck, J.; Pauly, S. Properties of Cu-Based Shape-Memory Alloys Prepared by Selective Laser Melting. *Shape Mem. Superelast.* **2016**, *3*, 24–36. [CrossRef]
20. Kolb, T.; Huber, F.; Akbulut, B.; Donocik, C.; Urban, N.; Maurer, D.; Franke, J. Laser Beam Melting of NdFeB for the production of rare-earth magnets. In Proceedings of the 2016 6th International Electric Drives Production Conference (EDPC), Nuremberg, Germany, 30 November–1 December 2016; pp. 34–40. [CrossRef]
21. Kagathara, J.; Wieland, S.; Gärtner, E.; Uhlenwinkel, V.; Steinbacher, M. Heat Treatment and Formation of Magnetocaloric 1:13 Phase in $LaFe_{11.4}Si_{1.2}Co_{0.4}$ Processed by Laser Beam Melting. *Materials* **2020**, *13*, 773. [CrossRef]
22. Westermann, J.; Raatz, G. Improving quality control through effective particle characterisation of metal powders. *Powder Metall. Rev.* **2016**, *5*, 95–99. Available online: www.pm-review.com (accessed on 15 October 2020).

23. Thiede, T.; Mishurova, T.; Evsevleev, S.; Serrano-Munoz, I.; Gollwitzer, C.; Bruno, G. 3D Shape Analysis of Powder for Laser Beam Melting by Synchrotron X-ray CT. *Quantum Beam Sci.* **2019**, *3*, 3. [CrossRef]
24. Russo, P. (Ed.) *Handbook of X-ray Imaging: Physics and Technology*, 1st ed.; Series in Medical Physics and Biomedical Engineering; CRC Press Taylor & Francis Group: Boca Raton, FL, USA, 2018; ISBN 9781498741521.
25. Camsizer Website. Microtrac Retsch GmbH: Haan, Germany. Available online: www.microtrac.de (accessed on 15 October 2020).
26. Gustmann, T.; Gutmann, F.; Wenz, F.; Koch, P.; Stelzer, R.; Drossel, W.G.; Korn, H. Properties of a superelastic NiTi shape memory alloy using laser powder bed fusion and adaptive scanning strategies. *Prog. Addit. Manuf.* **2020**, *5*, 11–18. [CrossRef]
27. Miao, X.; Wang, W.; Liang, H.; Qian, F.; Cong, M.; Zhang, Y.; Muhammad, A.; Tian, Z.; Xu, F. Printing $(Mn,Fe)_2(P,Si)$ magnetocaloric alloys for magnetic refrigeration applications. *J. Mater. Sci.* **2020**, *55*, 6660–6668. [CrossRef]
28. Waygate Technologies Website. Waygate Technologies (Baker Hughes Digital Solutions GmbH): Wunstorf, Germany. Available online: www.bakerhughesds.com (accessed on 15 October 2020).
29. Thermo Fischer Scientific Avizo Software Website. Thermo Fischer Scientific Inc.: Waltham, MA, USA. Available online: www.thermofisher.com (accessed on 15 October 2020).
30. Ewert, U.; Zscherpel, U.; Heyne, K.; Jechow, M.; Bavendiek, K. Image quality in digital industrial radiology. *Mater. Eval.* **2012**, *70*, 955–964.
31. Jaenisch, G.-R.; Ewert, U. On the radiographic detectability of pores in spherically shaped matter by digital detector arrays. *Nucl. Instrum. Methods Phys. Res. Sec. A* **2021**, submitted.
32. ISee! Software Website. BAM: Berlin, Germany. Available online: http://www.zscherpel.info/ic/ (accessed on 15 October 2020).
33. ASTM. *ASTM E2446-16—Standard Practice for Manufacturing Characterization of Computed Radiography Systems*; ASTM: West Conshohocken, PA, USA, 2014.
34. Bellon, C.; Deresch, A.; Gollwitzer, C.; Jaenisch, G.-R. Radiographic Simulator *aRTist*: Version 2. In Proceedings of the 18th World Conference on Nondestructive Testing, Durban, South Africa, 16–20 April 2012.
35. Bellon, C.; Jaenisch, G.-R. *aRTist*—Analytical RT Inspection Simulation Tool. In Proceedings of the International Symposium on Digital Industrial Radiology and Computed Tomography DIR 2007, Lyon, France, 25–27 June 2007.
36. aRTist Software Website. BAM: Berlin, Germany. Available online: www.aRTist.bam.de (accessed on 15 October 2020).
37. Gustmann, T.; Schwab, H.; Kühn, U.; Pauly, S. Selective laser remelting of an additively manufactured Cu-Al-Ni-Mn shape-memory alloy. *Mater. Des.* **2018**, *153*, 129–138. [CrossRef]
38. Du Plessis, A.; Yadroitsava, I.; Yadroitsev, I. Effects of defects on mechanical properties in metal additive manufacturing: A review focusing on X-ray tomography insights. *Mater. Des.* **2020**, *187*, 108385. [CrossRef]
39. Jennings, B.R.; Parslow, K. Particle Size Measurement: The Equivalent Spherical Diameter. *Proc. R. Soc. Lond. A* **1988**, *419*, 137–149. [CrossRef]
40. DIN. *DIN ISO 9276-6:2012-01—Darstellung der Ergebnisse von Partikelgrößenanalysen—Teil 6: Deskriptive und quantitative Darstellung der Form und Morphologie von Partikeln*; DIN: Berlin, Germany, 2012.

metals

MDPI

Article

Experimental Determination of the Emissivity of Powder Layers and Bulk Material in Laser Powder Bed Fusion Using Infrared Thermography and Thermocouples

Gunther Mohr [1,2,*], Susanna Nowakowski [3], Simon J. Altenburg [1], Christiane Maierhofer [1] and Kai Hilgenberg [1,2]

1 Federal Institute for Materials Research and Testing (BAM; Bundesanstalt für Materialforschung und Prüfung), Unter den Eichen 87, 12205 Berlin, Germany; Simon.Altenburg@bam.de (S.J.A.); Christiane.Maierhofer@bam.de (C.M.); Kai.Hilgenberg@bam.de (K.H.)

2 Institute of Machine Tools and Factory Management, Technische Universität Berlin, Straße des 17. Juni 135, 10623 Berlin, Germany

3 Faculty of Engineering and Natural Sciences, Technical University of Applied Sciences Wildau, Hochschulring 1, 15745 Wildau, Germany; SNowakowski@gmx.de

* Correspondence: gunther.mohr@bam.de; Tel.: +49-030-8104-4865

Received: 29 October 2020; Accepted: 18 November 2020; Published: 20 November 2020

Abstract: Recording the temperature distribution of the layer under construction during laser powder bed fusion (L-PBF) is of utmost interest for a deep process understanding as well as for quality assurance and in situ monitoring means. While having a notable number of thermal monitoring approaches in additive manufacturing (AM), attempts at temperature calibration and emissivity determination are relatively rare. This study aims for the experimental temperature adjustment of an off-axis infrared (IR) thermography setup used for in situ thermal data acquisition in L-PBF processes. The temperature adjustment was conducted by means of the so-called contact method using thermocouples at two different surface conditions and two different materials: AISI 316L L-PBF bulk surface, AISI 316L powder surface, and IN718 powder surface. The apparent emissivity values for the particular setup were determined. For the first time, also corrected, closer to real emissivity values of the bulk or powder surface condition are published. In the temperature region from approximately 150 °C to 580 °C, the corrected emissivity was determined in a range from 0.2 to 0.25 for a 316L L-PBF bulk surface, in a range from 0.37 to 0.45 for 316L powder layer, and in a range from 0.37 to 0.4 for IN718 powder layer.

Keywords: laser powder bed fusion (L-PBF); selective laser melting (SLM); laser beam melting (LBM); thermography; emissivity; calibration; thermocouples; 316L; IN718; process monitoring

1. Introduction

Additive manufacturing (AM) technologies comprise several different modern manufacturing methods. Within metallic production routes, laser powder bed fusion (L-PBF) is of outstanding interest [1]. Due to the layer-wise nature of the process, L-PBF offers unique opportunities to monitor the complete production of a part layer by layer. Numerous monitoring approaches using various technologies and diverse methodologies can be found in the relevant literature [2,3]. They are used to monitor several different objects in L-PBF, e.g., powder bed compaction [4], particle gas emissions [5], laser power [6], and thermal emissions [7].

In this introductory section, a brief overview is given on thermal monitoring in L-PBF and the theoretical background about emissivity of real metallic surfaces. Furthermore, a brief review on temperature adjustment attempts for thermography in AM and a short survey on oxidation and its influence on emissivity are given. In Section 1.5, the calibration approach of this study is introduced.

1.1. Thermal Monitoring in L-PBF

As melting, solidification and cooling are essential for L-PBF, thermal aspects of the component during manufacturing are of utmost interest. Transient heat flux and thermal history directly affect part quality and properties of L-PBF components. Therefore, the most applied monitoring approaches deal with the spatial and temporal monitoring of heat radiation of the built-up [8]. Among the approaches of thermal condition monitoring, contactless measurement techniques are most common. Passive infrared (IR) thermography is a technology which was used by several groups of authors for thermal in situ process monitoring means [7,9–14]. IR thermography can acquire data of thermal emissions of the process layer by layer with variations in spatial and temporal resolutions, depending on the particular equipment and setup [12]. Compared to highly localized pyrometric measurements, IR cameras allow for a relatively large field of view, as well as for the capability to capture different build parts or different sections of one part at the same time without the need for an implementation that is coaxial to the laser path.

However, without appropriate calibration or adjustment of the signal of the IR camera used in the particular conditions of a specific setup, the acquired data provide information on absolute or relative radiation intensity or apparent temperatures, but not on real surface temperature values. It is interesting to note that there are currently no standardized procedures and reference standards for the calibration of infrared cameras in additive manufacturing setups. Therefore, in the following, temperature adjustment and temperature calibration are used as synonyms. This is due to the fact that the computation of temperature from data recorded with an IR camera is not only based on measured radiation intensity but highly depends on the emissivity of the target object [15]. IR cameras deliver either calculated IR signal values or apparent temperature values [13] (in the case of a previous black body calibration of the camera itself), or they deliver IR signals proportional to the radiant flux absorbed by the camera detector [16] (in the case of no previous black body calibrations or no computational considerations of such a forgone calibration). Commercial IR thermography cameras are often calibrated for black body radiation by their manufacturer. In these cases, the IR signal values, delivered by the camera, are sometimes referred as apparent temperatures, as done by the authors in other work [12]. These apparent temperatures are well below the actual temperatures of the regarded surfaces. The discrepancy between real temperatures and these apparent temperatures is mainly a result of differences in emissivity of real surfaces and black body radiators [15] on the one hand, and additional attenuation effects by optical elements in the optical path between camera and object in the particular scenery on the other hand [9,12].

Depending on the monitoring goal, the use of IR thermography without determination of real temperatures due to a missing calibration of the setup can still be very valuable, as for many issues the relative comparison of apparent temperatures or cooling rates can already provide enough informative value for particular conclusions, e.g., defect detection [7,12,16] or detection of areas of heat accumulation [7,13,17]. Therefore, the knowledge of an absolute temperature is not always necessary, especially when IR thermography is used as comparative mean against a kind of standard condition. However, there are also several research questions in which the knowledge of a calibrated absolute temperature or at least a reliable approximation of the absolute temperature would be desirable, e.g., in the field of validations of numerical modelling [18] or for considerations and classifications of in situ heat treatments during the L-PBF process [13]. Hence, a calibration of an installed IR camera at a L-PBF machine is considered to be very useful.

1.2. Theoretical Background about Emissivity or Real Metallic Surfaces

According to Usamentiaga et al. [15], the most important calibration parameter for temperature measurement using IR thermography is emissivity. They reported a general procedure to measure emissivity, i.e., the so-called contact method. This method uses a thermocouple to acquire a reference temperature of the target object, which is heated up to a temperature of real working conditions. At these conditions, the apparent temperature of the IR camera can be calibrated against the temperature of the thermocouple.

Hereafter, a brief excursus into the theoretical background and definition of emissivity is given and the introduced equations will be used in subsequent sections: The radiosity $W_{\lambda b}$ of a black body (a black body is defined by transmittance $\tau = 0$ and reflectance $\rho = 0$; hence, absorptance $\alpha = 1$, and thus its radiosity equals its radiant exitance) is a function of temperature T and wavelength λ and can be described by Planck's law, where c_1 and c_2 are radiation constants [15,19]. The peak intensities for higher temperatures are shifted towards smaller wavelengths. The shift can be explained by Wien's displacement law. Both is graphically illustrated in Figure 1.

$$W_{\lambda b}(\lambda, T) = \frac{c_1}{\lambda^5} \frac{1}{\exp\left(\frac{c_2}{\lambda \cdot T}\right) - 1} \tag{1}$$

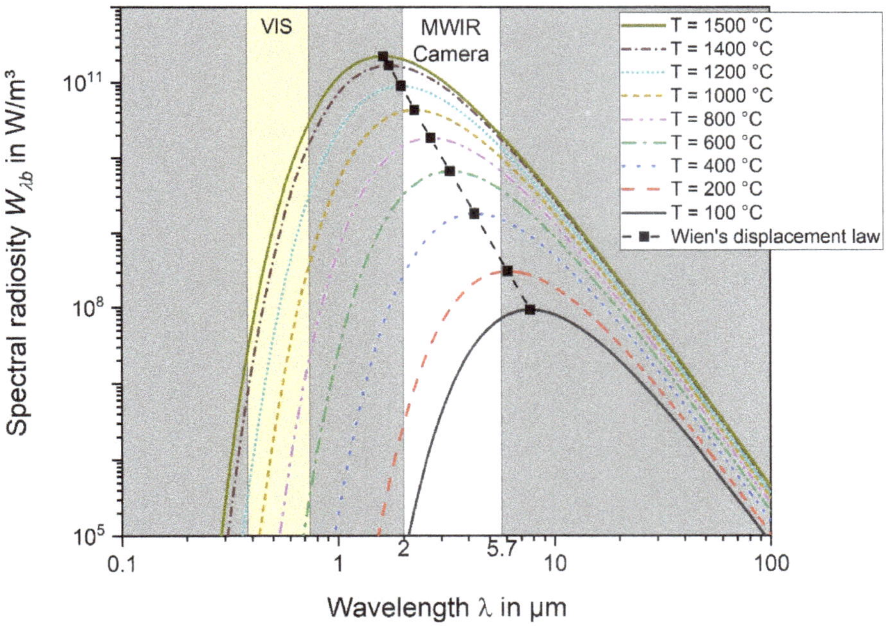

Figure 1. Planck's law and Wien's displacement law graphically expressed. The visible spectrum (VIS) and the mid-wavelength infrared (MWIR) region of the camera of this study are highlighted.

It is worth noticing from Figure 1 that the most relevant temperature regions of solidified surface temperatures of L-PBF parts during cooling down after laser exposure as well as in the pre-heating condition prior to laser exposure have their maximum intensity in the mid-wavelength infrared (MWIR) region. As also summarized by Usamentiaga et al. [15], the integration of Planck's law through all wavelengths leads to the radiant exitance of a blackbody and can be defined as in Equation (2), where σ is the Stefan-Boltzmann constant (σ = 5.6704 ... $\times 10^{-8}$ W/(m^2 K^4)).

$$W_{b,total} = \sigma T^4 \quad (2)$$

For the determination of the spectral emissivity, a computation of the spectral exitance is required. An integration over all wavelengths, leading to Equation (2), would be an ineligible simplification in the case of the restricted spectrum of wavelengths in thermographic applications. As IR sensors are always sensitive in a restricted spectral range only, an integration of Planck's law over the appropriate wavelengths is required. Figure 2 shows the difference in calculations for an integration over 2 µm to 5.7 µm (comparable to the spectral range of the sensor used in this study) compared to the Stefan-Boltzmann approach. This demonstrates clearly that the use of Stefan-Boltzmann would lead to significant errors.

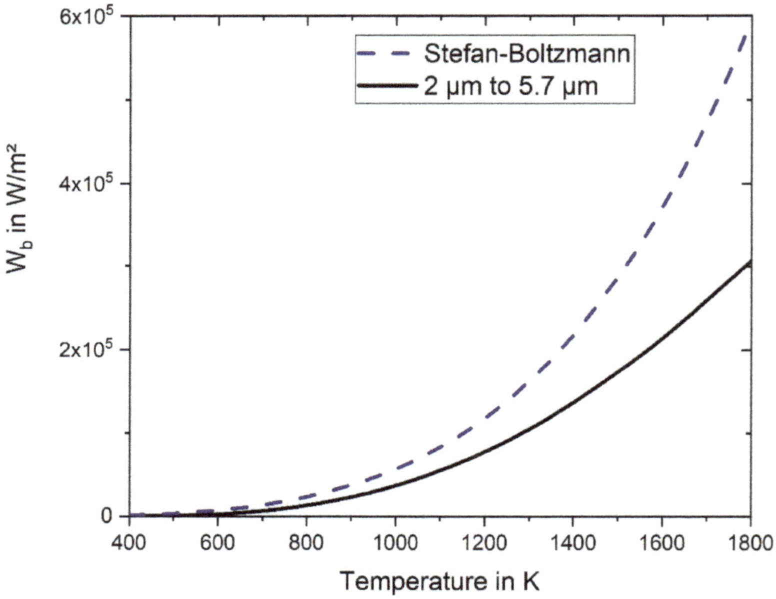

Figure 2. Theoretical calculations of radiant exitance W_b using the Stefan-Boltzmann approach and an integration through a limited spectrum of wavelengths.

Real surfaces do not fulfill the definition of a black body. They always emit less energy than a black body. The ratio of the spectral exitance of a real body $W_{\lambda r}$ to that of a black body $W_{\lambda b}$ at the same temperature is defined as spectral emissivity ε_λ, see Equation (3).

$$\varepsilon_\lambda(\lambda) = \frac{W_{\lambda r}(\lambda)}{W_{\lambda b}(\lambda)} \quad (3)$$

The emissivity of a black body is $\varepsilon = 1$, the emissivity of real bodies is smaller ($\varepsilon < 1$). For simplification reasons, the emissivity ε of solid objects is often treated as a constant and independent of the wavelength within short intervals, in which IR sensors work [15]. In doing so, real bodies are assumed to be grey bodies [15].

In the case of unknown transmission losses of the optical path between sensor and target object in a specific experimental setup, a calculation of the emissivity based on measured reference temperatures and apparent temperatures of an IR camera would result in an apparent emissivity ε_{app} of the target object in the particular scenery which is smaller than the actual emissivity of the target object ε_{real}.

As the aim of this work was the temperature adjustment of a specific MWIR camera setup used as thermal monitoring device at a particular L-PBF system, thereafter, ε_{app} was used in the further considerations in a first stage. This simplified the analysis, as the transmission losses of the optical elements were neglected under the assumption of being constant over the respective temperature regions. These have often been referred to as being influential factors on emissivity, e.g., by Lane et al. [9]. In a second stage, a correction of transmission losses and a consideration of radiation of the ambient temperature were conducted to calculate an approximation of the real emissivity of the target object ε_{corr}, although still within the grey body approximation.

The emissivity of a real target object depends on several factors: material, surface condition (surface roughness and oxidation state), viewing angle, temperature and wavelength [20]. The published reference values of emissivity of different materials in data sheets are usually considered as being captured perpendicular to the surface of the target object [21]. Metals and their alloys have considerably low emissivity values and undergo heavy variations due to their surface conditions [22], e.g., polished steel sheets have $\varepsilon = 0.1$ at a temperature of $T = 310$ °C, but in aged and oxidized condition, they show an emissivity of up to $\varepsilon = 0.8$. For stainless steel AISI 316 in polished condition, one can find emissivity values between 0.24 and 0.31 in a temperature range of 200 °C to 1040 °C [21].

1.3. Calibration Attempts for Thermography in Additive Manufacturing

Within the scope of IR thermography in AM, some work has been published that includes attempts to calibrate a particular IR camera setup or to evaluate emissivities for particular process conditions. The differences of the cooling rate of the transition between liquidus and solidus in the melt pool can be used as a kind of single-point calibration in cases where the temporal or spatial resolution of the camera allows for reliable capturing of this condition. This was done by Doubenskaia et al. [23] for laser metal deposition (LMD) using TiAl6V4. They used an IR camera sensitive in a spectral range from 3 µm to 5 µm and calculated an emissivity of $\varepsilon = 0.201$ at the transition temperature. Yadroitsev et al. [24] also used the liquidus solidus transition during a L-PBF process using TiAl6V4. They used a CCD camera setup coaxial to the laser path and measured at a wavelength of 0.8 µm. They calculated an emissivity of $\varepsilon = 0.35$ at the transition temperature. Heigel and Whitenton [25] determined the liquidus-solidus transition for the calibration of a SWIR camera monitoring the L-PBF process of the nickel-based alloy IN625. Additionally, they calculated the effective emissivity of a L-PBF IN625 surface as $\varepsilon = 0.168$, based on the transition temperatures [25]. Other work by Lane et al. [9] argued for using assumption-based estimations of uniform emissivity values of $\varepsilon = 0.5$ as long as no measured emissivity values are available and in order to still use temperature values instead of IR signal values.

As mentioned above, a possible approach for transforming IR signals or apparent temperatures of IR cameras into real temperatures is the synchronous use of a second but already calibrated temperature measuring technique during IR capturing, which measures the temperature of the object of investigation directly as a reference. The classic approach for this is the contact method, using thermocouples (TC) for reference temperature measurement. There is only limited work published on this for L-PBF. Heigel et al. [26] and Williams et al. [14] presented calibration results of the contact method for their specific camera setups.

Williams et al. [14] installed an IR camera of type A 35 (FLIR Systems Inc., Wilsonville, OR, USA) in the build chamber of a commercial L-PBF system of type AM250 (Renishaw plc, Wotton-under-Edge, UK) using an observation angle of 24° (detector plane to build plane). The resulting resolution of their setup was approximately 1 mm^2 per pixel, captured at a framerate of 60 Hz. They used a heated AM calibration component manufactured by L-PBF using 316L with and without powder on top of it. The real surface temperature was measured by one TC. The results were directly used for a determination of in-process surface temperatures of built parts, which were monitored afterwards in the same study. They abstained from any calculation of apparent or real emissivity values.

Heigel et al. [26] calibrated an IR camera sensitive in a reduced spectral range of 1.35 μm to 1.6 μm by using a purposely built calibration setup outside of a L-PBF system. Their calibration setup ensured similar conditions to their monitoring setup at a L-PBF system of type M270 (EOS GmbH, Krailing, Germany) presented by Lane et al. [9], i.e., using an observation angle of 45° to 43.7°. They used a heated AM calibration component manufactured by L-PBF using IN625. They calculated an emissivity $\varepsilon = 0.680$ of a rather smooth surface (Sa approximately 12 μm) and an emissivity $\varepsilon = 0.761$ of a rather rough surface (Sa approximately 27 μm). Oxidation of IN625 was also considered, but resulted in being negligible as a factor of emissivity in their study. However, the question as to whether the not purposely oxidized specimens (post process oxidation was conducted for some specimens) might have already been oxidized during the built process remained unclear.

1.4. Oxidation and Its Influence on Emissivity

In the frame of this study, knowledge regarding oxidation growth at steel surfaces and its effect on emissivity in general is needed for the discussion section. Hence, a brief excursus is given hereafter, based on the relevant literature, summarizing published results on oxide layer thickness of steels and its influence on emissivity. Thickness growth of material dependent oxide layers depends on atmosphere, temperature and time. When assuming a constant atmosphere, temperature and time play important roles in the evolution of an oxide layer.

Hakiki et al. [27] measured the oxide film thickness of austenitic stainless steel AISI 304 (1.4301) tempered in the temperature range between 50 °C and 450 °C in air for 2 h. The film thickness varied between 8 nm (50 °C) and 30 nm (450 °C) [27].

Kämmerer [28] measured the thickness of oxide layers of cold-rolled plates of the ferritic steel AISI 441 (1.4509) tempered in air. She showed the exponential relationship between tempering temperature and oxide layer thickness. She examined a change in oxide layer growth rate between 400 °C and 500 °C towards higher growth rates for higher temperatures, at a tempering time of 10 h. At a tempering temperature of 400 °C, no significant change of oxidation thickness could be revealed between holding times of 1 h and 10 h. The measured oxide layer thickness was between 3 nm and 3.5 nm at a tempering temperature of 400 °C and approximately 4 nm at 500 °C (holding time 10 h). However, repeated tempering with interim cooling down (10 times 1 h at 400 °C) resulted in doubling of the oxidation layer thickness [28].

According to Janssen [29], who studied oxidation of austenitic stainless steel AISI 304 (1.4301) in air, the literature values about the thickness of a passive oxidation layer of steels varied between 1.5 nm and 8 nm (mostly in the area 2 nm +/− 0.6 nm). During his oxidation experiments, a slight yellow annealing color started to be visible by eye at 550 °C (5 min holding time) changing to gold-yellow at 600 °C (9 min holding time) and darkened after a subsequent 10 min holding time at 595 °C. During the temperature rise from 410 °C to 550 °C, he observed a distinct increase in oxidation of chromium and iron. From approximately 580 °C and especially from 600 °C, the oxidation rate of iron increased strongly, while the chromium oxidation rate increased slightly. This was in accordance with his measurements of the oxidation layer thickness: A first distinct growth of the oxidation layer was measurable after a temperature rise from 410 °C to 550 °C, followed by a strong increase in the temperature region from 550 °C to 600 °C. At 900 °C, the oxidation layer thickness was around 35 nm to 65 nm with an additional 2 μm to 3 μm thick scale layer.

Iuchi et al. [30] investigated the modelling of an emissivity change during the growth of oxide layers for a wavelength of $\lambda = 1.5$ μm at cold rolled steel tempered at 500 °C. They found that the emissivity behavior of oxide films thinner than 5.8 nm was almost identical to that of non-oxidized surfaces. For oxide films thicker than 39.1 nm, the emissivity behavior changed drastically from approximately $\varepsilon = 0.3$ (at 5.8 nm) to approximately $\varepsilon = 0.7$ (at 39.1 nm); $\varepsilon = 0.8$ (at 54.2 nm); $\varepsilon = 0.87$ (at 82.7 nm) [30].

According to Zauner et al. [31], the change of emissivity of steel surfaces due to the growth of an oxide layer reached a first maximum above $\varepsilon > 0.8$ for an oxide film thickness of about 100 nm.

While the emissivity fluctuated in the beginning of the growth phase, it started to stabilize after reaching approximately 500 nm film thickness at a value of $\varepsilon = 0.85$. These results were based on theoretical calculations. The fluctuations stem from interference phenomena of reflectance at oxide films at metallic surfaces which depended on oxide film thickness [31].

Del Campo et al. [32] studied the oxidation kinetics of iron below 570 °C and conducted emissivity measurements at four different temperatures: 415 °C, 480 °C, 535 °C, 570 °C. Their results for thin oxide films were not contradicting to the afore mentioned. However, they had a stronger focus on various spectral wavelengths and on thicker oxide layers resulting from longer holding times.

1.5. Calibration Approach of This Study

The literature basis regarding emissivity determination and IR camera calibration for L-PBF is limited, also due to the requirement of specificity of each setup. Thus, this study aimed for the experimental temperature adjustment of an off-axis MWIR thermography setup, which was used for in situ thermal data acquisition of L-PBF processes in other studies of the authors [13,17]. The contact method was applied for this purpose. Furthermore, a correction of apparent emissivity values of the specific setup was pursued in in order to receive setup independent emissivity values of L-PBF bulk material and powder layers, which might be useful for, e.g., numerical simulations. A third goal of this work was the transferability of the technical equipment used during the calibration for the later use in other specific setups.

2. Materials and Methods

This section describes the experimental setup, the used equipment and materials, as well as the experimental variations.

2.1. Thermographic Measurement Setup and L-PBF System

A MWIR thermographic camera of type ImageIR8300 (InfraTec GmbH, Dresden, Germany) was mounted on top of a commercial L-PBF system of type SLM280HL (SLM Solutions Group AG, Lübeck, Germany). It had optical access to the build chamber through a purposely installed sapphire window in the ceiling of the chamber. The optical path was deflected by two gold mirrors to shift the observation field in the direction of the center of the build plate while keeping the angle of observation at approximately 0° (angle between detector plane and build plate plane). Compared to similar off-axis IR camera setups used as thermal monitoring device in L-PBF (cf. Krauss et al. [33], Lane et al. [9]), the nearly perpendicular view onto the build plane is beneficial in terms of having a large focus area. Tilted systems are always faced with very limited areas which are in focus and usually have large defocused areas. A schematic and two photographs illustrate the setup in Figure 3. The same setup was also used in other work of the authors [13,17] in the same configuration for in situ process monitoring means. The schematic also includes the heating device, which is described in Section 2.2.

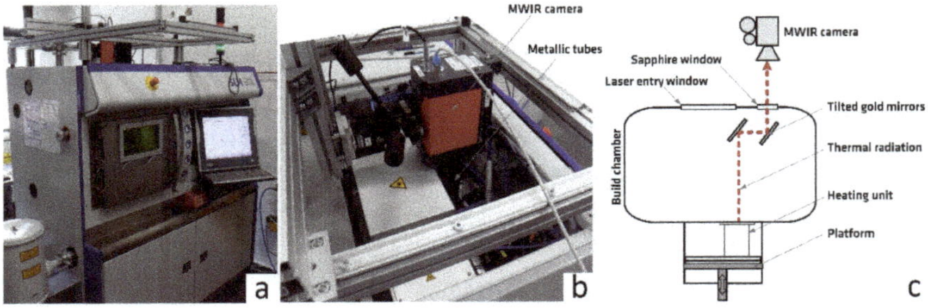

Figure 3. MWIR camera setup: installed camera on top of L-PBF machine ((**a**) side view, dimensions of the process chamber: (280 × 280 × 360) mm^3; (**b**) top view, dimensions of the camera body; (244 × 120 × 160) mm^3; (**c**) schematic, not to scale).

The camera used a cooled InSb-focal-plane-array of size 640 pixel × 512 pixel and a bit resolution of 14 bits was used. A 25 mm objective lens was used, resulting in a spatial resolution of the setup of approximately 420 µm per pixel length. No additional external filters were used. The camera was sensitive in a spectral range from 2 µm to 5.7 µm. The optical path outside of the build chamber was shielded by blackened metallic tubes, it could be interrupted manually by a shutter. According to the manufacturer's specifications, the reflectivity of the gold mirrors was above 99% in the spectral range of the camera, the transmissivity of the window above 83%. Typical spectral reflectivity (mirrors), transmissivity (window, objective, camera internal filters) and sensitivity (camera detector) were known to be at least in the relevant spectral range between 2 µm and 5.7 µm and were used for the emissivity calculation in Section 3.3.

The camera was calibrated using a black body radiator by the manufacturer in different calibration ranges that could be chosen for the specific experiment. The following camera calibration ranges were used:

- 60–200: valid for black body temperature of 60 °C–200 °C (integration time: 89 µs),
- 200–400: valid for black body temperature of 200 °C–400 °C (integration time: 193 µs; internal attenuation filter A within the camera),
- 300–600: valid for black body calibration temperature 300 °C–600 °C (integration time: 45 µs; internal attenuation filter B within the camera).

Frame capturing of the camera was conducted at 100 Hz. To reduce the amount of data, a subframe was used for capturing a size of 224 pixel × 160 pixel, similar to the cited in situ measurements.

2.2. Heated Reference Device

A specimen was manufactured by L-PBF using 316L, which was used as the heated AM reference part. The manufacturing parameters were in accordance with the standard parameters used in [13] (laser power of 275 W; scanning velocity of 700 mm/s; hatch distance of 0.12 mm; layer thickness of 0.05 mm; bi-directional scanning parallel to the edges of the specimen). No additional up-skin parameter was applied. The surface area roughness of the top surface was approximately Sa = 7 µm, determined at two areas of 0.8 mm × 10 mm using a coherence scanning interferometry profilometer of the type Nexview (Zygo Corp., Middlefield, CT, USA). The cuboid specimen had the dimensions 13 mm × 20 mm × 140 mm. In the middle of the upper surface of the specimen, a cross-like artefact with a depth of 0.5 mm was manufactured to help finding the focus level during the calibration experiments. The rim of the specimen was elevated by 0.5 mm at a width of 1 mm.

A fixture was constructed that ensured the upright standing of the specimen, as well as thermal insulation between the specimen and the fixture by ceramic plates. A heating mat and a heating inverter of type ST11 (Sokol-Therm Deutschland GmbH, Eisenhüttenstadt, Germany) were used to heat up the specimen. The specimen was wrapped by the heating mat and placed within a sheet metal enclosure. For insulation reasons, an insulation wool was placed between the heating mat and the enclosure. To shield heat radiation towards the camera which did not come from the heated AM specimen, high-temperature fiberboard was used to cover the top of the heating device. Four K-type thermocouples (TC, TC1 to TC4) of 0.127 mm diameter were spot welded at the four sides of the specimen approximately 1 mm below the upper surface of the rim. At one side, an additional TC (TC Inv.) was installed to act as target measuring point for the heating inverter. The responses from the thermocouples were acquired through a measuring amplifier device of type MX1609 (HBM GmbH, Darmstadt, Germany) at a sampling rate of 10 Hz. The heating device was positioned by lowering the build plate lift of the machine in such a way that the surface of the reference specimen was in the usual build plane of the L-PBF setup. A schematic of the specimen surface and the clamping are shown in Figure 4. This also includes the positioning of the thermocouples. Figure 5 shows photographs of the heating device. Two separate TCs were spot welded at the sheet metal and the substrate, respectively, in order to monitor the heat development at the device.

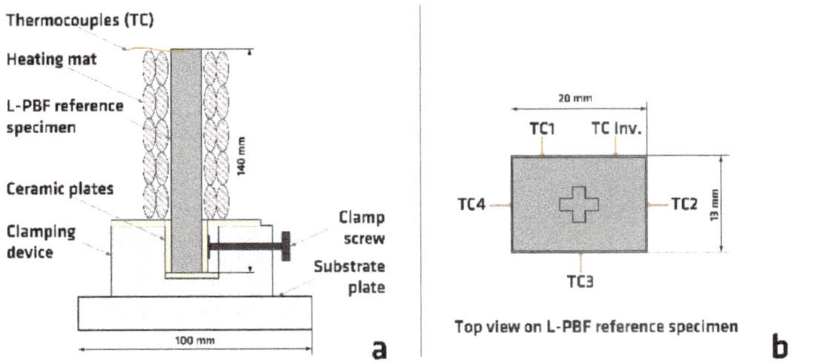

Figure 4. (**a**) schematic of heating device without outer insulation; (**b**) schematic of specimen surface indicating TC positions.

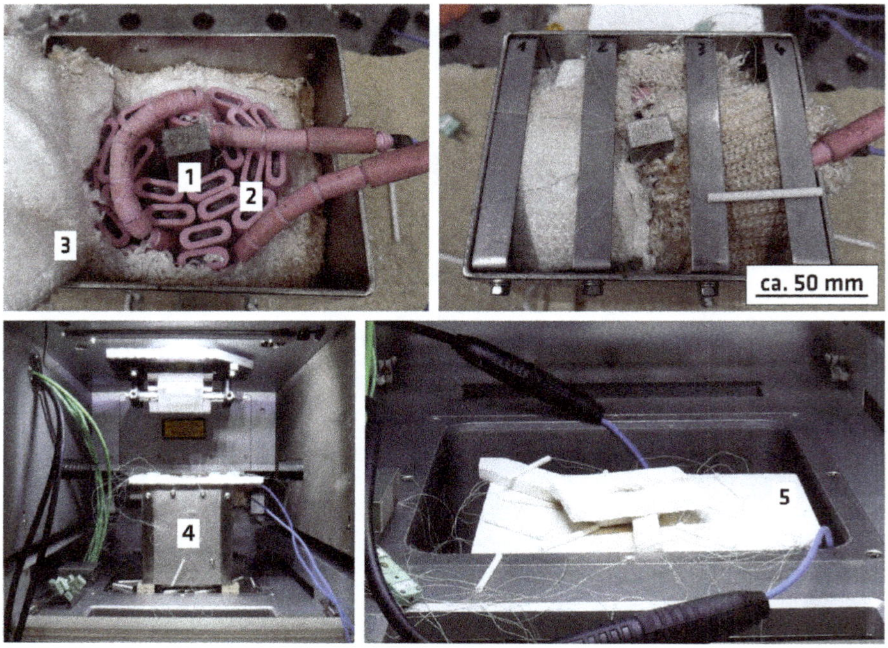

Figure 5. Photographs of mounted heating device: L-PBF reference specimen (**1**); heating mat (**2**); insulation wool (**3**); sheet metal enclosure (**4**); high-temperature fiberboard on top of the device (**5**); the heating device is placed at its measuring position (position 1, as described in Section 2.4.3).

2.3. Examination Methodology

Recording of MWIR camera data and TC data was started synchronously by the experimental conductors at different times during the heating cycle. The apparent temperature data of the MWIR camera were gained by using the software IRBIS3 professional (InfraTech GmbH, Dresden, Germany). Four regions of interest (ROIs) were defined across the top surface of the specimen, excluding the cross in the middle of the surface as well as the rim in order to exclude additional surface property effects, cf. Figure 6. The mean temperatures of these four ROIs were exported as ASCII files. Subsequent analysis was conducted using the software Origin 2019 (OriginLab Corporation, Northampton, MA, USA). The total mean of all pixels within the four ROIs was taken as apparent temperature value of the camera. Considerations of standard deviations and fluctuations are discussed in Section 3.5. The TC temperature data were also analyzed using Origin 2019.

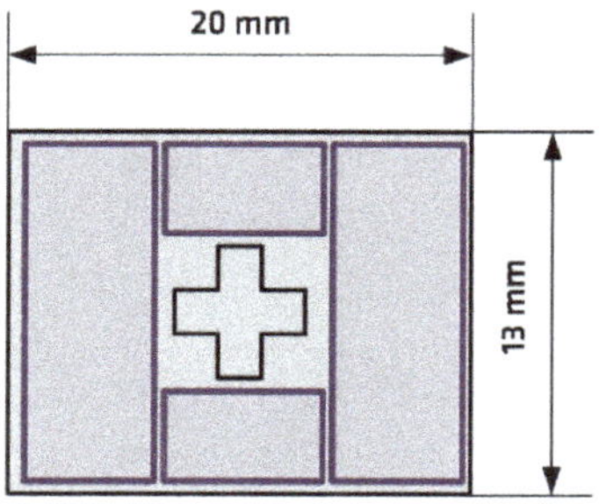

Figure 6. Schematic of the ROIs (purple color) for data capture.

2.4. Experimental Measurment Variations

The temperature calibration experiments were conducted at different conditions, which are introduced in this section. Unless stated otherwise, all experiments were conducted under typical L-PBF working conditions. Thus, the experiments were conducted inside the build chamber and under argon gas atmosphere at a gas flow velocity of approximately 21.5 m/s (measuring sensor in the gas circulation pipes), with a gas flow flushing the chamber from the right to the left. A one-point nonuniformity correction (NUC) was always conducted before the beginning of any new capturing of the MWIR camera. This was done by placing a shutter in the optical path outside of the build chamber and using the NUC function of the camera. The optical shutter was at room temperature. A manual refocusing of the objective lens was conducted whenever a blackbody calibration range of the camera was changed. This was necessary, since the additional camera internal filters A and B in the calibration ranges 200–400 and 300–600 changed the optical path length.

The following experimental variations were investigated using different camera calibration ranges:

- 316L L-PBF surface at temperatures between 100 °C and 700 °C,
- 316L powder layer at temperatures between 100 °C and 650 °C,
- IN718 powder layer at temperatures between 100 °C and 650 °C, and
- 316L oxidized L-PBF surface at three different positions.

The individual experiments are explained in detail in the respective sections.

2.4.1. Temperature Variations at a Non-Oxidized L-PBF Surface (316L)

During the temperature variation experiments, the virgin 316L L-PBF specimen was heated up to a temperature of 750 °C in several stages in the L-PBF build chamber. Data couples of TC temperatures and apparent temperatures were acquired. The heat up was paused at different temperature stages to reach temperature plateaus. Different camera calibration ranges were used during the data acquisition (see Section 2.1). The sheet metal enclosure of the heating device reached a maximum temperature of approximately 350 °C and the substrate of the fixture a maximum temperature of approximately 375 °C during the heating experiments. All temperatures measured by the temperature sensors of the L-PBF system stayed within the narrow rage of the specifications of the manufacturer during the experiments, e.g., the build chamber temperature was between 31 °C and 36 °C.

2.4.2. Temperature Variations at Powder Surface (316L and IN718)

Two different powder materials were spread on top of the specimen in two separate experiments before the heating up. The powder was manually placed using a spatula in the middle of the top surface of the specimen and then spread manually over its entire surface by using a razor blade. Thereby, a powder bed of approximately 500 μm thickness was generated according to the dimensions of the rim. Table 1 contains information of the powder properties according to supplier's information (SLM Solutions Group AG, Lübeck, Germany). The heating device with the powder bed on top was heated up to 650 °C in several stages.

Table 1. Powder properties according to supplier's information.

Powder Properties	Powder 1 AISI 316L (1.4404)	Powder 2 Inconel 718
D_{10} in μm	18.22	25.50
D_{50} in μm	30.50	37.60
D_{90} in μm	55.87	57.07
Mean diameter in μm	34.69	39.49
Apparent density in g/cm^3	4.58	4.56

2.4.3. Positioning

To evaluate a potential influence of the x-y position of the target object within the build chamber, measurements were conducted at three different positions as illustrated in Figure 7. The heating device was placed manually. The measurements were carried out at a specimen surface temperature of around 400 °C. The center of the specimen was approximately located at the following coordinates (reference 0 as in Figure 7, coordinates in mm): x140, y66 (Position 1); x113, y81 (Position 2); x143, y140 (Position 3). Position 1 and position 2 are approximately at the positions where two of the in situ monitored specimens of [13,17] were located. Position 3 was representative of a specimen in the center of the L-PBF substrate plate. The tilt of the gold mirrors as well as the focus of the camera had to be adjusted to capture position 3. Position 1 and position 2 were located within the same chosen field of view. These measurements were conducted using an oxidized specimen, which was oxidized in pre-tests of the heating device up to a temperature of 600 °C outside of the L-PBF machine at ambient atmosphere beforehand. Apart from the position shifts described here, all other experiments were conducted at approximately position 1.

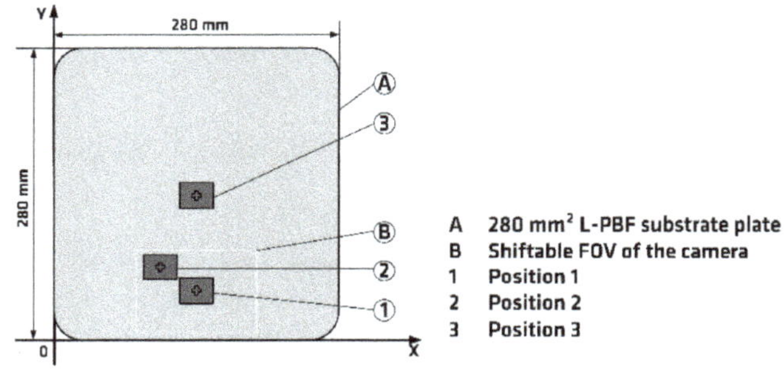

Figure 7. Positioning of the heated specimen on top of the (lowered) build plate.

2.5. Determination of Emissivity Values

Two different ways of determining emissivity values were implemented, resulting in non-corrected, so-called apparent emissivity values and corrected, closer-to-real emissivity values. The values were calculated for the distinct measurement points, which comprised the temperature of the thermocouples T_{TC}, the apparent temperature of the MWIR camera T_{app}, and the temperature in the build chamber T_0.

2.5.1. Determination of Apparent Emissivity Values

The computation of emissivity values without consideration of transmission losses and thermal stray radiation leads to the so-called apparent emissivity ε_{app}. The apparent emissivity ε_{app} can be computed using Equation (3) (neglecting the wavelength dependency, grey body approximation). Thereby, an integration of Equation (1) (Planck's law) in the spectral range of the MWIR camera (λ_1 = 2 µm to λ_2 = 5.7 µm) for the reference TC temperature ($W_{\lambda b}$) and for the apparent temperature of the camera of the respective calibration range ($W_{\lambda r}$) must be conducted:

$$\varepsilon_\lambda(T = T_{TC}) = \frac{\int_{\lambda_2}^{\lambda_1} W_{\lambda r}\left(\lambda, T_{app}\right) d\lambda}{\int_{\lambda_2}^{\lambda_1} W_{\lambda b}(\lambda, T_{TC}) d\lambda} \tag{4}$$

This simplified analysis was performed, since it is a widely used way to estimate emissivity values.

2.5.2. Determination of Corrected Emissivity Values

The calculation of the apparent emissivity by Equation (4) is a rough estimate. This analysis neglects the radiation that was reflected from the surroundings at the surface as well as all the spectral characteristics of all optical elements and the camera detector. Assuming a homogeneous thermal background radiation of a black body at temperature T_0, the emitted spectral radiosity of a grey body is:

$$W_{\lambda r}(\lambda, T, \varepsilon, T_0) = \varepsilon(T) \cdot W_{\lambda b}(\lambda, T) + (1 - \varepsilon(T)) \cdot W_{\lambda b}(\lambda, T_0) \tag{5}$$

Here, the spectral and angular dependence of ε are still neglected; thus, the determined values are still effective values for the spectral range of the camera. To better estimate the surface emissivity, first, the total irradiance measured by the camera during calibration at a black body was calculated, considering the spectral transmissivity of the optics present during the calibration $\tau_{opt,cal}(\lambda)$ and the spectral responsivity of the detector $S(\lambda)$:

$$E_{cal}(T) = \int_{-\infty}^{\infty} W_{\lambda b}(\lambda, T) \cdot \tau_{opt,cal}(\lambda) \cdot S(\lambda) \cdot d\lambda \tag{6}$$

Please note that the influence of the atmosphere (absorption and emission) is neglected here as well. Knowing the needed optical properties (at least typical values), $E_{cal}(T_{app})$ can be calculated for all measured apparent temperatures T_{app}, using Equation (6). Then, in a next step, the radiance of the surface that was measured during the experiment by the camera can be calculated as follows:

$$E_{meas}(T, \varepsilon, T_0) = \int_{-\infty}^{\infty} W_{\lambda r}(\lambda, T, \varepsilon, T_0) \cdot \tau_{opt,meas}(\lambda) \cdot S(\lambda) \cdot d\lambda \tag{7}$$

For each measurement point above, the temperature of the inner ceiling of the build chamber was monitored by the L-PBF system's sensors. This temperature was used as surroundings temperature T_0 here. Please note that the experimental transmissivity $\tau_{opt,meas}(\lambda)$ was dependent on the selected calibration range of the camera, since the camera internal absorptive filters differed.

As the camera outputs the same values at the same irradiance rather than at the same temperature of the object observed by the camera, the emissivity can be reconstructed by setting:

$$E_{cal}\left(T_{app}\right) = E_{meas}(T_{TC}, \varepsilon, T_0) \tag{8}$$

for each measurement point (T_{TC}, T_{app}, T_0), where ε was the only unknown variable. Thus, ε_{corr} was determined by:

$$\varepsilon_{corr} = \underset{0<\varepsilon<1}{\arg\min} \left| E_{meas}(T_{TC}, \varepsilon, T_0) - E_{cal}\left(T_{app}\right) \right| \tag{9}$$

3. Results and Discussion

3.1. Selection of Thermocouples

The TCs showed temperature deviations depending on their position with respect to the sample surface and the gas flow, as shown in Figure 4. Figure 8 shows exemplarily a sequence of TC measurements during a cooling down phase of the specimen. While deviations of the measured temperature were small between TC1, TC3 and TC4, the temperature measured at TC2 was remarkably lower than at the other TCs. TC2 was directly placed in the gas flow. This deviation could be associated with the gas flow, as the difference decreased when reducing the gas flow velocity at the times of 103.5 s (21.5 m/s before 103.5 s), 217 s (reduction to 14.9 m/s until 217 s), 330 s (reduction to 8.9 m/s until 330 s) and 344 s (reduction to 0 m/s until 344 s, no gas flow after 344 s) in the presented example. Thus, TC2 values were excluded from further examinations. The mean of TC1, TC3 and TC4 was taken as surface reference temperature in all subsequent considerations. A constant surface temperature was assumed. A consideration of measurement uncertainty is given in Section 3.5.

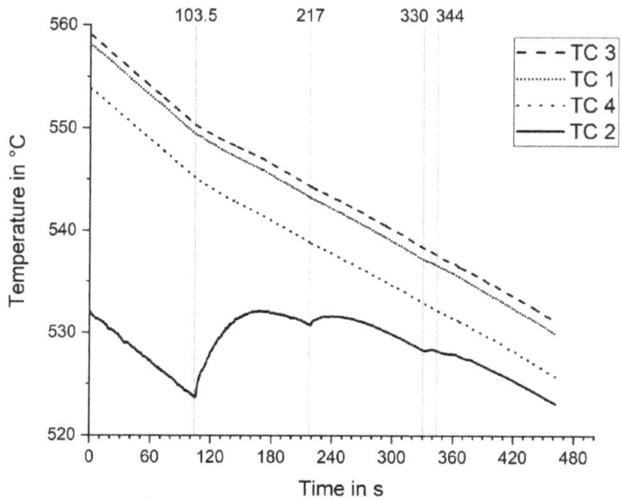

Figure 8. Effect of gas flow on TC temperature. The gas flow velocity decreases at the highlighted times.

3.2. Apparent Emissivity

The apparent emissivity values are presented in the following subsections for the 316L L-PBF surface, the 316L powder layer and the IN718 powder layer.

3.2.1. Apparent Emissivity of the 316L L-PBF Surface

The comparison of the reference temperature values of the 316L L-PBF surface, the mean value of three TCs (see Section 3.1), and the respective apparent temperature values, calculated as the mean apparent temperature of the four ROIs (see Section 2.3) of the MWIR camera data, clearly indicated that the apparent temperature values of the black body calibrated MWIR camera underestimated the reference temperatures. The difference between reference temperature and apparent temperature increased from approximately 64 °C to 181 °C at reference temperatures of approximately 134 °C and 579 °C, respectively. This underestimation was not surprising, as the emissivity of the real L-PBF surface was expected to be much smaller than unity, which was assumed by the apparent temperature computation of the camera. Figure 9 displays data couples of TC reference temperatures and apparent temperatures of the MWIR camera (black hollow symbols). Three different calibration ranges of the camera were used to measure the large variance of surface temperatures during the experiment (see Section 2.1). The measured data for each calibration range are distinguished by the distinct symbols (circle, square, triangle) in the plot. The connected data points of the calibration ranges 60–200 and 200–400 follow a linear trend, whereas a non-linear trend is revealed for the calibration range 300–600. In addition, steps between the curves of the distinct calibration ranges are noticeable. Both are discussed in the course of the emissivity determination, below.

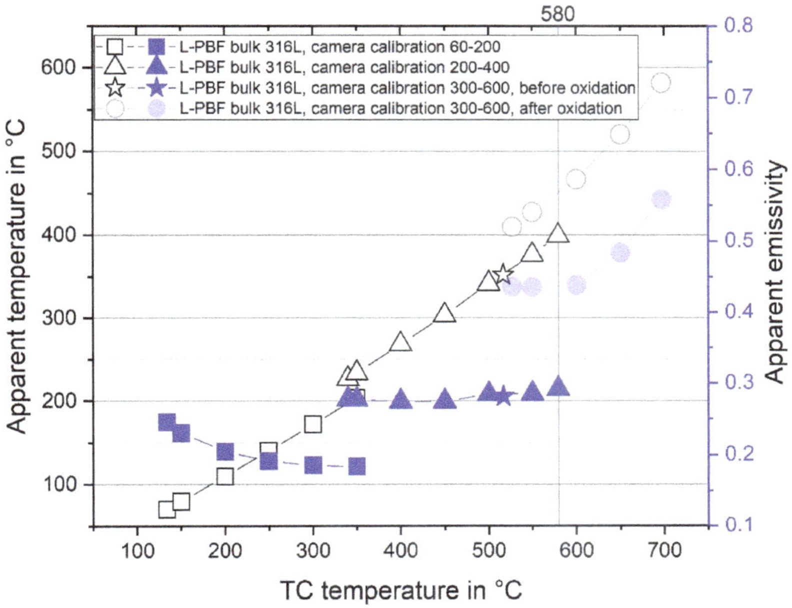

Figure 9. Apparent temperature (black hollow symbols) and apparent emissivity (blue full symbols) of 316L L-PBF bulk surface over measured TC temperature. In the camera calibration range 300–600, measurements were conducted before the onset of increased oxidation (star symbols) and after oxidation (circle symbols).

The apparent emissivity ε_{app} of the 316L L-PBF surface was computed using Equation (4) (neglecting the wavelength dependency, grey body approximation) for all data couples presented in Figure 9 (reference temperatures between 134 °C and 700 °C). In addition to the temperature data couples (black hollow symbols), Figure 9 displays the respective apparent emissivity ε_{app} (blue full symbols).

The apparent emissivity ε_{app} of the surface of the L-PBF specimen in the particular setup showed a decrease in the camera calibration range 60–200 with increasing temperature, starting at $\varepsilon_{app} = 0.25$ at a temperature of 134 °C, leveling off to $\varepsilon_{app} = 0.18$ at temperatures above 300 °C. However, the radiation of the ambient temperature was not considered for this calculation of the apparent emissivity. Especially for lower temperatures, which were of a similar magnitude as the ambient temperature, the radiation of ambient temperature is expected to lead to significant falsifications if not considered. A correction analysis including data of measured ambient temperatures of the build chamber of around 33 °C is presented in Section 3.3.

When switching the calibration range from 60–200 to 200–400 at constant TC temperature, there was a jump in the apparent temperature of more than 25 K. This led to an increased apparent emissivity of the surface of the L-PBF specimen of approximately $\varepsilon_{app} = 0.28$ at temperatures between 350 °C and 580 °C. To examine this peculiar effect, additional experiments were conducted outside of the L-PBF setup: Firstly, experiments were performed using a black body radiator (Fluke 4181, Fluke Corporation, Norwich, UK) set to a temperature of 200 °C. Changes of the camera calibration range between 60–200 and 200–400 showed only a small deviation of the measured temperature by the camera of below 1 K, indicating a correct black body calibration. Secondly, further investigations using 316L samples produced in the described L-PBF machine, heated on a hot plate to 200 °C were conducted. Here, the jump in the apparent temperature observed in the calibration experiment at the L-PBF setup was reproduced. Thus, this effect was in fact caused by the non-unity emissivity of the material. A consultation of the camera manufacturer revealed that the absorptive filter elements that were introduced to the optical path within the camera in the calibration ranges at elevated temperatures (200–400, 300–600) have a transmissivity that strongly depends on the wavelength. Therefore, their spectral transmissivity had to be considered for a correct emissivity determination and correction. A further analysis is presented in Section 3.3. The big step between the apparent emissivity values of distinct calibration ranges obtained by the simple analysis approach can be seen as an imposing example of the risk of data misinterpretation when using commercial thermography cameras with various calibration ranges. Unless clearly stated by the vendor, one has to be very careful when transferring experimentally determined values to slightly other conditions of the setup, in this case to another camera calibration range.

3.2.2. Oxidation Effects on the Apparent Emissivity of the 316L L-PBF Surface

At TC temperatures above approximately 500 °C, tempering colors could be recognized by the human eye, beginning with a slightly brownish appearance, which darkened until approximately 580 °C and then turned into a bluish appearance, inspected through the green UV protection window of the process chamber door. Thus, despite the low oxygen content in the build chamber during the experiments, oxidation of the heated surface was still occurring. Oxidation layers can drastically change emissivity values. This effect is well known in the literature [20], and an excursus into this matter is therefore given in the introduction Section 1.4.

Such an oxidation-driven change of emissivity values was clearly revealed at temperatures above 580 °C (compare Figure 9). Additionally, a slight increase in apparent emissivity could be seen for temperatures above 500 °C (increase from 0.28 to 0.29), which might be attributed to the onset of oxidation as visually noticed during the heat up. While the apparent emissivity change between 500 °C and 580 °C was very small, the emissivity values changed drastically above 580 °C. The oxidation of the surface got too strong to present reliable emissivity values of an unoxidized or only slightly oxidized 316L L-PBF surface above 580 °C. The measurement data which are displayed transparent in Figure 9 and their respective computations of emissivity values were heavily influenced by a change of emissivity due to oxidation. This was in good agreement with the visually noticeable tempering colors and the literature review. For example, apparent emissivity values of 0.58 were determined using the same camera in the laser metal deposition of AISI 316L, where stronger oxidation is expected to occur due to a less efficient shielding of oxygen by a local shielding gas flow in surrounding air conditions [34].

Remarkably, there are measurement points between 500 °C and 580 °C TC temperature, which were obtained at different times and different camera calibration ranges, i.e., calibration ranges 200–400 and 300–600, showing huge differences in apparent emissivity for the same temperatures. The measurement of the data points symbolized by full circles (calibration range 300–600) was conducted approximately 70 min after the measurement series symbolized by triangles (calibration range 200–400). The temperature of the specimen in between these two measurement series was constantly higher than 530 °C, most of the time around 600 °C. Apparently, the oxidation layer thickness was still very small during the measurements in the calibration range 200–400, but had enough time and temperature to grow before the measurements at a calibration range of 300–600 were conducted. This would explain the discrepancy between apparent emissivity for the same TC temperatures. Interestingly, the apparent emissivity step in Figure 9 between the calibration ranges 200–400 and 300–600 can be explained by oxidation without further correction for the change of the internal filter of the camera: The measurement of the extra dot (star symbol) was taken using the calibration range 300–600 during the measurement series represented by triangles (200–400), and thus before the strong oxidation started. It shows a very similar apparent emissivity to the results obtained in the 200–400 calibration range. This observation might be explained by a similar spectral dependence of the transmissivity of filter A and B (at lower amplitudes for filter B). This is important to know, for any later use of these calibration ranges, which were identified as most relevant for in situ L-PBF monitoring means (cf. [13]).

3.2.3. Apparent Emissivity of the 316L Powder Layer

The determination of the apparent emissivity and the comparison of TC reference temperatures and thermographically acquired apparent temperatures of the 316L powder layer followed the same procedure as for the solid 316L L-PBF surface, described and discussed in the previous section. Figure 10 displays the respective temperature data couples and computed apparent emissivity values of the 316L powder for the three different camera calibration ranges. For comparison reasons, it also contains the respective values of the L-PBF 316L surface at similar conditions in transparent colors.

The non-corrected emissivity values of the powder layer leveled to approximately ε_{app} = 0.33 in the camera calibration range 60–200. Before the onset of increased oxidation above 580 °C, apparent emissivity was calculated to approximately ε_{app} = 0.43 for the camera calibration ranges 200–400 and 300–600. In comparison to the solid L-PBF surface of the same material, the 316L powder layer showed significantly increased emissivity values. This was already mentioned in the literature, e.g., in [14], and originates from the strongly increased surface roughness of a powder layer compared to a solid L-PBF surface. It also explains the potential occurrence of apparent temperatures of new recoated powder layers which could be higher than the apparent temperatures at the same position prior to recoating in L-PBF real manufacturing, monitored by in situ thermography, as in [17].

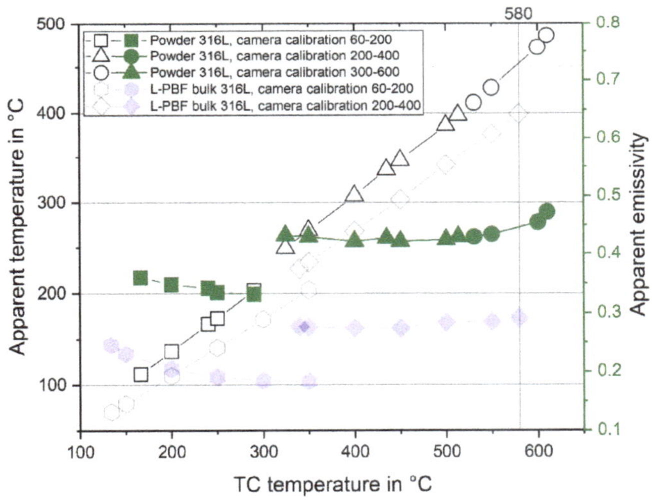

Figure 10. Apparent temperature (black hollow symbols) and apparent emissivity (green full symbols) of the 316L powder layer over measured TC temperature. For comparison, apparent temperature (gray hollow symbols) and apparent emissivity (blue full symbols) of the 316L L-PBF bulk surface over TC temperature are also depicted.

3.2.4. Apparent Emissivity of the IN718 Powder Layer

The determination of the apparent emissivity and the comparison of TC reference temperatures and thermographically acquired apparent temperatures of the IN718 powder layer followed the same procedure as for the 316L L-PBF surface, described and discussed in Section 3.2.1. The results are plotted in Figure 11. The non-corrected emissivity values of the IN718 powder layer leveled to approximately ε_{app} = 0.34 in the camera calibration range 60–200. The apparent emissivity was calculated to approximately ε_{app} = 0.41 to 0.42 for the camera calibration ranges 200–400 and 300–600. The determined apparent emissivity values of the IN718 powder layer were at a similar level as the values of the 316L powder layer, depicted in Figure 12. In contrast to the 316L powder, the rise of apparent emissivity values above 580 °C is significantly smaller, which is assumed to be attributed to other oxidation kinetics for the nickel-based IN718 as compared to the stainless steel 316L. This is in good agreement with experiments by del Campo et al. [35], who showed a rather small influence of short-term oxidation at 700 °C of IN718 on emissivity.

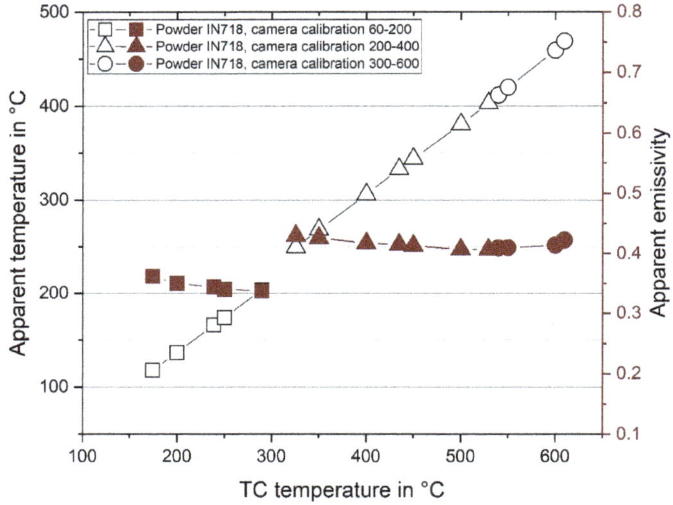

Figure 11. Apparent temperature (black hollow symbols) and apparent emissivity (brown full symbols) of IN718 powder layer over measured TC temperature.

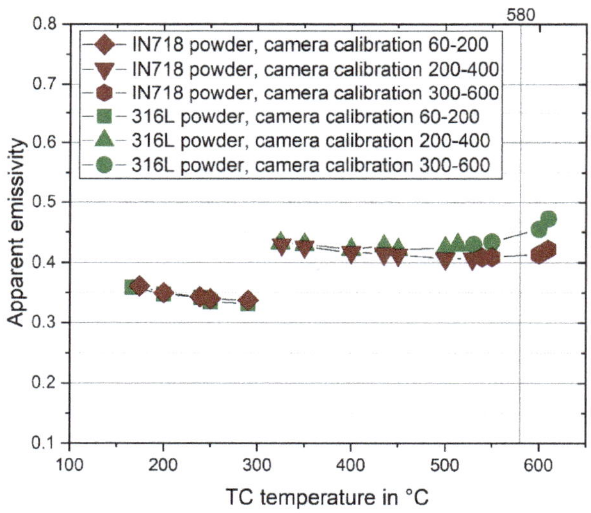

Figure 12. Comparison of apparent emissivity results of IN718 powder layer (brown full symbols) and 316L powder layer (green symbols) over measured TC temperature.

3.3. Determination of Corrected Emissivity

The corrected emissivity values ε_{corr} are presented in the following subsections for the 316L L-PBF surface, the 316L powder layer and the IN718 powder layer. The correction analysis is described in Section 2.5.2.

3.3.1. Corrected Emissivity of 316L L-PBF Surface

When considering the radiation that was reflected from the surroundings at the surface as well as all the spectral characteristics of the optical elements including the different internal filters as described above, some fundamental changes of the corrected emissivity curves compared to the apparent emissivity curves (discussed in Section 3.2.1) can be observed. Figure 13 compares the corrected emissivity values with the non-corrected apparent emissivity values. The following discussion focuses on the emissivity values of the non-oxidized surface, i.e., below 580 °C.

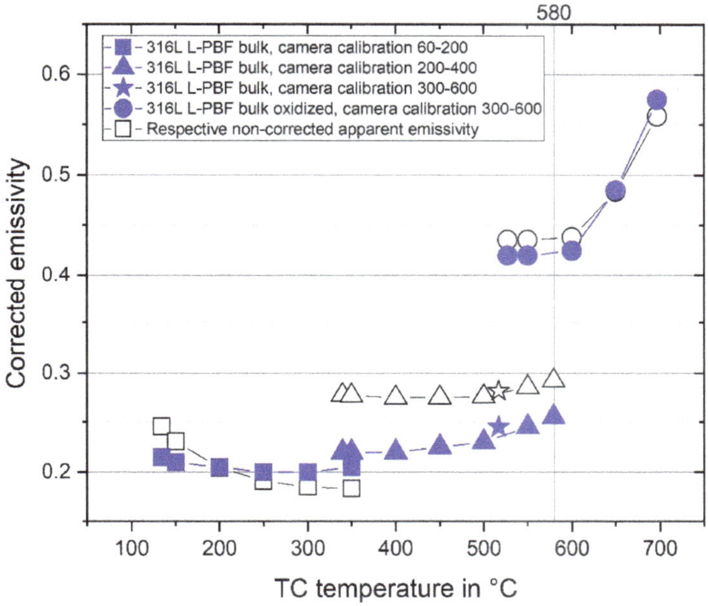

Figure 13. Comparison of corrected (blue symbols) and non-corrected apparent (black hollow symbols) emissivity values of 316L L-PBF bulk surface over measured TC temperatures.

First of all, the significant effect of apparently increasing emissivity values in the lower temperature region decreased drastically. The radiation of the surroundings (build chamber temperature around 33 °C) had a strong influence at relatively low temperatures of the target object, which resulted in the apparent increase of the non-corrected emissivity values with decreasing TC temperatures. Therefore, the corrective analysis flattened the curve in this region. However, there is still a slight increase below a TC temperature of 250 °C. This was assumed to be a result of the position of the TC of the L-PBF system within the build chamber, which was located toward the front of the chamber ceiling, rather than close to the optical path of the IR camera. As a result of the TC position, a slight underestimation of the radiation of the surroundings T_0 could be assumed to be responsible for this, as, e.g., a value of $T_0 = 38$ °C leads to a complete flattening of the curve.

Secondly, the huge jump of the apparent emissivity values connected to the change of the calibration range of the camera from 60–200 to 200–400 almost disappeared by the correction analysis as a result of the consideration of the distinct spectral transmissivities of the respective internal filters, which changed with the changing calibration ranges. The small remaining jump between the real emissivity values of the different calibration ranges is assumed to be attributed to possible slightly differing optical parameters of the actual optical elements from the typical values used for the calculations (see Section 3.5).

It was revealed that the emissivity of the L-PBF bulk surface used in this experimental setup increased with increasing temperatures. The computed emissivity of 316L L-PBF bulk surface varied between $\varepsilon_{corr} = 0.2$ and $\varepsilon_{corr} = 0.23$ in the temperature region from 200 °C to 500 °C and between $\varepsilon_{corr} = 0.23$ and $\varepsilon_{corr} = 0.25$ in the temperature region between 500 °C and 580 °C, where slight oxidation effects could not be excluded, as discussed in Section 3.2.1. These values are in good agreement with literature values of stainless steel [36], as presented in Figure 14.

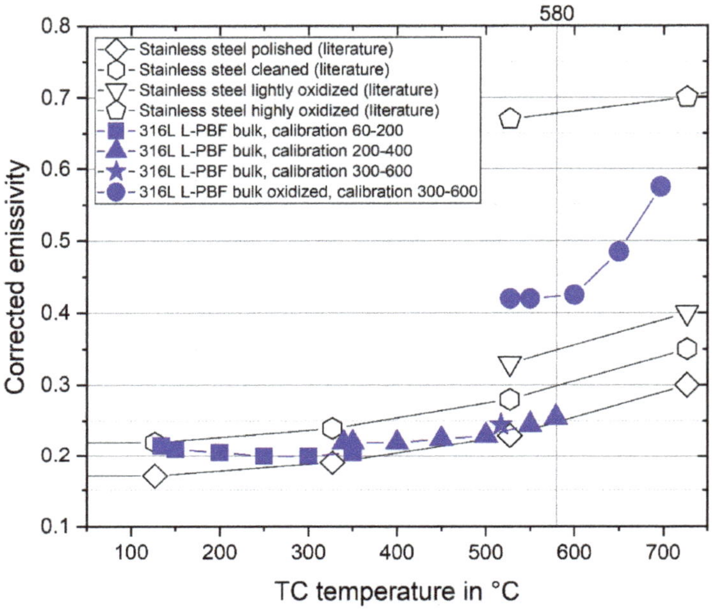

Figure 14. Corrected emissivity of 316L L-PBF bulk surface (blue symbols) over measured TC temperature. For comparison, temperature depend literature values of emissivity of stainless steel in different conditions (black hollow symbols) are added from [36].

3.3.2. Corrected Emissivity of Powder Layers

The same general changes between corrected emissivity values and apparent emissivity values, as discussed in the previous section for a 316L L-PBF bulk surface, also applied to the corrective analysis of the emissivity values of the two different powders. Figure 15 contains a comparison of the corrected emissivity values of 316L powder layer, IN718 powder layer and 316L L-PBF bulk surface.

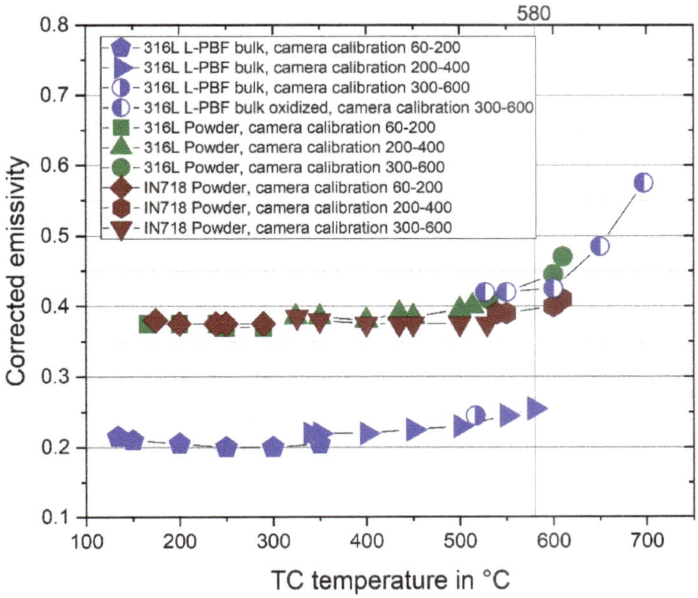

Figure 15. Corrected emissivity of 316L powder layer (green symbols) and of IN718 powder layer (brown symbols) and of 316L L-PBF bulk (blue symbols) over measured TC temperature.

The computed corrected emissivity of the 316L powder layer varied between ε_{corr} = 0.37 and ε_{corr} = 0.4 in the temperature region from 200 °C to 500 °C and between ε_{corr} = 0.4 and ε_{corr} = 0.45 in the temperature region between 500 °C and 600 °C. The increase at higher temperatures was attributed to oxidation effects.

The computed corrected emissivity of the IN718 powder layer varied between ε_{corr} = 0.37 and ε_{corr} = 0.38 in the temperature region from 200 °C to 500 °C and between ε_{corr} = 0.38 and ε_{corr} = 0.4 in the temperature region between 500 °C and 600 °C.

3.4. *Influence of Measurement Position*

No significant differences between the three positions of the heated sample could be found, as can be seen in Figure 16, which shows single temperature couples and the respective apparent emissivity values at a similar temperature, determined for the same calibration range. A comparison of the different positions was conducted using the camera calibration 200–400 measuring around a reference temperature of about 400 °C. Please note that the calculated emissivity values did not contain any corrections, as discussed in Section 3.3, since a correction was not necessary for this relative comparison. It is also interesting to note that these measurements were conducted at a specimen which was heated up to 604 °C in air outside of the build chamber prior to these measurements. The surface temperature was above 570 °C for about 480 s. Therefore, the presented measurements stem from a slightly oxidized surface condition. The apparent emissivities under these conditions were: at position 1 ε_{app} = 0.28; at position 2 ε_{app} = 0.28; at position 3 ε_{app} = 0.29. The deviations of the emissivity values with regard to the different positions were within the measurement accuracy of the camera (see Section 3.5). According to published results on the angular dependence of emissivity [20,35], this result was not surprising, as the angular tilt between the particular measurement sceneries for the different positions was small. However, the confirmation of comparable results irrespective of the position of the target object was important for further monitoring tasks.

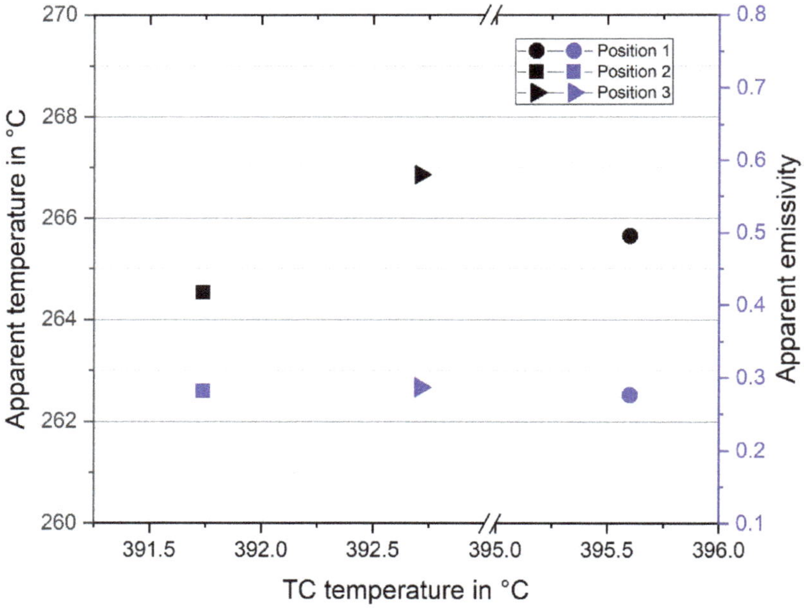

Figure 16. Apparent temperature and apparent emissivity of a 316L L-PBF bulk surface at three different positions within the build chamber according to the positions shown in Figure 7.

3.5. Measurement Uncertainty

As discussed above, unnoticed changes of the surface condition of the target object, e.g., an onset of oxidation layer growth, could lead to inaccuracies or misinterpretation. In addition, potential systematic measurement errors of the applied setup contributed to the measurement uncertainty. A rough quantification of the main contributing factors is given hereafter. The main factors contributing to the measurement uncertainty were identified as follows: accuracy of thermocouples, accuracy of MWIR camera, temperature heterogeneities over the target surface (with respect to TC and to IR camera values).

Accuracy of thermocouples: The standard limit of error of the used thermocouples was specified according to DIN EN 60594-1: +/−2.5 °C or +/−0.75% [37].

Accuracy of the camera: The manufacturers' specifications of the MWIR camera allow for a deviation of up to 1% in the determination of apparent temperature in °C or 1 °C, whichever is larger.

Temperature heterogeneities over the target surface: Figure 8 shows a temperature plot of the single thermocouples over a short period of time at temperatures between 520 °C and 560 °C. It was proposed to define the mean of TC1, TC3 and TC4 as the surface temperature. This mean was taken in the discussion section without consideration of its standard deviation. The standard deviation of the temperatures of the three thermocouples was either smaller than 2.5 °C or smaller than 0.75% of the measured temperature over the entire region of examined temperatures (130 °C–700 °C). Only in the temperature region between 300 °C and 470 °C were the standard deviations slightly higher, resulting in standard deviations of up to 1.1%. To compare the temperature of the top surface and the temperature at the described positions TC1-TC4, one test specimen was heated up outside of the build chamber, which had two TCs on top of the upper surface, replacing TC2 and TC4. The temperature differences between these two TCs on top and TC1 and TC2 at the side surface were below the measuring errors mentioned above.

For a conservative estimation of the measuring error of the apparent temperature, the standard deviations of the max. and min. apparent temperature values of the four ROIs (see Figure 6) were calculated for the L-PBF bulk surface. They were up to 7.9% of the respective mean value in the calibration range 60–200; up to 4.2% in the calibration range 200–400; and up to 6.5% in the calibration range 300-600. It was interesting to note that the standard deviation decreased drastically above an apparent temperature of approximately 500 °C to 2% in the measuring data of the calibration range 300–600. This corresponds well with the onset of oxidation and thus an increase of the emissivity above this value. Table 2 summarizes the resulting deviations per temperature regime. This results in a potential uncertainty of the emissivity determination of approximately 0.05. Apart from these mentioned potential measuring errors, the separate measurements at three different positions (Sections 2.4.3 and 3.4) without significant deviations in apparent emissivity results demonstrate the good repeatability of the conducted measurements.

Table 2. Measurement uncertainties of temperature determination.

Target Object Temperature in °C	Deviations of TC Values	Standard Deviation of the Apparent Temperature Over the Target Surface
130–300	+/−2.5 °C	calibration 60–200: +/−7.9%
300–470	+/−1.1%	calibration 60–200: +/−7.9%
		calibration 200–400: +/−4.2%
470–700	+/−0.75%	calibration 200–400: +/−4.2%
		calibration 300–600: +/−6.5%

Soldan [38] (p. 26) pointed out that potential measurement errors of thermographic measurements can occur due to incorrect focusing of the camera with regard to the target object. This is problematic in the context of unknown target objects, as there is no absolute measure for image sharpness [38]. In the frame of the thermographic setup of this study, the focusing of the IR camera was conducted by manual adjustment of the objective lens until the operator had the subjective impression of a sharp image in the live view mode of the software. This procedure had to be repeated when a calibration range of the camera was changed. Although the camera had a nearly perpendicular view of the target object and, therefore, a large lateral area at the same focus position, deviations from the ideal focus plane could not be completely precluded. However, the effect of defocused measurements is negligible when the region of interest does not contain edges, i.e., large temperature gradients. Here, only a plane surface area was taken into account for the measurements (compare Figure 6). A step-wise change of the z-position of the heating device of up to 10 mm difference in z-height revealed no differences in the mean apparent temperature. Thus, the manual focusing seemed to be reliable for the measuring procedure of this study.

4. Conclusions

An experimental temperature adjustment of an off-axis MWIR thermography setup, which was installed at a L-PBF machine, was conducted using the usual L-PBF working conditions. The apparent emissivity values for the specific setup were determined for two materials at two different conditions using the contact method: 316L L-PBF bulk material, 316L powder layer and IN718 powder layer. For this purpose, a heated reference device was placed inside a L-PBF build chamber. A corrective analysis considering transmission losses due to optical elements within the optical path as well as the affecting radiation of the surroundings revealed corrected emissivity values for the spectral range of 2 µm to 5.7 µm. In the temperature region from approximately 150 °C to 580 °C, where oxidation did not strongly effect the measurements, the corrected emissivity is in a range from 0.2 to 0.25 for a 316L L-PBF bulk surface, in a range from 0.37 to 0.45 for 316L powder layer, and in a range from 0.37 to 0.4 for IN718 powder layer. With the knowledge of these emissivity values, a real temperature determination for in situ thermographic measurements can be conducted. The findings will also be very useful for numerical simulations. Additionally, the heated reference device can be used for temperature adjustments of other thermographic setups that show differences, e.g., in the spectral sensitivity of the camera.

Author Contributions: Conceptualization, G.M.; methodology, G.M., S.N., and S.J.A.; validation, G.M., S.N., and S.J.A.; formal analysis, G.M., S.N. and S.J.A.; investigation, G.M., S.N., and S.J.A.; data curation, G.M., S.N.; writing—original draft preparation, G.M.; writing—review and editing, G.M., S.N., S.J.A., C.M. and K.H.; visualization, G.M., S.N.; supervision, K.H.; project administration, S.J.A., C.M. All authors have read and agreed to the published version of the manuscript.

Funding: This research was funded by BAM within the focus area Materials.

Acknowledgments: The authors would like to thank Mathias Röllig for his support in the verification of the camera calibration ranges against a black body radiator. The authors would also like to thank Matthias Weise for the measurement of the surface roughness.

Conflicts of Interest: The authors declare no conflict of interest. The funders had no role in the design of the study; in the collection, analyses, or interpretation of data; in the writing of the manuscript, or in the decision to publish the results.

References

1. Herzog, D.; Seyda, V.; Wycisk, E.; Emmelmann, C. Additive manufacturing of metals. *Acta Mater.* **2016**, *117*, 371–392. [CrossRef]
2. Grasso, M.; Colosimo, B.M. Process defects and in situ monitoring methods in metal powder bed fusion: A review. *Meas. Sci. Technol.* **2017**, *28*, 044005. [CrossRef]
3. Colosimo, B.M.; Grasso, M. In-situ monitoring in L-PBF: Opportunities and challenges. *Procedia CIRP* **2020**, *94*, 388–391. [CrossRef]
4. Ali, U.; Mahmoodkhani, Y.; Imani Shahabad, S.; Esmaeilizadeh, R.; Liravi, F.; Sheydaeian, E.; Huang, K.Y.; Marzbanrad, E.; Vlasea, M.; Toyserkani, E. On the measurement of relative powder-bed compaction density in powder-bed additive manufacturing processes. *Mater. Des.* **2018**, *155*, 495–501. [CrossRef]
5. Mohr, G.; Seeger, S.; Hilgenberg, K. Measurement of particle emissions in Laser Powder Bed Fusion (L-PBF) processes and its potential for in-situ process monitoring. In Proceedings of the Euro PM 2019, EPMA, Maastricht, The Netherlands, 13–16 October 2019.
6. Alberts, D.; Schwarze, D.; Witt, G. High speed melt pool & laser power monitoring for selective laser melting (SLM®). In Proceedings of the 9th International Conference on Photonic Technologies LANE, Fürth, Germany, 19–22 September 2016.
7. Krauss, H.; Zeugner, T.; Zaeh, M.F. Layerwise Monitoring of the Selective Laser Melting Process by Thermography. *Phys. Procedia* **2014**, *56*, 64–71. [CrossRef]
8. Schmidt, M.; Merklein, M.; Bourell, D.; Dimitrov, D.; Hausotte, T.; Wegener, K.; Overmeyer, L.; Vollertsen, F.; Levy, G.N. Laser based additive manufacturing in industry and academia. *CIRP Ann.* **2017**, *66*, 561–583. [CrossRef]
9. Lane, B.; Moylan, S.; Whitenton, E.; Ma, L. Thermographic Measurements of the Commercial Laser Powder Bed Fusion Process at NIST. *Rapid Prototyp. J.* **2016**, *22*, 778–787. [CrossRef] [PubMed]
10. Heigel, J.C.; Lane, B.M. Measurement of the Melt Pool Length during Single Scan Tracks in a Commercial Laser Powder Bed Fusion Process. *J. Manuf. Sci. Eng.* **2018**, *140*, 051012. [CrossRef]
11. Bartlett, J.L.; Heim, F.M.; Murty, Y.V.; Li, X. In situ defect detection in selective laser melting via full-field infrared thermography. *Addit. Manuf.* **2018**, *24*, 595–605. [CrossRef]
12. Mohr, G.; Altenburg, S.J.; Ulbricht, A.; Heinrich, P.; Baum, D.; Maierhofer, C.; Hilgenberg, K. In-Situ Defect Detection in Laser Powder Bed Fusion by Using Thermography and Optical Tomography—Comparison to Computed Tomography. *Metals* **2020**, *10*, 103. [CrossRef]
13. Mohr, G.; Altenburg, S.J.; Hilgenberg, K. Effects of inter layer time and build height on resulting properties of 316L stainless steel processed by laser powder bed fusion. *Addit. Manuf.* **2020**, *32*, 101080. [CrossRef]
14. Williams, R.J.; Piglione, A.; Rønneberg, T.; Jones, C.; Pham, M.S.; Davies, C.M.; Hooper, P.A. In situ thermography for laser powder bed fusion: Effects of layer temperature on porosity, microstructure and mechanical properties. *Addit. Manuf.* **2019**, *30*, 100880. [CrossRef]

15. Usamentiaga, R.; Venegas, P.; Guerediaga, J.; Vega, L.; Molleda, J.; Bulnes, F.G. Infrared thermography for temperature measurement and non-destructive testing. *Sensors* **2014**, *14*, 12305–12348. [CrossRef] [PubMed]
16. Lough, C.S.; Wang, X.; Smith, C.C.; Adeniji, O.; Landers, R.G.; Bristow, D.A.; Kinzel, E.C. Use of SWIR Imaging to Monitor Layer-to-Layer Part Quality during SLM of 304L Stainless Steel. In Proceedings of the 29th Annual International Solid Freeform Fabrication Symposium, Austin, TX, USA, 13–15 August 2018.
17. Mohr, G.; Scheuschner, N.; Hilgenberg, K. In situ heat accumulation by geometrical features obstructing heat flux and by reduced inter layer times in laser powder bed fusion of AISI 316L stainless steel. *CIRP Proceedia* **2020**, *94*, 155–160. [CrossRef]
18. Khan, K.; Mohr, G.; Hilgenberg, K.; De, A. Probing a novel heat source model and adaptive remeshing technique to simulate laser powder bed fusion with experimental validation. *Comput. Mater. Sci.* **2020**, *181*, 109752. [CrossRef]
19. Schuster, N.; Kolobrodov, V.G. *Infrarotthermographie*; Wiley-VCH: Weinheim, Germany, 2004.
20. Vollmer, M.; Möllmann, K.P. *Infrared Thermal Imaging: Fundamentals, Research and Applications*; Wiley-VCH: Weinheim, Germany, 2017.
21. Siegel, R.; Howell, J.R.; Lohrengel, J. *Wärme- und Stoffübertragung. Teil 1: Grundlagen und Materialeigenschaften*; Springer: Berlin, Germany, 1988.
22. Glückert, U. *Erfassung und Messung von Wärmestrahlung. Eine Praktische Einführung in die Pyrometrie und Thermographie*; Franzis: Munich, Germany, 1992.
23. Doubenskaia, M.; Pavlov, M.; Grigoriev, S.; Smurov, I. Definition of brightness temperature and restoration of true temperature in laser cladding using infrared camera. *Surf. Coat. Technol.* **2013**, *220*, 244–247. [CrossRef]
24. Yadroitsev, I.; Krakhmalev, P.; Yadroitsava, I. Selective laser melting of Ti6Al4V alloy for biomedical applications: Temperature monitoring and microstructural evolution. *J. Alloys Compd.* **2014**, *583*, 404–409. [CrossRef]
25. Heigel, J.C.; Whitenton, E. Measurement of thermal processing variability in powder bed fusion. In Proceedings of the 2018 ASPE and Euspen Summer Topical Meeting-Advancing Precision in Additive Manufacturing, Berkeley, CA, USA, 22–25 July 2018.
26. Heigel, J.; Lane, B.; Moylan, S. Variation of Emittivity with Powder Bed Fusion Build Parameters. In Proceedings of the 2016 Annual International SFF Symposium, Austin, TX, USA, 8–10 August 2016.
27. Hakiki, N.; Montemor, M.; Ferreira, M.; da Cunha Belo, M. Semiconducting properties of thermally grown oxide films on AISI 304 stainless steel. *Corros. Sci.* **2000**, *42*, 687–702. [CrossRef]
28. Kämmerer, B. *Abhängigkeit der Korrosionsbeständigkeit von der Chemischen Oberflächenzusammensetzung von Chromstählen*; Universität Augsburg: Augsburg, Germany, 2012.
29. Joachim, J. *Untersuchung von Oxidationsprozessen an Oberflächen von FeCr-Legierungen und Austenitstahl mittels Röntgenabsorptionsspektroskopie unter Streifendem Einfall*; Rheinische Friedrich-Wilhelms-Universität Bonn: Bonn, Germany, 2003.
30. Iuchi, T.; Furukawa, T.; Wada, S. Emissivity modeling of metals during the growth of oxide film and comparison of the model with experimental results. *Appl. Opt.* **2003**, *42*, 2317–2326. [CrossRef]
31. Zauner, G.; Mayrhofer, F.; Hendorfer, G. Optical characterization of growing thin films at high temperatures by analysis of near infrared emissivity variations using CCD thermal imaging. In Proceedings of the 2010 International Conference on Quantitative InfraRed Thermography, Quebec, QC, Canada, 24–29 June 2010
32. Del Campo, L.; Pérez-Sáez, R.B.; Tello, M.J. Iron oxidation kinetics study by using infrared spectral emissivity measurements below 570 °C. *Corros. Sci.* **2008**, *50*, 194–199. [CrossRef]
33. Krauss, H.; Eschey, C.; Zaeh, M.F. Thermography for Monitoring the Selective Laser Metling Process. In *Proceedings of the Solid Freeform Fabrication Symposium*; University of Texas: Austin, TX, USA, 2012; pp. 999–1014.
34. Altenburg, S.J.; Straße, A.; Gumenyuk, A.; Maierhofer, C. In-situ monitoring of a laser metal deposition (LMD) process: Comparison of MWIR, SWIR and high-speed NIR thermography. *Quant. InfraRed Thermogr. J.* **2020**, 1–18. [CrossRef]
35. Del Campo, L.; Pérez-Sáez, R.B.; González-Fernández, L.; Esquisabel, X.; Fernández, I.; González-Martín, P.; Tello, M.J. Emissivity measurements on aeronautical alloys. *J. Alloys Compd.* **2010**, *489*, 482–487. [CrossRef]
36. Incropera, F.P.; DeWitt, D.P.; Bergman, T.L.; Lavine, A.S. *Fundamentals of Heat and Mass Transfer*, 6th ed.; John Wiley & Sons: Hoboken, NJ, USA, 2007.

37. Deutsches Institut für Normung (DIN). *DIN EN 60584-1: Thermocouples—Part 1: EMF Specifications and Tolerances*; DIN: Berlin, Germany, 2014.
38. Soldan, S. *Sensordatenfusionsansätze in der Thermografie zur Verbesserung der Messergebnisse*; Universität Kassel: Kassel, Germany, 2014.

metals

MDPI

Article

Use of X-ray Computed Tomography for Assessing Defects in Ti Grade 5 Parts Produced by Laser Melting Deposition

Diana Chioibasu [1,2], Sabin Mihai [1,3], Muhammad Arif Mahmood [1,4], Mihail Lungu [1], Ioana Porosnicu [1,4], Adrian Sima [1], Cosmin Dobrea [1], Ion Tiseanu [1] and Andrei C. Popescu [1,*]

1 National Institute for Laser, Plasma and Radiation Physics (INFLPR), Magurele, 077125 Ilfov, Romania; diana.chioibasu@inflpr.ro (D.C.); sabin.mihai@inflpr.ro (S.M.); arif.mahmood@inflpr.ro (M.A.M.); mihail.lungu@inflpr.ro (M.L.); ioana.porosnicu@inflpr.ro (I.P.); adrian.sima@inflpr.ro (A.S.); cosmin.dobrea@inflpr.ro (C.D.); ion.tiseanu@inflpr.ro (I.T.)

2 Faculty of Applied Science, University Politehnica of Bucharest, 060042 Bucharest, Romania

3 Faculty of Industrial Engineering and Robotics, University Politehnica of Bucharest, 060042 Bucharest, Romania

4 Faculty of Physics, University of Bucharest, Magurele, 077125 Ilfov, Romania

* Correspondence: andrei.popescu@inflpr.ro; Tel.: +40-21-4574550 (ext. 2414/2423)

Received: 30 September 2020; Accepted: 21 October 2020; Published: 23 October 2020

Abstract: Laser Melting Deposition (LMD) is a metal printing technique that allows for the manufacturing of large objects by Directed Energy Deposition. Due to its versatility in variation of parameters, the possibility to use two or more materials, to create alloys in situ or produce multi-layer structures, LMD is still being scientifically researched and is still far from industrial maturity. The structural testing of obtained samples can be time consuming and solutions that can decrease the samples analysis time are constantly proposed in the scientific literature. In this manuscript we present a quality improvement study for obtaining defect-free bulk samples of Ti6Al4V under X-Ray Computed Tomography (XCT) by varying the hatch spacing and distance between planes. Based on information provided by XCT, the experimental conditions were changed until complete elimination of porosity. Information on the defects in the bulk of the samples by XCT was used for feedback during parameters tuning in view of complete removal of pores. The research time was reduced to days instead of weeks or months of samples preparation and analysis by destructive metallographic techniques.

Keywords: laser melting deposition; 3D printing; X-ray computed tomography; porosity; non-destructive analysis

1. Introduction

Additive manufacturing (AM) with titanium alloys revealed several possible breakthroughs in producing complex metal parts that would have been impossible to achieve via traditional manufacturing methods (e.g., casting, milling, forging) [1–3]. One of them is the possibility to create lattices instead of bulk allowing for mass reduction of parts [4] and increasing the cost efficiency [5]. Another is production of trabecular structures for components of metallic prostheses. For the success of AM a comprehensive understating of the relationship between manufacturing parameters (e.g., laser power, scanning speed, powder feed rate, scanning strategy), powder quality and overall built integrity of the final part (e.g., defects, voids, inclusions) is essential.

One of the common AM techniques for 3D printing of metallic parts is Laser Melting Deposition (LMD) [1,6]. In this technique, the laser irradiates a substrate and locally melts it, while powder is injected into the molten pool. The path of the laser beam also represents the contour of the part to be printed. Layer by layer, a 3D part can be constructed in this manner. Another common 3D printing technique for metallic materials is Selective Laser Melting (SLM). In this method, a bed of powder is irradiated by a laser beam. The powder is melted in the areas scanned by the laser beam and is rapidly solidified after the action of the laser beam stops. A new layer of powder is applied on the top via a leveling set-up, followed by a new laser irradiation step. Thus, layer by layer a 3D shape is produced [7,8]. LMD allows for fabrication of large size near-net-shape components and high versatility in composition (as it can feed various types of metallic powders by using powder distributors with two or more hopers), while SLM is usually dedicated to smaller parts which require high resolution. Consequently, this method is suitable for building complex parts with topological optimization design [9] made of composite materials [10] or compositionally graded materials [11,12]. Furthermore, LMD technique is ideal for repairing high-valued metallic parts [13] due to its capability to add material on top of worn or damaged regions with minimum waste [14–16] and preserving or even improving the mechanical properties of the initial components [17]. LMD manufacturing has been applied for personalized metallic parts with major applications in medical field (implants or medical instruments) [18], automotive [19], aerospace [20] and aeronautics (turbine blades, gas turbine-blisks) [21].

Ti6Al4V is widely used in industrial applications for the manufacturing of high value components, due to its high corrosion resistance [20], high strength [22], low weight ratio [23] and high-temperature performance [24,25]. Aerospace and aircraft [26], electronics [6], automotive [27] and medical devices [22,28,29] are only a few examples of the industries where Ti6Al4V is used. This Ti alloy is suitable for aerospace parts production such as flaps, engine subassemblies or air-frames [30] because of its weight saving, high level of fatigue resistance, significant volume reduction and high-temperature mechanical properties [31]. There are a few reports which demonstrate the superior properties of the Ti6Al4V parts produced by LMD compared to traditional manufacturing procedures [32–34]. Even though it is well-known that mechanical properties are strongly influenced by thermal load cycles, Ti6Al4V parts produced by LMD show a great balance between mechanical strength and ductility.

In theory, LMD can produce fully dense structures, but the powder quality, optimized processing parameters and scanning strategies can influence the fabrication resolution and integrity of the final part. It was reported in several papers that the size, shape, and defects of the starting powder are affecting deposited track thickness, interlayer porosity [35], surface roughness, hardness and mechanical properties of the final LMD components [36,37]. The process parameters that can influence the overall quality of the LMD Ti6Al4V parts are laser power, scanning speed, powder feed rate, laser spot size, nozzle type and correct alignment of the focus spot with the powder stream [38]. Besides these parameters, the scanning strategy that includes hatch interspace and scan pattern, can drastically affect the width, height, and surface quality of the final LMD product. A properly determined scanning strategy might decrease the amount of residual stress and thermal distortions [38].

Internal defects in LMD parts have substantial negative effect on their mechanical reliability. These defects are inaccessible to surface investigation methods based on tactile, light-optical sensors, and even destructive methods; thus, a non-destructive tool is highly recommended in order to obtain an in-depth structural overview. The X-ray computed tomography (XCT) is one of the most suitable non-destructive methods for sample analysis with tolerance in the range of a few microns [39]. In case of long time exposure to X-Rays materials can suffer surface or internal modifications in terms of texturing or plastic deformations [40,41]. In our case such modifications are not envisageable due to the short irradiation times (minutes to tens of minutes). This method is widely applied in various industrial applications due to its major advantages in metrological inspection independent to surface, shape, or material [42–44]. The XCT is becoming a reliable non-destructive inspection tool of metallic AM parts, due to its possibility to provide in-depth volumetric and structural analysis with high resolution scanning, versatility, and post-processing algorithms for statistical interpretation. Recent studies have confirmed the time to cost efficiency of the XCT applied in large volume investigation, in comparison with destructive methods such as metallographic examination [45]. While XCT has been tested for characterization of small samples produced by SLM, no optimization study on XCT monitoring was found in the scientific literature for LMD, which usually produces large samples.

In this paper, we report LMD scanning strategy improvement using XCT as an analysis tool. Its feedback can offer valuable information in a relatively short amount of time on the defects' formation in various deposition conditions. The aim of this study is to obtain final products fabricated by LMD without internal defects, such as pores, cracks, or non-melted inner regions. We will show that the scanning strategy for producing a part can be optimized in a matter of hours instead of days, as would be the case when using destructive analyses techniques.

2. Materials and Methods

The starting material used in this study was commercially available Ti6Al4V metallic powders (Carpenter Additive, UK) with spherical shape and size diameter of 45–106 μm. The powder was obtained using gas atomization technique. In order for the particles to have the desired diameters range, the powders were sieved after the production process. The alloy contains Al (min 5.5% and max 6.5%), V (min 3.5% and max 4.5%), and the rest up to 100%Ti. The density of this alloy is the range of 4.43–4.5 g/cm^3 [46].

2.1. LMD Set-Up

The LMD experiments were performed using an Yb:YAG laser source (TruDisk 3001, Trumpf, Ditzingen, Germany) with wavelength λ = 1030 nm emitting in continuous wave, which was connected to a 6-axis robotic arm (Kuka, Augsburg, Germany) equipped with a three beam nozzle for powder flow and a metallic powder distributor (Figure 1). The focused spot of the laser beam was 800 μm with a top-hat energy distribution. The metallic powder was transported from the feeder to the substrate via a mix of He and Ar gases.

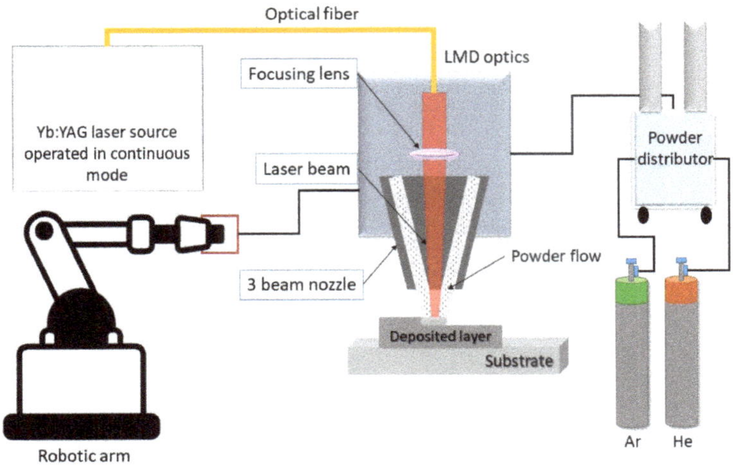

Figure 1. Schematic representation of the LMD system used in our experiments.

The processing parameters, such as laser power, scanning speed and powder feed rate were previously optimized in order to obtain high density depositions free of defects such as pores, cracks or non-fusion powder particles [47]. They are provided in Table 1 and were preserved for all experiments. However, if depositions are conducted on larger areas or when using different scanning strategies, these conditions no longer ensure defects free depositions and a new study is necessary, this time taking into account hatch spacing and the distance between layers. The defects can be due to the higher temperature of the substrate caused by prolonged irradiation, which can cause higher rate evaporation, larger heat affected zones, deposition with non-uniform thickness, change of metallographic structure or repeated dilatation-contraction cycles.

Table 1. Optimized process parameters for tracing a clearly defined, parallel borders, single line of Ti6Al4V with the least residual material.

Process Parameter	Value
Laser power	700 W
Scanning speed	15 mm/s
Powder flow rate	3 gr/min
Laser spot size	800 μm
Layer thickness	2.5 mm
Nozzle stand off	16 mm
Ar shielding gas flow rate	10 L/min
He shielding gas flow rate	3 L/min

During the fabrication process, the distance between the nozzle and the deposited surface was kept constant at 16 mm. A pure titanium plate with a thickness of 10 mm and a diameter of 100 mm was used as a substrate material.

Bulk structures in shape of parallelepipeds with size of 30 mm × 15 mm × ΔZ mm (ΔZ varying function of the scanning strategy) were performed using eight different scanning patterns (Table 2). Three parameters were considered essential during scanning: laser beam path trajectory, hatch spacing and offset between meander planes (ΔZ). Laser beam path trajectory presented in Figure 2a was defined as a meander in a Computer-Aided Manufacturing (CAM) software, TruTops Cell® (Trumpf, Ditzingen, Germany). The hatch spacing and offset between planes were varied with a 0.25 mm incremental step starting from 0.5 mm up to 2 mm for the two parameters. The hatch spacing establishes the overlapping (Ov) percentage on XY-plane, while the layer height of a meander establishes the overlapping on XZ-plane, which in this case is equal with the overlapping value on YZ-plane. For all the experiments, the meander was multiplied 10 times on YZ-plane (Figure 2b). To improve the roughness of the final part, a supplemental contour trajectory was traced after each meander. The contour is enhancing the final roughness of the parts fabricated by LMD [47]. Each samples was produced in triplicate in order to provide statistical relevant results.

Table 2. Scanning strategy parameters.

Sample	Offset between Meander Planes ΔZ [mm]	Hatch Spacing [mm]	Overlap [%]
S1	0.5	1	33
S2	0.5	1.25	20
S3	0.5	1.5	0
S4	0.75	0.5	66
S5	0.75	0.75	50
S6	1	1	33
S7	1	1.25	20
S8	1	1.5	0

The Computer-Aided Design (CAD) models of the samples were designed in SolidWorks software (Dassault Systemes, Vélizy-Villacoublay, France) and subsequently imported in the CAM program, where the surfaces were vectored and the robot paths established.

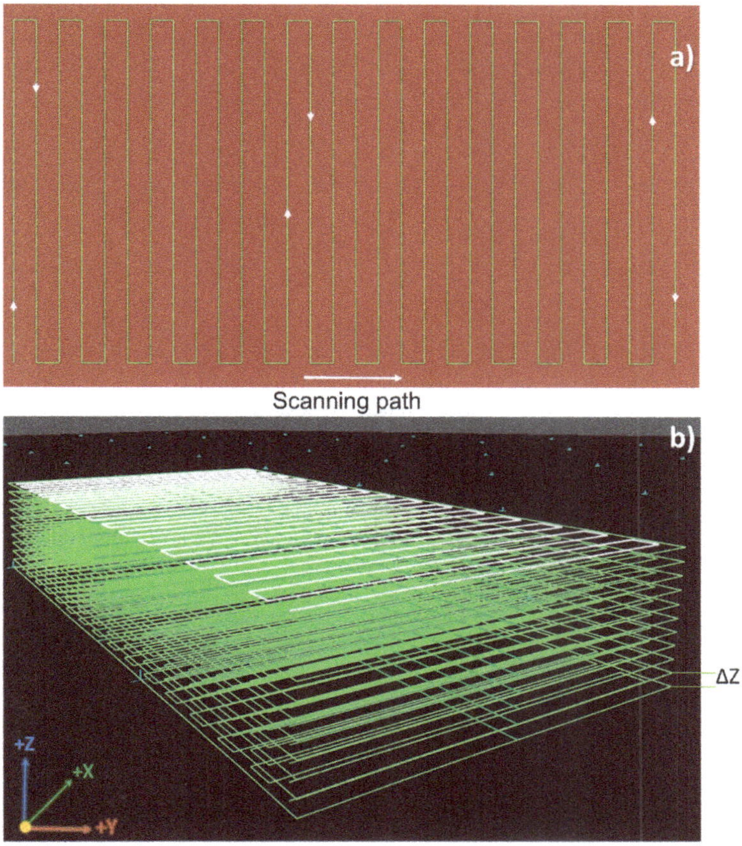

Figure 2. Laser beam path trajectory in form of a meander (**a**), Superposed meanders with offset between meander planes ΔZ (**b**).

2.2. *Quantitative Analysis of High Resolution Computed Tomography Scans*

The XCT method was applied for characterization of Ti6Al4V powder size distribution and 3D printed samples fabricated by LMD method. The XCT equipment is a custom-made system [48] and the experimental configurations were optimized in order to achieve certain requirements such as high spatial resolution for metallic powders and bulk sample analysis.

The microtomography configuration (μXCT) integrates a transmission nano focus X-ray source, a large flat panel detector with a detection matrix of 2 K by 2 K pixels and a motorized stage that allows sample manipulation on four coordinate axes (XYZθ) [49]. Several optimization protocols were imposed in order to achieve a steady resolution of ~2 μm, validated with a micro resolution calibration grid (JIMA-Japan Inspection Instruments Manufacturers Association).

The high-power XCT configuration integrates a reflection target energetic source (up to 320 kV), a high load motorized axis, using a similar flat panel detector as presented for the μXCT configuration.

The 3D reconstructed model of the samples was obtained within the cone-beam configuration using a modified Feldkamp (FDK) algorithm. Appropriate beam hardening corrections and adaptive Gauss filtration were applied to the reconstructions. All data processing applied on the obtained 3D reconstructions were performed with Volume Graphics software (VGStudio Max version 3.4 Volume Graphics GmbH, Heidelberg, Germany).

In this paper, for in-depth powder analysis and sample validation, we applied the Porosity/Foam Analysis modules using VGDefX/Only threshold algorithm. Advanced function for surface determination that separates the material from the air with healing parameter set as "remove all particles and voids" was applied before pore analysis [50]. The program automatically assesses the gray value variation to identify pores using threshold mode that was set manually based on the maximum gray value assigned for pores. The threshold value was determined by looking in the top view section for the preview analyzed area, assuring that all pores with a size higher than eight voxels are detected. The maximum size for pore detection was set at 15 mm, in order to ensure that all voids inside the sample were identified.

The Ti6Al4V metallic powder was inserted into a carbon fiber tube and scanned at a voltage of 90 kV and a current of 110 μA. For a well-balanced reconstruction the number of acquired projections were 2400 with an equidistant rotation step of 0.15°. At a magnification of ~130 the X-ray effective beam width of 1.5 μm is well matched with the source focus size of ~2 μm.

The detection of the closed pore in the Ti6Al4V particles using the Porosity/Inclusions Analysis module and the volume classification of the Ti6Al4V particles by Foam Analysis module was performed following the same procedure described in the literature [50].

The LMD produced samples were measured at 110 kV in order to obtain optimal penetrating X-ray beam. The current intensity was set at 250 μA, thus providing the best signal to noise ratio for the implemented scanning configuration. The scanning parameters for a well-balanced reconstruction were: 1800 projections (0.2° increment) and a magnification of X40 resulting in a ~13.8 μm effective beam width. For an easier identification of XCT parameters in case of powder and bulk they are gathered in Table 3.

Table 3. X-ray measurement parameters.

Parameter	Powder	Bulk
Voltage [kV]	90	110
Current [μA]	110	250
Acquired projections	2400	1800
Rotation step [°]	0.15	0.2
Beam width [μm]	1.5	13.8
Magnification	130	40

In both types of measurements, the Ti6Al4V metallic powder and the LMD produced samples, the X-ray beam was filtered by a 0.5 mm thickness Cu foil. In all XCT scanning experiments the total measuring time was kept around one hour by collecting images of two averaged frames of maximum one second integration time.

3. Results

3.1. *Micro-Scale Characterization of Ti6Al4V Metallic Powder*

After conducting the XCT reconstruction, the metallic powder was analyzed using the pore module analysis of the Volume Graphics software. The pore processing module is only suitable for particle embedded pores. The gas including pores are easily trackable due to their high sphericity.

A preliminary analysis of the closed pores inside of the Ti6Al4V particles from the commercial powder was performed. One could observe in the front view slice (Figure 3a) the shape variation and the presence of fully embedded pores with diameters of up to 50 μm. In the partially transparent 3D reconstructed image (Figure 3b) one highlights the presence of particles with closed pores in the Ti6Al4V metallic powder.

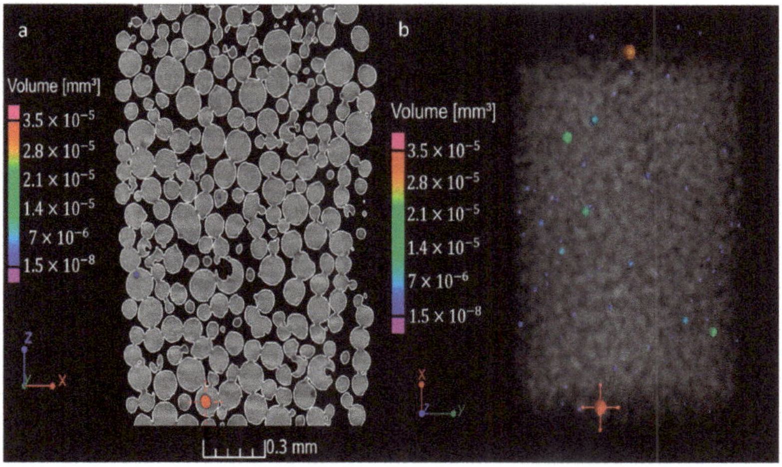

Figure 3. Embedded pores identification in Ti6Al4V powder by means of porosity module: front view slice (**a**) and 3D reconstruction of Ti6Al4V powder showing the identified embedded pores in the particles (**b**).

We mention that besides the fully embedded pores, there are also partially embedded ones and all of them can induce gases when molten, with significant probability of causing visible defects in the structure of the final product [50].

However, our analysis revealed that porosity in the powder material was negligible. From a total of 5104 analyzed particles only 91 displayed internal pores (counting for less than 2% of the particles).

3.2. *Scanning Strategy for Porosity Removal*

Figure 4 shows an optical microscopy image of a single line deposited by LMD using the optimized conditions specified in Table 1. The deposition is uniform, the borders of the line are parallel and the residual material deposited around the line is low. More details on the process of scanning and laser parameters can be found in Reference [47].

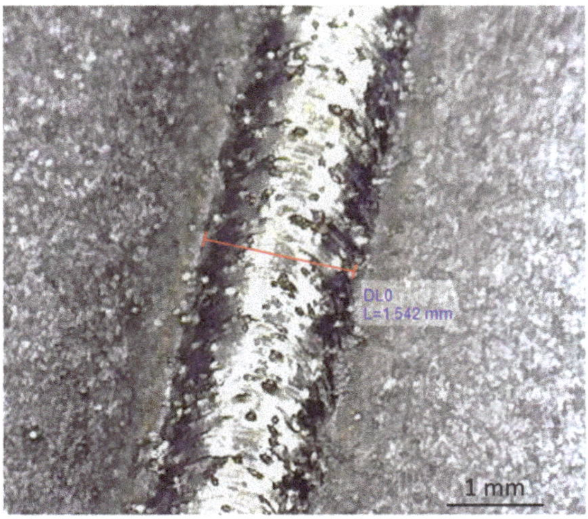

Figure 4. Single line traced by LMD using the optimized conditions from Table 1.

Two different quality samples obtained with the conditions given in Table 2 and with a changed meander trajectory between them, are shown in Figure 5. It can be observed from the picture that the area around the corners is higher for one of the samples. This excess of material is due to the decreased scanning speed around the sharp corners, which leads to increased heat input per unit time. The slower deposition speed due to deceleration around the sharp edges was investigated and explained by D. Thakkar et al. [51]. They found that by adding a small rounding on the corners, the scanning speed remains constant, which leads to uniform energy distribution and homogenous microstructure.

Figure 5. Bulk samples manufactured by LMD using different scanning parameters.

After experiments, the height of the bulk structures was measured in three different points, then the mean of these values was calculated and compared with the theoretical height (Table 4) obtained from the offset between meander planes (set in TruTops Cell) multiplied by the number of planes. In addition, we calculated the overlapping on XY-plane and error percentage (Equation (1)) to evaluate how the scanning strategy affects the height of the final samples.

$$\delta = \left| \frac{v_E - v_T}{v_T} \right| \times 100\% \quad (1)$$

where δ = percent error, v_E = experimental value measured, v_T = theoretical value.

Table 4. Experimental height variation as compared to the theoretical value in case of LMD deposited samples.

Sample	Experimental Height [mm]	Theoretical Height [mm]	Error [%]
S1	8.69 ± 0.3	5	74
S2	6.42 ± 0.1	5	28
S3	5.96 ± 0.05	5	19
S4	12.32 ± 0.15	7.5	64
S5	11.39 ± 0.05	7.5	52
S6	8.98 ± 0.01	10	10
S7	5.89 ± 0.05	10	41
S8	5.48 ± 0.02	10	45

The minimum error was obtained using 1 mm hatch space and 1 mm offset between meander planes. In this case, the theoretical value was 10 mm and the experimental value was 9.02 mm. It is important to keep the standing off distance constant, otherwise the laser beam and the powder spot will be defocused, thus directly affecting the quality and dimensions of the final part.

3.3. XCT Examination of The Bulk Samples Manufactured by LMD

Samples produced with the scanning strategies described in Table 2 were XCT scanned in order to identify defects that occurred during deposition. We added a paragraph on page line. The samples were aligned with respect to the X-ray tomography system by placing the substrate interface parallel to the detector and the long side along direction of rotation. In other words, the Z axis from the sample coordinates is perpendicular to the X-ray detector.

The appearance of pore clusters is more probable among first several deposited layers that are nearest to the substrate (low ΔZ value). The thermo-dynamic stability is improving as the layer grows on the Z axis. Figure 6 shows the preferential pores positioning in a LMD deposited sample, scanned by XCT.

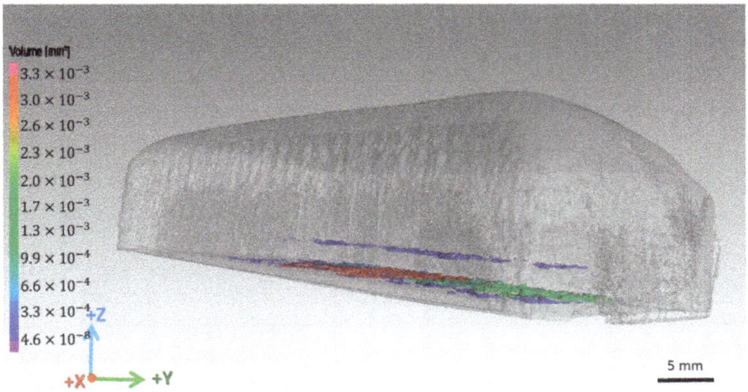

Figure 6. 3D rendering on sample 4 highlighting the appearance of pore clusters, were the most predominant are near the substrate region.

For a better overview regarding the present defects, we extracted a XCT slice from each sample reconstruction in the XY-plane, where the isolated or cluster pores are the most visible (Figure 7).

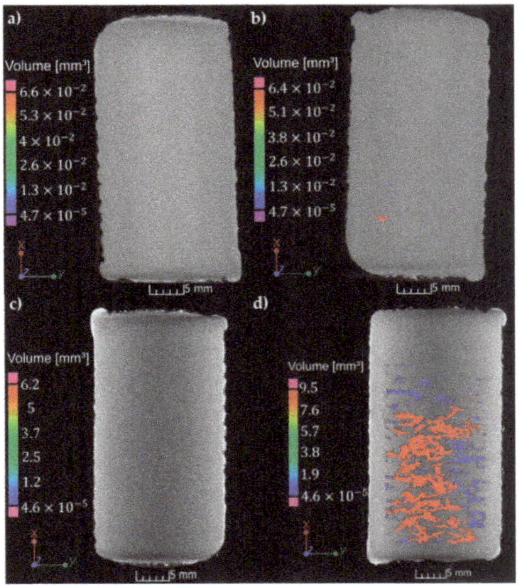

Figure 7. *Cont.*

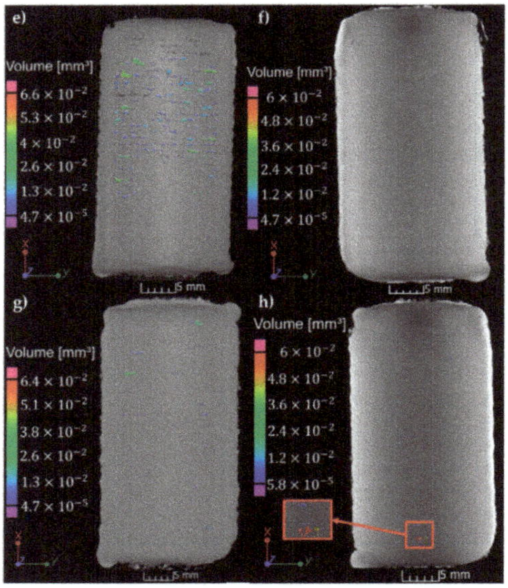

Figure 7. XCT slices in the areas with the most pronounced porosity of samples (**a**) S1, (**b**) S2, (**c**) S3, (**d**) S4, (**e**) S5, (**f**) S6, (**g**) S7, (**h**) S8.

The voids were the only defects identified in the bulk of the deposited samples. No inclusions of higher density were detected as expected since they are absent in the powder as well. Therefore, the porosity was calculated based on Equation (2) and the determined values are presented in Table 5.

$$Porosity = \frac{Total\ detected\ defect\ volume}{Total\ scanned\ volume} \quad (2)$$

Table 5. Determined porosity for samples produced by LMD.

Sample No.	S1	S2	S3	S4	S5	S6	S7	S8
Volume analyzed [mm^3]	1500	1500	1400	2400	2200	1200	1800	1700
Defect volume [mm^3]	0.0037	0.66	0.073	13	2.5	0.87	1.3	0.21
Porosity percentage [%]	0.00024	0.044	0.0052	0.54	0.12	0.0725	0.072	0.012
Slice Position relative to substrate interface [mm]	1.64	2.04	2.51	1.66	2.24	1.64	2.04	2.77

Because the samples were uneven in terms of thickness the porosity values were normalized and expressed in percentages in order to be able to compare them.

Besides samples S4 and S5 which displayed 0.54% and 0.12% porosity, all other scanning strategies with optimized deposition parameters proved efficient for deposition of structures with low number of defects. The porosity in case of samples S1–S8 (except S4 and S5) was less than 0.1%. We note that specifically for the analyzed planes in which the defects are concentrated, the porosity has higher values that are specified in the last line of Table 4. Thus, for the most porous samples (S4 and S5) even though the porosity of the whole samples is ~0.5%, locally, in the areas where the pores are concentrated, it reaches ~50%. 3D rendering for the most defective sample (S4) provides us the distribution of pore clusters that are parallel with the XY-plane (Figure 6). Analyzing the µXCT of sample S4 (Figure 7d) it becomes clear that the defects align along the scanning direction. The pores interconnect and form elongated voids of ~10 µm to 1 mm. By comparing the µXCT of samples S4 and S5, one can see that sample S5 (Figure 7e) displays significantly lower number of defects as compared to sample S4

(Figure 7d). The defects maintain the same tendency to align along the scanning direction, but they are generally spherical and their number is lower. The only difference between the scanning strategy of samples S4 and S5 is the hatch spacing. In case of S4 it is 0.5 mm and for S5 it is 0.75 mm. Indeed, when increasing even more the hatch spacing to 1 up to 1.5 mm the defects are reduced even more. A suggestive example is sample S7 (Figure 7g) that was produced with a hatch space of 1.25 mm between lines and the distance between planes of 1 mm. The sample displays defects that are aligned to the scanning direction in form of parallel lines made of interconnected pores. The defects volume is significantly reduced as compared to sample S5 (Figure 7e) for a small increase of hatch spacing of 0.5 mm and offset between meander planes of 0.25 mm. The next step was to increase the hatch space to 1.5 mm (Figure 7h). Again the sample's quality improved. Only two remaining areas of interconnected pores were visible, with much diminished length of ~300 μm. Besides these lines, only a few isolated pores of spherical shape and ~30 μm diameters were identified in the bulk of the deposited sample. If the hatch spacing is maintained to 1.5 mm and the offset between planes is set to 0.5 mm (Figure 7c) the pores disappear almost completely from the bulk of the sample. There are a few isolated polyhedral pores with length of ~20 μm at the borders of the sample in the area were the supplemental lines for contour were traced. In Figure 8a we present a sketch that explains the most probable cause of porosity at the borders. Basically, the "U" turns of the meander at the borders of the sample leaves some small voids, as shown in Figure 8a. If the contour is traced according to the technical drawing, these voids will remain unfilled or will be partially filled. The solution in this case is to translate the contour line by 0.5 mm in order to completely cover the voids (Figure 8b). μXCT was extremely useful in assessing non-destructively the pores location and in indicating the necessary offset of the contour line toward the sample bulk. Figure 8c,d present in detail the area close to the border of the sample without and with offset of the contour line. One can clearly see that when offsetting the contour line, the porosity is completely eliminated.

Table 4 shows the variations of the deposited the sample's height function of the hatch spacing. The closer the lines of the meander in respect with each other the thicker is the deposited layer. Thus, for a sample produced of 10 layers with a hatch spacing of 0.5 mm the sample's height was of ~12 mm, while a similar sample produced with a hatch spacing of 1.5 mm had a height of ~6 mm. The orange line corresponds to the theoretical height of the samples. It does not vary with the hatch spacing, as the offset between meander planes was selected to be in all cases of 1 mm. However, in practice the sample's height drastically varies with the hatch spacing. One can deduce that no sample's height matches the theoretical height value of 10 mm. Therefore, knowing that hatch spacing of 1.5 mm produces samples without internal defects, we kept this parameter intact and started to vary the distance between planes in order to match the experimental height with the theoretical one. We produced samples with plane distances of 1, 0.75 and 0.5 mm. It became obvious that the distance between planes in the range of 0.5 mm does not affect the quality of the produced samples. In all cases, the samples were free of pores or other internal defects. However, the offset between meander planes affected the height of the manufactured samples. A very close match was obtained for a hatch spacing of 1.5 mm and an offset between meander planes of 0.5 mm (Figure 7c).

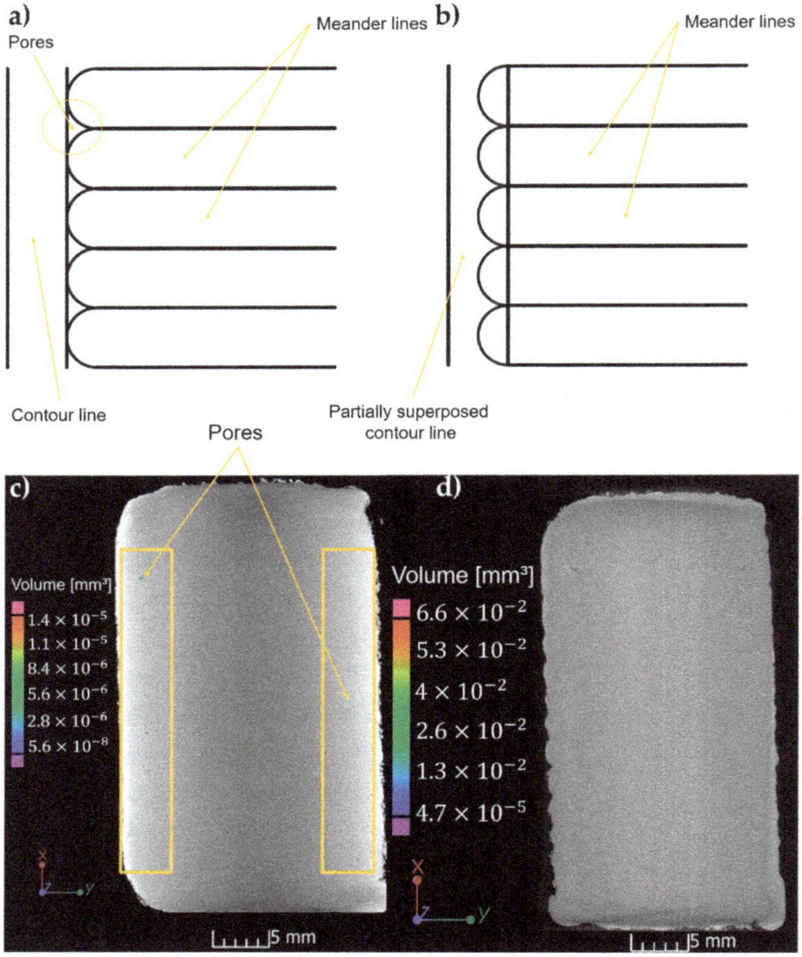

Figure 8. Schematic representation of scanning strategy with offset between contour and meander (**a**) and the corresponding XCT micrograph (**b**), schematic representation of scanning strategy with overlap between contour and meander (**c**) and the corresponding XCT micrograph (**d**).

4. Discussion

Ti6Al4V was selected for this study due to its applications in medicine and aerospace industry. Implants and prostheses are presently produced in several fixed sizes that are adopted for various types of patients and wounds. 3D printing can revolutionize implantology by manufacturing of personalized medical devices specifically designed for treating a patient's particular wound. Both for implants and aerospace industry the reduced weight and adequate mechanical properties are of high importance. Even if titanium has a reduced weight compared to other industrial metals and alloys, this can be further improved by replacing parts of the bulk via a lattice structure that can be easily obtained by 3D printing.

3D printing of metals can be the most commonly achieved by SLM and LMD. The technique are complementary as SLM produces finer parts with larger resolution, but reduced size as compared to LMD. Moreover, LMD is a more versatile technique, allowing for manufacturing of multilayer structures or parts with gradient composition.

As shown in the results section, the scanning strategy plays a determinant role on the quality of the LMD deposited sample.

In all cases, using a mix of experimental conditions that we optimized in time and changing the scanning strategy, allowed for a small degree of porosity, of maximum 0.5%. In all cases, porosity was concentrated close to the interface between deposition and substrate.

This preferential positioning of the pores has been also observed by other research groups and was attributed to the incomplete stacking of the layers, caused by non-molten powders. After deposition of a few superposed layers, the substrate temperature is of the order of hundreds of degrees and the scanning laser beam adds more heat that will completely melt the powder [34,52,53]. After the critical area is surpassed, the growth of material is smooth and free of defects.

Two types of pores have been identified:

(a) The pores in the bulk of the material, which are due to the use of a high energy density and slow scanning speed that create local vaporization. During rapid solidification of the material, the gases are trapped into the bulk in form of circular pores. The reason for the pores to be oriented along the laser scanning direction is that the liquid tends to flow oppositely to the laser movement and the gases will be trapped along this direction when the liquid solidifies. The pores can be isolated or they can form a necklace of adjacent voids. In case of high overlaps, the necklace of pores can be remelted and the gases gather in large polyhedral voids. During LMD deposition with high overlaps, real interconnected networks of cavities can be created by successive melting and solidification of material. A relevant example is Figure 7d. Overlaps of 70% or more between the meander lines are therefore not suitable for depositing quality samples by LMD. Apparently, a hatching superposition of 60% or less influence only the built height, but not the quality of the deposition. In our case, overlaps less than 60% between the hatching lines produced samples with very low porosity (0.0002–0.005%). However, the possibility of pores with size less than 13.8 μm should not be ruled out as the XCT cannot identify them.

(b) Pores caused by incomplete seam between the bulk layers and the contour layer. The fact that we use a meander trajectory to deposit our layers leaves at the borders some non-uniformities. The contour layer aims to fix them; however, if it is not superposed over the initial border, some voids can remain after contour solidification. The solution to remove them is to fill them with material by superposing part of the contour line with the border of the meander (Figure 8c).

Pores trapped in the volume of material can create stress concentration centers that in time might generate fractures in case of highly solicited parts. Therefore, the elimination of pores or their drastically reduction is e prerequisite in numerous quality control procedures.

XCT is an excellent non-destructive tool for assessment of pores, geometry and location. The custom proprietary built XCT set-up integrates a high penetration power X-Ray source and is applicable as metrological analysis tool for large and thick bulk samples from elements with high atomic number. The drawback as compared with metallographic techniques is that it is limited by the voxel size in identification of pores and it cannot provide quantitative data on their size. In our case, the minimum voxel was a cube with the side's length of 13.8 μm. Therefore, pores with smaller diameters than 13.8 μm will be hardly visible by XCT. For our study, the technique was however extremely useful, as it drastically reduced the analysis time to hours instead of days or even weeks of samples' preparation and analysis by metallographic techniques. It has to be mentioned that by reducing the analyzed area the voxel size is decreased and thus the resolution of the method increases. A compromised has to be reached between the analysis time and the resolution to be achieved.

The best conditions to obtain defect-free bulk samples are not necessarily the best conditions for printing a 3D part. It is highly possible that with the best conditions for defect-free samples to obtain a mismatch in dimensions (especially height) between the CAD model and the actual printed part. Our solution was to keep constant the optimal process parameters for a single line and the hatch superposition, in order to obtain defect-free samples. Next, we varied the offset between deposited planes in order to achieve the match with the CAD file.

Alternative solution to non-destructive monitoring and parameters tuning would be to accept samples with certain degree of porosity and to apply a post-processing heat treatment with hot isostatic pressing (HIP). Using HIP post-treatment, the porosity percentage decreased from 0.08% to 0.01% [54,55]. However, with our solution, the percentage of porosity was an order of magnitude lower.

The parts quality monitoring technique by XCT is not limited to the LMD method or to the material used in this manuscript. It can be useful for assessing defects in metallic samples with sizes of tens of centimeters obtained by casting, powder metallurgy or 3D printing by laser or electron beam [56]. The technique is also compatible with the most common metals used in metallurgy, so it can cover a wide range of applications.

5. Conclusions

A quality improvement study for Laser Melting Deposition of Ti6Al4V parts was undertaken in order to completely eliminate internal defects. X-Ray Computed Tomography proved to be an invaluable tool for non-destructive characterization of the bulk for defects assessment. We started from a set of laser and scanning parameters that were considered the best for tracing single lines (good definition, uniform deposition, minimal unmolten powder around deposition) and conducted a study on the influence of hatch spacing and distance between deposition planes on internal defects in the obtained samples. All obtained samples were scanned by XCT and the porosity was evaluated. The main results can be summarized as follows:

a. Two types of pores were identified by XCT: some spherical caused by gases produced by local evaporation and some polyhedral ones caused by lack of superposition between the lines that produced the samples and a supplemental contour line traced or correction of the sample borders.
b. The circular pores were affected mainly by the hatch spacing, while the distance between planes had no major effect on the samples' quality.
c. The distance between planes affected the build height in corroboration with the hatch spacing. Distance between planes of 0.5 mm produced samples with height very close to the theoretical one, while increasing the offset between meander planes to 1 mm caused lower heights as compared to the theoretical ones.
d. The circular pores were concentrated close to the interface with the substrate. They were aligned along scanning direction forming parallel rows of voids. The larger the hatch spacing the lower the number of pores. By increasing the hatch spacing to 1.5 mm the pores disappeared completely. For a hatch spacing of 0.5 mm we recorded the highest level of porosity of 0.0002%.
e. For complete elimination of polyhedral pores from the borders of the samples, the contour line was superposed in various ratio of its thickness. XCT revealed that an overlap of 50% between the contour line and the meander was sufficient to eliminate the pores.
f. XCT was an excellent assisting tool during the tuning of the scanning parameters, as it succeeded to provide valuable information on the pores' arrangement, location, and size for every varied parameter. The alternative analysis would have been cutting of each sample in various locations, polishing and microscopic analysis that could slow the manufacturing process by days or even weeks. The alternative with non-destructive control with ultrasounds is also unreliable due to the lack of visual information.

Author Contributions: Conceptualization, D.C., I.P. and A.C.P.; software and methodology, S.M. and M.A.M.; investigation, I.P., M.L. and I.T.; software, A.S. and C.D.; writing—original draft preparation, D.C., S.M., M.L. and A.C.P.; supervision, I.T.; funding acquisition, A.C.P. and I.T.; writing—review and editing, D.C., A.C.P. and I.T. All authors have read and agreed to the published version of the manuscript.

Funding: This research was funded by the National Core Program, contract numbers 16N/2019. A.C.P. and D.C. have received the funding of the PN-III-P1-1.1-TE-2016-2015 (TE136/2018) Project. S.M. was financed from the Project 25PCCDI/2018. M.A.M. has received financial support from the European Union's Horizon 2020 (H2020) research and innovation program under the Marie Skłodowska-Curie, grant agreement No. 764935.

Conflicts of Interest: The authors declare no conflict of interest.

References

1. Herzog, D.; Seyda, V.; Wycisk, E.; Emmelmann, C. Additive manufacturing of metals. *Acta Mater.* **2016**, *117*, 371–392. [CrossRef]
2. Frazier, W.E. Metal additive manufacturing: A review. *J. Mater. Eng. Perform.* **2014**, *23*, 1917–1928. [CrossRef]
3. Zhang, D.; Wang, W.; Guo, Y.; Hu, S.; Dong, D.; Poprawe, R.; Schleifenbaum, J.H.; Ziegler, S. Numerical simulation in the absorption behavior of Ti6Al4V powder materials to laser energy during SLM. *J. Mater. Process. Technol.* **2019**, *268*, 25–36. [CrossRef]
4. Diegel, O.; Nordin, A.; Motte, D. Additive manufacturing technologies. In *A Practical Guide to Design for Additive Manufacturing*; Springer: Singapore, 2019. [CrossRef]
5. Galarraga, H.; Warren, R.J.; Lados, D.A.; Dehoff, R.R.; Kirka, M.M.; Nandwana, P. Effects of heat treatments on microstructure and properties of Ti-6Al-4V ELI alloy fabricated by electron beam melting (EBM). *Mater. Sci. Eng. A* **2017**, *685*, 417–428. [CrossRef]
6. Raju, R.; Duraiselvam, M.; Petley, V.; Verma, S.; Rajendran, R. Microstructural and mechanical characterization of Ti6Al4V refurbished parts obtained by laser metal deposition. *Mater. Sci. Eng. A* **2015**, *643*, 64–71. [CrossRef]
7. Kozior, T.; Bochnia, J. The influence of printing orientation on surface texture parameters in powder bed fusion technology with 316L steel. *Micromachines* **2020**, *11*, 639. [CrossRef] [PubMed]
8. Kozior, T.; Bochnia, J.; Zmarzły, P.; Gogolewski, D.; Mathia, T.G. Waviness of Freeform Surface Characterizations from Austenitic Stainless Steel (316L) Manufactured by 3D Printing-Selective Laser Melting (SLM) Technology. *Materials* **2020**, *13*, 4372. [CrossRef]
9. Korsmik, R.S.; Rodionov, A.A.; Korshunov, V.A.; Ponomarev, D.A.; Prosychev, I.S.; Promakhov, V.V. Topological optimization and manufacturing of vessel propeller via LMD-method. *Mater. Today Proc.* **2020**, 1–7. [CrossRef]
10. Liu, S.; Shin, Y.C. The influences of melting degree of TiC reinforcements on microstructure and mechanical properties of laser direct deposited Ti6Al4V-TiC composites. *Mater. Des.* **2017**, *136*, 185–195. [CrossRef]
11. Liu, S.; Shin, Y.C. Simulation and experimental studies on microstructure evolution of resolidified dendritic TiCx in laser direct deposited Ti-TiC composite. *Mater. Des.* **2018**, *159*, 212–223. [CrossRef]
12. Wilson, J.M.; Shin, Y.C. Microstructure and wear properties of laser-deposited functionally graded Inconel 690 reinforced with TiC. *Surf. Coat. Technol.* **2012**, *207*, 517–522. [CrossRef]
13. Wilson, J.M.; Piya, C.; Shin, Y.C.; Zhao, F.; Ramani, K. Remanufacturing of turbine blades by laser direct deposition with its energy and environmental impact analysis. *J. Clean. Prod.* **2014**, *80*, 170–178. [CrossRef]
14. Thompson, S.M.; Bian, L.; Shamsaei, N.; Yadollahi, A. An overview of Direct Laser Deposition for additive manufacturing; Part I: Transport phenomena, modeling and diagnostics. *Addit. Manuf.* **2015**, *8*, 36–62. [CrossRef]
15. Gasser, A.; Backes, G.; Kelbassa, I.; Weisheit, A.; Wissenbach, K. Laser Metal Deposition (LMD) and Selective Laser Melting (SLM) in Turbo-Engine Applications Laser Additive Manufacturing. *Laser Tech. J.* **2010**, *7*, 58–63. [CrossRef]
16. Petrat, T.; Graf, B.; Gumenyuk, A.; Rethmeier, M. Laser metal deposition as repair technology for a gas turbine burner made of inconel 718. *Phys. Procedia* **2016**, *18*, 761–768. [CrossRef]
17. Kumar, L.J.; Nair, C.G.K. Laser metal deposition repair applications for Inconel 718 alloy. *Mater. Today Proc.* **2017**, *4*, 11068–11077. [CrossRef]
18. Culmone, C.; Smit, G.; Breedveld, P. Additive manufacturing of medical instruments: A state-of-the-art review. *Addit. Manuf.* **2019**, *27*, 461–473. [CrossRef]
19. Everton, S.K.; Hirsch, M.; Stravroulakis, P.; Leach, R.K.; Clare, A.T. Review of in-situ process monitoring and in-situ metrology for metal additive manufacturing. *Mater. Des.* **2016**, *95*, 431–445. [CrossRef]
20. Vaithilingam, J.; Goodridge, R.D.; Hague, R.J.; Christie, S.D.; Edmondson, S. The effect of laser remelting on the surface chemistry of Ti6al4V components fabricated by selective laser melting. *J. Mater. Process. Technol.* **2016**, *232*, 1–8. [CrossRef]
21. Yasa, E.; Deckers, J.; Kruth, J.P. The investigation of the influence of laser re-melting on density, surface quality and microstructure of selective laser melting parts. *Rapid Prototyp. J.* **2011**, *17*, 312–327. [CrossRef]

22. Wang, X.; Xu, S.; Zhou, S.; Xu, W.; Leary, M.; Choong, P.; Qian, M.; Brandt, M.; Xie, Y.M. Topological design and additive manufacturing of porous metals for bone scaffolds and orthopaedic implants: A review. *Biomaterials* **2016**, *83*, 127–141. [CrossRef] [PubMed]
23. Hrabe, N.; Gnäupel-Herold, T.; Quinn, T. Fatigue properties of a titanium alloy (Ti–6Al–4V) fabricated via electron beam melting (EBM): Effects of internal defects and residual stress. *Int. J. Fatigue* **2017**, *94*, 202–210. [CrossRef]
24. Majorell, A.; Srivatsa, S.; Picu, R.C. Mechanical behavior of Ti-6Al-4V at high and moderate temperatures-Part I: Experimental results. *Mater. Sci. Eng. A* **2002**, *326*, 297–305. [CrossRef]
25. Picu, R.C.; Majorell, A. Mechanical behavior of Ti-6Al-4V at high and moderate temperatures-Part II: Constitutive modeling. *Mater. Sci. Eng. A* **2002**, *326*, 306–316. [CrossRef]
26. Kasperovich, G.; Haubrich, J.; Gussone, J.; Requena, G. Correlation between porosity and processing parameters in TiAl6V4 produced by selective laser melting. *Mater. Des.* **2016**, *105*, 160–170. [CrossRef]
27. Shunmugavel, M.; Polishetty, A.; Littlefair, G. Microstructure and Mechanical Properties of Wrought and Additive Manufactured Ti-6Al-4V Cylindrical Bars. *Procedia Technol.* **2015**, *20*, 231–236. [CrossRef]
28. Moiduddin, K.; Mian, S.H.; Alkhalefah, H.; Umer, U. Digital design, analysis and 3D printing of prosthesis scaffolds for mandibular reconstruction. *Metals* **2019**, *9*, 569. [CrossRef]
29. Weißmann, V.; Bader, R.; Hansmann, H.; Laufer, N. Influence of the structural orientation on the mechanical properties of selective laser melted Ti6Al4V open-porous scaffolds. *Mater. Des.* **2016**, *95*, 188–197. [CrossRef]
30. Liu, Q.; Wang, Y.; Zheng, H.; Tang, K.; Ding, L.; Li, H.; Gong, S. Microstructure and mechanical properties of LMD-SLM hybrid forming Ti6Al4V alloy. *Mater. Sci. Eng. A* **2016**, *660*, 24–33. [CrossRef]
31. Sterling, A.J.; Torries, B.; Shamsaei, N.; Thompson, S.M.; Seely, D.W. Fatigue behavior and failure mechanisms of direct laser deposited Ti-6Al-4V. *Mater. Sci. Eng. A* **2016**, *655*, 100–112. [CrossRef]
32. Amsterdam, E.; Kool, G.A. High cycle fatigue of laser beam deposited Ti-6Al-4V and inconel 718. Bridging the gap between theory and operational practice. In Proceedings of the 25th Symposium of the International Committee on Aeronautical Fatigue, Rotterdam, The Netherlands, 27–29 May 2009; pp. 1261–1274.
33. Keist, J.S.; Palmer, T.A. Role of geometry on properties of additively manufactured Ti-6Al-4V structures fabricated using laser based directed energy deposition. *Mater. Des.* **2016**, *106*, 482–494. [CrossRef]
34. Carroll, B.E.; Palmer, T.A.; Beese, A.M. Anisotropic tensile behavior of Ti-6Al-4V components fabricated with directed energy deposition additive manufacturing. *Acta Mater.* **2015**, *87*, 309–320. [CrossRef]
35. Cunningham, R.; Nicolas, A.; Madsen, J.; Fodran, E.; Anagnostou, E.; Sangid, M.D.; Rollett, A.D. Analyzing the effects of powder and post-processing on porosity and properties of electron beam melted Ti-6Al-4V. *Mater. Res. Lett.* **2017**, *5*, 516–525. [CrossRef]
36. Ahsan, M.N.; Pinkerton, A.J.; Moat, R.J.; Shackleton, J. A comparative study of laser direct metal deposition characteristics using gas and plasma-atomized Ti-6Al-4V powders. *Mater. Sci. Eng. A* **2011**, *528*, 7648–7657. [CrossRef]
37. Liou, F.W. Rapid prototyping processes. In *Rapid Prototyping and Engineering Applications*; CRC Press: Boka Raton, FL, USA, 2007; Chapter 6; p. 84. [CrossRef]
38. Bian, L.; Thompson, S.M.; Shamsaei, N. Mechanical Properties and Microstructural Features of Direct Laser-Deposited Ti-6Al-4V. *JOM* **2015**, *67*, 629–638. [CrossRef]
39. Reimers, P.; Goebbels, J. New Possibilities of Nondestructive Evaluation by X-ray Computed Tomography. *Mater. Eval.* **1983**, *41*, 732–737, ISSN 0025-5327.
40. Krztońa, H.; Muchab, J.; Witkowskib, W. The application of laboratory X-ray micro-diffraction to study the effects of clinching process in steel sheets. *Acta Phys. Pol. A* **2016**, *130*, 985. [CrossRef]
41. Gajjar, P.; Styliari, I.D.; Nguyen, T.T.H.; Carr, J.; Chen, X.; Elliott, J.A.; Hammond, R.B.; Burnett, T.L.; Roberts, K.; Withers, P.J.; et al. 3D characterization of dry powder inhaler formulations: Developing X-ray micro computed tomography approaches. *Materials* **2020**, *151*, 32–44. [CrossRef]
42. De Chiffre, L.; Carmignato, S.; Kruth, J.P.; Schmitt, R.; Weckenmann, A. Industrial applications of computed tomography. *CIRP Ann. Manuf. Technol.* **2014**, *63*, 655–677. [CrossRef]
43. Svintsitskiy, D.A.; Yu Kardash, T.; Lazareva, E.V.; Bondareva, V.M. X-ray-induced transformations on the surface of MoVNbTe mixed oxide catalyst: An XPS study. *Appl. Surf. Sci.* **2021**, *535*, 147676. [CrossRef]
44. Chen, L.; Chang, J.; Ding, D.; Feng, H. Preparation, luminescent properties and X-ray imaging result of Lu2O3: Eu structured scintillation film on sapphire substrate by LCVD method. *Mater. Lett.* **2021**, *282*, 128689. [CrossRef]

45. Romano, S.; Abel, A.; Gumpinger, J.; Brandão, A.D.; Beretta, S. Quality control of AlSi10Mg produced by SLM: Metallography versus CT scans for critical defect size assessment. *Addit. Manuf.* **2019**, *28*, 394–405. [CrossRef]
46. Brandt, D.A.; Warner, J.C. *Metallurgy Fundamentals: Ferrous and Nonferrous*, 6th ed.; Goodheart-Willcox: Tinley Park, IL, USA, 2019; ISBN 10 1635638747.
47. Chioibasu, D.; Achim, A.; Popescu, C.; Stan, G.E.; Pasuk, I.; Enculescu, M.; Iosub, S.; Duta, L.; Popescu, A. Prototype Orthopedic Bone Plates 3D Printed by Laser Melting Deposition. *Materials* **2019**, *12*, 906. [CrossRef] [PubMed]
48. Tiseanu, I.; Zani, L.; Craciunescu, T.; Cotorobai, F.; Dobrea, C.; Sima, A. Characterization of superconducting wires and cables by X-ray micro-tomography. *Fusion Eng. Des.* **2013**, 1613–1618. [CrossRef]
49. Tiseanu, I.; Craciunescu, T.; Petrisor, T.; Corte A Della. 3D X-ray micro-tomography for modeling of NB3SN multifilamentary superconducting wires. *Fusion Eng. Des.* **2007**, *82*, 1447–1453. [CrossRef]
50. Du Plessis, A.; Sperling, P.; Beerlink, A.; Tshabalala, L.; Hoosain, S.; Mathe, N.; Le Roux, S.G. Standard method for microCT-based additive manufacturing quality control 1: Porosity analysis. *MethodsX* **2018**, *5*, 1102–1110. [CrossRef]
51. Thakkar, D.; Sahasrabudhe, H. Investigating microstructure and defects evolution in laser deposited single-walled Ti6Al4V structures with sharp and non-sharp features. *J. Manuf. Process.* **2020**, *56*, 928–940. [CrossRef]
52. Li, P.H.; Guo, W.G.; Huang, W.D.; Su, Y.; Lin, X.; Yuan, K.B. Thermomechanical response of 3D laser-deposited Ti-6Al-4V alloy over a wide range of strain rates and temperatures. *Mater. Sci. Eng. A* **2015**, *647*, 34–42. [CrossRef]
53. Zhang, K.; Wang, S.; Liu, W.; Long, R. Effects of substrate preheating on the thin-wall part built by laser metal deposition shaping. *Appl. Surf. Sci.* **2014**, *317*, 839–855. [CrossRef]
54. Liu, S.; Shin, Y.C. Additive manufacturing of Ti6Al4V alloy: A review. *Mater. Des.* **2019**, *164*, 107552. [CrossRef]
55. Kasperovich, G.; Hausmann, J. Improvement of fatigue resistance and ductility of TiAl6V4 processed by selective laser melting. *J. Mater. Process. Technol.* **2015**, *220*, 202–214. [CrossRef]
56. Abdullah, M.F.; Anwar, S.; Al-Ahmari, A. Thermomechanical Simulations of Residual Stresses and Distortion in Electron Beam Melting with Experimental Validation for Ti-6Al-4V. *Metals* **2020**, *10*, 1151. [CrossRef]

metals

MDPI

Article

Separation of the Formation Mechanisms of Residual Stresses in LPBF 316L

Alexander Ulbricht [1,*], Simon J. Altenburg [1], Maximilian Sprengel [1], Konstantin Sommer [1], Gunther Mohr [1,2], Tobias Fritsch [1], Tatiana Mishurova [1], Itziar Serrano-Munoz [1], Alexander Evans [1], Michael Hofmann [3] and Giovanni Bruno [1,4]

1 Bundesanstalt für Materialforschung und -prüfung (BAM), Unter den Eichen 87, 12205 Berlin, Germany; Simon.Altenburg@bam.de (S.J.A.); Maximilian.Sprengel@bam.de (M.S.); Konstantin.Sommer@bam.de (K.S.); Gunther.Mohr@bam.de (G.M.); tobias.thiede@bam.de (T.F.); tatiana.mishurova@bam.de (T.M.); Itziar.Serrano-Munoz@bam.de (I.S.-M.); alexander.evans@bam.de (A.E.); giovanni.bruno@bam.de (G.B.)

2 Institute of Machine Tools and Factory Management, Chair of Processes and Technologies for Highly Loaded Welds, Technische Universität Berlin, Straße des 17. Juni 135, 10623 Berlin, Germany

3 Heinz Maier-Leibnitz Zentrum (MLZ), Technische Universität München, Lichtenbergstraße 1, 85747 Garching, Germany; Michael.Hofmann@frm2.tum.de

4 Institute of Physics and Astronomy, University of Potsdam, Karl-Liebknecht-Straße 24/25, 14476 Potsdam, Germany

* Correspondence: alexander.ulbricht@bam.de; Tel.: +49-308-104-4140

Received: 29 July 2020; Accepted: 3 September 2020; Published: 14 September 2020

Abstract: Rapid cooling rates and steep temperature gradients are characteristic of additively manufactured parts and important factors for the residual stress formation. This study examined the influence of heat accumulation on the distribution of residual stress in two prisms produced by Laser Powder Bed Fusion (LPBF) of austenitic stainless steel 316L. The layers of the prisms were exposed using two different border fill scan strategies: one scanned from the centre to the perimeter and the other from the perimeter to the centre. The goal was to reveal the effect of different heat inputs on samples featuring the same solidification shrinkage. Residual stress was characterised in one plane perpendicular to the building direction at the mid height using Neutron and Lab X-ray diffraction. Thermography data obtained during the build process were analysed in order to correlate the cooling rates and apparent surface temperatures with the residual stress results. Optical microscopy and micro computed tomography were used to correlate defect populations with the residual stress distribution. The two scanning strategies led to residual stress distributions that were typical for additively manufactured components: compressive stresses in the bulk and tensile stresses at the surface. However, due to the different heat accumulation, the maximum residual stress levels differed. We concluded that solidification shrinkage plays a major role in determining the shape of the residual stress distribution, while the temperature gradient mechanism appears to determine the magnitude of peak residual stresses.

Keywords: additive manufacturing; Laser Powder Bed Fusion; LPBF; AISI 316L; online process monitoring; thermography; residual stress; neutron diffraction; X-ray diffraction; computed tomography

1. Introduction

In recent years, Additive Manufacturing (AM) has evolved from a method for rapid prototyping to a mature production process for certain parts in industries, such as the aerospace industry [1]. Among the different AM manufacturing techniques, Laser Powder Bed Fusion (LPBF) is an important technique for the production of net shaped metallic parts [2]. Early research conducted by Mercelis and

Kruth [3] showed that metallic parts made by LPBF inherently contain residual stresses (RS). They had described two driving mechanisms for the formation of RS: the Temperature Gradient Mechanism (TGM) and the Solidification Shrinkage Mechanism (SSM). The two mechanisms are interlinked and their combined effect on RS in AM 316L is a topic of current research [4–7]. Wang et al. [6] showed within long bars of LPBF 316L that scan strategies using shorter scan tracks reduced RS and attributed this to lower solidification shrinkage. Roehling et al. [7] observed a decrease of RS in samples with a bridge geometry manufactured by LPBF of 316L due to post-solidification heating of each layer during the build job using selective large-area diode surface heating. This method aimed to decrease the cooling rate. Each of the two publications had mainly utilized one of the two mechanisms to reduce RS: Wang et al. [6] mainly utilized the SSM, whereas Roehling et al. [7] mainly utilized the TGM. In both cases, a reduction of RS was observed. However, there is still a level of uncertainty on the magnitude of the influence of each mechanism onto the shape of the resulting RS field.

Diffraction is a well-known non-destructive method to evaluate RS [8–10]. Determining elastic strains by measuring the variation of lattice spacing provides a powerful method to identify RS. This is achieved at the surface by Lab X-ray Diffraction (XRD), up to a depth of about 5 µm in metals, as well as in the bulk by Neutron Diffraction (ND) up to a depth of about a few mm to a few cm [11–13]. In this work, the bulk triaxial RS state was determined using ND and was combined with the Lab XRD biaxial RS state at the surface. This methodology allows for the mapping of the RS distribution across the complete cross-sectional plane. Such residual stress tends to be compressive in the bulk and tensile near the surface [14]. ND enables the non-destructive determination of the triaxial RS state over a complete two-dimensional (2D) plane or three-dimensional (3D) volume. Destructive methods, such as incremental or deep hole drilling, slitting, or contour method, would also yield stress depth profiles, but it would be extremely difficult to determine triaxial stress states over a complete cross-section.

To exploit the benefits of lightweight, load driven structural designs for LPBF metallic parts, it is necessary to understand RS in those parts, since their effect on fatigue life can be significant [15]. In order to understand RS, it is necessary to decouple the contributing mechanisms, especially if we aim at modelling the manufacturing process.

Therefore, this study aims at unravelling the contributions of the two mechanisms, to provide a basis for discussion on the length scale of RS introduced into parts by the TGM and the SSM. Therefore, the specimen design was chosen to provide similar solidification shrinkage, but at the same time different cooling rates without changing the volumetric energy density (VED) of specimens. Based on this design, similar RS results should be assigned to equal solidification shrinkage, whereas differences should be caused by the different cooling rates. Online monitoring by thermography during the build process was used in order to assess these cooling rates and their effect on RS formation.

The TGM is mainly related to process parameters (e.g., VED) and the SSM is mainly related to the length of shrinking scan tracks. Therefore, the results of this study might help to decide which approach is more suitable to reduce RS for a specific part design and its expected load profile.

Additionally, the results from Micro Computed Tomography (µCT) and Optical Microscopy (OM) were evaluated to link RS fields and defect distributions, with the aim to produce a holistic approach towards the analysis of the interconnection of TGM and SSM.

2. Materials and Methods

2.1. Material and LPBF Processing Conditions

Austenitic stainless steel 316L powder was processed by the commercial LPBF system SLM280 HL (SLM Solutions Group AG, Lübeck, Germany). The powder was characterised by its supplier: it has an apparent density of 4.58 g cm^{-3} and a mean diameter of 34.69 µm. The cumulative mass values of the particle size distribution are: $D_{10} = 18.22$ µm, $D_{50} = 30.50$ µm, $D_{90} = 55.87$ µm. The LPBF system uses a single 400 W continuous wave ytterbium fibre laser with a spot size of approx. 80 µm in a focal position. The processes were conducted in an argon gas atmosphere with an oxygen content of less than 0.1%. The parts were manufactured on a stainless steel substrate plate, which was heated up to 100 °C as a preheating temperature before the start of the build process. Two prismatic specimens of the dimension 24 mm × 36 mm × 24.5 mm were manufactured in two separate built processes. In order to remove specimens from the base plate a band saw was used. This reduced the height to a final value of 21 mm. A specimen design of low aspect ratio was chosen for this experiment in order to prevent significant RS relaxation due to distortion after the removal from the substrate plate. Such a distortion had been reported in literature for LPBF 316L [16,17]. Although the removal from the substrate plate may have caused a degree of RS relaxation, the overall relative trend between the specimens was considered to be mainly unaffected. The specimens were placed close to the border of the base plate to fit within the field of view of the thermography camera setup. The specimens were manufactured using the following process parameters: layer thickness $t = 50\,\mu\text{m}$, scanning velocity $v = 700\,\text{mm}\,\text{s}^{-1}$, laser power $P = 275\,\text{W}$, and hatch distance $h = 0.12\,\text{mm}$. Two different so-called border fill scanning strategies were applied, which scan along the edges of the rectangular cross sections of the parts: for one the scanning sequence starts in the centre of the part with growing rectangles towards the perimeter and the other has a converse scanning sequence (see Figure 1). The interlayer time (according to Mohr et al. [18]) was approximately 27.6 ± 1.0 s due to time variations between re-coating forwards and backwards. Therefore, the total time for each build process of 490 layers added up to 3.76 h.

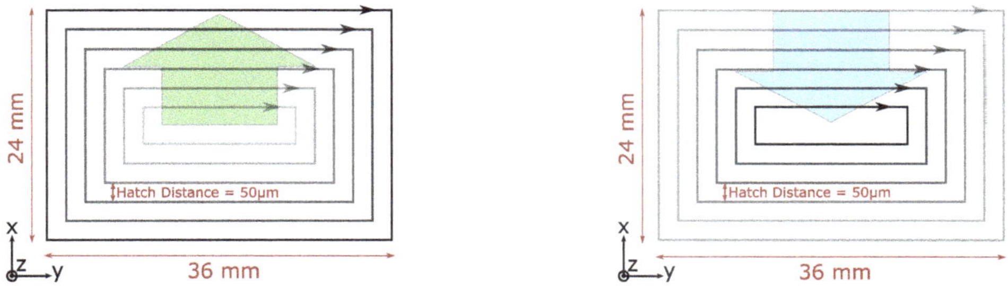

(**a**) ▲ Centre to Perimeter (CtP) (**b**) ▼ Perimeter to Centre (PtC)

Figure 1. Schematics of both border fill scan strategies. (**a**) Describes the Centre to Parameter (CtP) strategy indicated by the green arrow, while (**b**) shows the Parameter to Centre (PtC) scan strategy indicated by the blue arrow. The black arrows show the direction of scan of the laser around each border fill scan.

2.2. Thermography

An ImageIR 8300 hp camera (Infratec GmbH, Dresden, Germany) working in the spectral range of 2–5.7 µm was used for thermographic measurements. It was mounted on top of the SLM280 HL machine's build chamber, observing the build plate through a sapphire window. The chosen subframe image had a size of 160 px × 114 px featuring a geometric resolution on the build plate of $360\,\mu\text{m}\,\text{px}^{-1}$. The acquisition frame rate was set to 1000 Hz. The camera was calibrated for black body radiation. Due to the fact that the emissivity of the used material is well below unity [19] and the process was observed through optical elements, the calibration is not valid for quantitative evaluation of the obtained thermography data. Nonetheless, assuming that the emissivity remains (approximately) constant during the build process, the obtained apparent temperatures enable comparisons within a single build process and between the two different build processes. The thermography data for the two specimens were obtained during the build process while using two different calibration ranges: 673 K to 1073 K for the CtP specimen (▲) and 623 K to 973 K for PtC specimen (▼). Several overlapping ranges were being tested during this experiment to find the optimum for these specimens. Values within the overlapping apparent temperature range of 673 K to 973 K can be compared between the two build processes. For quantitative evaluation of the process, it would also be necessary to address the additional error in temperature estimation introduced by the limited spatial resolution of 360 µm px^{-1}. This error diminishes once the spatial temperature gradients have decreased due to lateral heat flow. These limitations are discussed in more detail by Mohr et al. [20]. Despite these limitations, qualitative analysis of the thermography data revealed results that contribute to the understanding of the presented RS results.

2.3. Lab X-ray Diffraction

A StressTech Xstress G3 X-ray diffraction instrument (Stresstech GmbH, Rennerod, Germany) was used in order to determine the RS distribution at the surface of the specimens according to the $\sin^2\psi$-method. Based on the assumption that that principal stresses are aligned with the geometrical axes of the specimens and the normal stress component can be neglected at the surface, the RS could be calculated from the slope of the linear fit of the lattice spacings over $\sin^2\psi$-plot [9,21]. The ψ-tilt was carried out in the angular range of $\psi = -45°$ to $\psi = 45°$ in 19 steps. The specimens were tilted around two perpendicular axes, to yield two perpendicular stress components. On the 36 mm surface, this corresponds to seven measurement positions of the prisms' normal and longitudinal stress component (see blue circles in Figure 2b). On the 24 mm surface, five measurement positions correspond to the normal and transversal component of the RS distribution of the prisms. The exposure time for each acquisition was 5 s. The 311 diffraction line at a corresponding 2ϑ angle of 152.26° was acquired using a Mn K_α radiation source and a 2 mm diameter collimator. Further details of the setup were described by Thiede et al. [13]. The software Xtronic (Stresstech GmbH, Rennerod, Germany) was used for data processing. The peak fitting process was performed using the Pearson VII function and the background was fitted with a parabolic function. The diffraction elastic constants (DEC) were calculated for austenitic steel 316L based on the Eshelby–Kroener model [22]. The calculated Young's modulus of $E^{311} = 184$ GPa and Poisson's ratio $\nu^{311} = 0.294$ agree with values reported by Rangaswamy et al. [23], as well as with results from measurements and simulations of the DEC values of LPBF 316L reported by Chen et al. [24].

2.4. Neutron Diffraction

Stress determination by neutron diffraction (ND) was carried out at the STRESS-SPEC diffraction instrument [25] at the neutron facility FRM II in Munich, Germany (Figure 2a).

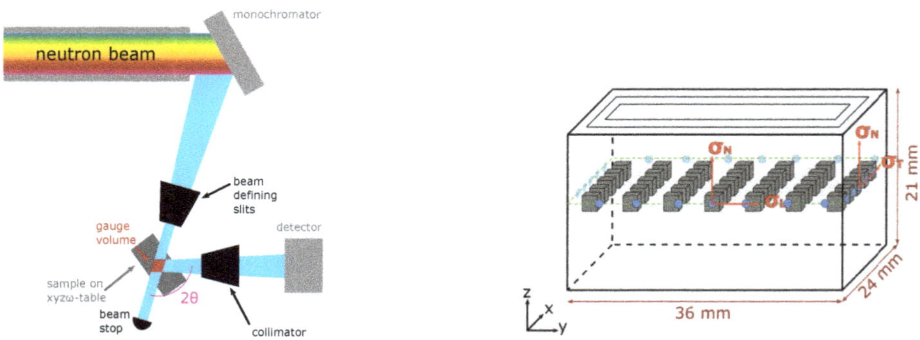

(**a**) principle sketch of the STRESS-SPEC beamline at FRM II, Munich

(**b**) map of measurement positions: blue circles (surface), grey cubes (bulk)

Figure 2. Beamline setup and measurement positions.

A bent Si_{400} single crystal monochromator was used to select the wavelength of 1.550 Å. The Fe_{311}-peak was selected in order to characterise the RS distribution due to the low accumulation of inter-granular stresses reported for this reflection in conventional face-cubic-centred (fcc) iron materials [10].

To map the RS in the cross sectional plane at the mid build height of the specimen, a gauge volume of 2 mm × 2 mm × 2 mm was used in a grid of 7 × 5 measurement points. (see grey cubes in Figure 2b). The coordinate system is depicted in Figure 2b. σ_L represents the RS along the y-direction, σ_T along the x-direction and σ_N along the z-direction, which was also the build direction.

A stress-free reference was needed to calculate strains from the measured d^{311} lattice-spacing. A small cube with the size of 3 mm × 3 mm × 3 mm was sectioned from the bottom corner of a separate test build job of the PtC specimen (▼). This cube was regarded as free of Type I macro-stresses [9], due to the mechanical relaxation during sectioning. The strain can subsequently be derived from the measured ϑ angles using Bragg's law [9].

$$\varepsilon = \frac{d^{311} - d_0}{d_0} = \frac{\sin \vartheta^{311}}{\sin \vartheta_0} - 1 \tag{1}$$

Assuming that the principal geometric directions correspond with the principal stress directions Hooke's law reads as the following, as described by Holden et al. [26]:

$$\sigma_{L,T,N} = \frac{E^{311}}{(1+\nu^{311})(1-2\nu^{311})}\left[\left(1-\nu^{311}\right)\varepsilon_{L,T,N} + \nu^{311}\left(\varepsilon_{T,N,L} + \varepsilon_{N,L,T}\right)\right] \tag{2}$$

The d_0 value was derived from the average of the ϑ^{311} measurements of the cube in longitudinal (L), transversal (T) and normal (N) direction, where the normal direction corresponds to the build direction (as depicted in Figure 2b). The same DECs that were derived from the Eshelby–Kroener model were applied to both Lab XRD and ND results (see Section 2.3).

2.5. Micro Computed Tomography

The small reference cube for ND was studied using Micro Computed Tomography (µCT) to obtain a detailed dataset of the internal defect structure. The µCT measurements were performed at a GE v|tome| × 180/300 CT scanner (GE Sensing & Inspection Technologies GmbH, Wunstorf, Germany) using the 180 kV source at a voltage of 150 kV and a current of 40 µA without any metal pre-filter. A voxel size of $(3\,\mu m)^3$ was achieved. The analysis of the data was performed using the commercial software VG Studio MAX version 3.2.1 (Volume Graphics GmbH, Heidelberg, Germany). A lower threshold limit of 8 voxels was used for pore detection.

2.6. Optical Microscopy

For Optical Microscopy (OM) investigations of the microstructure, the bottom faces of the samples were ground, polished, and etched. Emery papers with 180, 320, 600 and 1200 grits followed by clothes with 3 µm and 1 µm were used. For etching the Bloech & Wedl II method [27] (a solution of 50 mL H_2O, 50 mL HCl, 0.6 g $K_2S_2O_5$) was applied. The microstructure was captured using a Olympus BX53M microscope with a DP74 camera module (Olympus Corporation, Tokyo, Japan). The analysis was performed using the software Olympus Stream Essentials (Olympus Corporation, Tokyo, Japan).

3. Results

3.1. In-Situ Thermography

The build jobs of both samples were supervised in-situ by thermography in order to receive more detailed information on the local variation of the temperature gradient and cooling rates. Figure 3a,b display the maximum apparent temperatures at the mid-height layer.

The thermography data were averaged over 40 layers at the mid build-height to reduce noise and the influence of smoulder and spatter. In relation to the height of the specimens of 21 mm, these 40 layers represent an average of the height range from 9.35 mm to 11.35 mm. The number of 40 layers (=2 mm) was chosen, since this corresponds to the size of the used gauge volume for ND (2 mm × 2 mm × 2 mm).

The scan strategy of the PtC specimen (▼) resulted in an increased heat accumulation in the center, compared to the CtP specimen (▲). This could be observed as an increase in the maximum temperature at the centre, when comparing Figure 3d,c.

Figure 3d shows the different apparent temperature values for the four sections of the plane. These differences were caused by a combination of the different surface roughness of each section and of the shadowing effects from the smoulder.

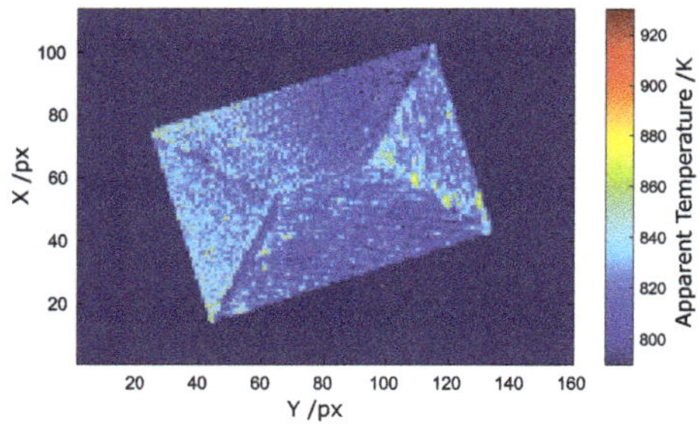

(**a**) ▲ CtP specimen's mid-height layer

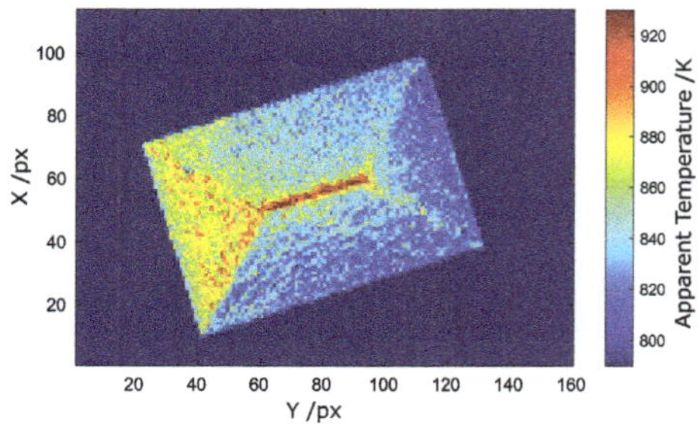

(**b**) ▼ PtC specimen's mid-height layer

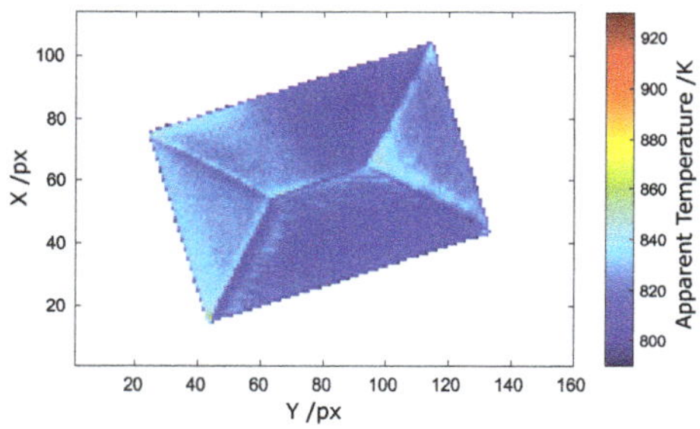

(**c**) ▲ CtP specimen: 40 layers at mid-height averaged

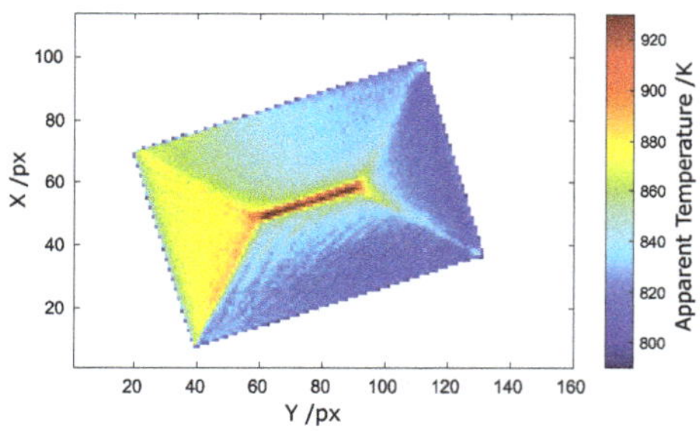

(**d**) ▼ PtC specimen: 40 layers at mid-height averaged

Figure 3. Apparent (uncalibrated) maximum temperature of the two specimens obtained from thermography data acquired during the build process at the sample's mid build height. Each border fill scan started and ended at the bottom right hand laser turn position of the images. (**a**,**b**) display the mid-height layer, (**c**,**d**) display an average of 40 layers at the mid-height.

Figure 4 shows the cooling rate of both specimens. The cooling rate—$\mathrm{d}T/\mathrm{d}t$ was obtained by comparing two images at $t = 1$ ms and $t = 2$ ms after an apparent temperature of 700 K was reached for the last time at the surface. Typical times for cooling from the maximum temperature to 700 K were between 2 ms and up to 15 ms. The cooling time of 15 ms was observed at the centre of the PtC specimen (▼).

As displayed in Figure 4d, the PtC specimen (▼) also featured a lower cooling rate in the centre of the plane in addition to the higher maximum temperature depicted in Figure 3d. The CtP specimen (▲) shows a low cooling rate at the edges of the specimen indicating that the surrounding metal powder served as an heat insulator for conduction, as assumed in modelling [28].

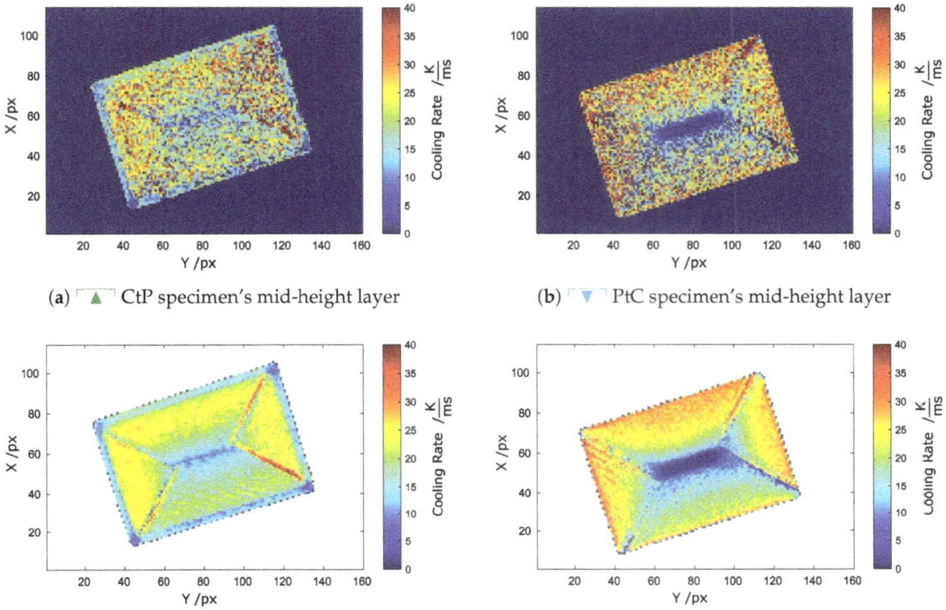

(**a**) CtP specimen's mid-height layer

(**b**) PtC specimen's mid-height layer

(**c**) CtP specimen: 40 layers at mid-height averaged

(**d**) PtC specimen: 40 layers at mid-height averaged

Figure 4. Cooling rates measured by the different between images taken at $t = 1\,\text{ms}$ and $t = 2\,\text{ms}$ after an apparent temperature of 700 K was observed for the last time. Thermography data for both samples was obtained during the build process at the middle of the total build height. (**a**,**b**) display the mid-height layer, (**c**,**d**) display an average of 40 layers around the middle of the total build height.

3.2. Combined Neutron and Lab X-ray Diffraction

The ND results (bulk RS) were combined with Lab XRD results (surface RS) to show the complete stress distribution across the full cross section of the specimens' middle plane. The contour plot function of the commercial software Origin 2018 (OriginLab Corporation, Northampton, USA), which is based on the Delaunay triangulation, was used to visualise the combined results. Figure 5 shows the combination of ND and XRD results. The RS distribution is visualised in longitudinal, transversal, and normal direction, where the normal direction corresponds to the build direction (Figure 2b). The XRD technique used only allows for the determination of stress components which are parallel to the surface (i.e., in-plane). Therefore, to visualise the third orthogonal stress component (perpendicular to the surface) in Figure 5 the following boundary condition was used: at the surfaces of the specimens corresponding to positions at $y = 0\,\text{mm}$ and $y = 36\,\text{mm}$, the value of the longitudinal stress component was assumed to $\sigma_L = 0\,\text{MPa}$. For the value of the transversal stress component at $x = 0\,\text{mm}$ and $x = 24\,\text{mm}$ it was assumed $\sigma_T = 0\,\text{MPa}$, since these are free surfaces in these corresponding stress directions. In general, for the two scan strategies, a similar RS distribution was observed in the longitudinal, transversal and normal direction. This distribution is characterised by compressive RS within the bulk, balanced by tensile RS at the surface. Instead, the RS distribution in the longitudinal and transversal direction of each specimen is similar in terms of shape and magnitude, the RS distribution in the normal (i.e., building) direction differed from the other directions in terms of shape and magnitude.

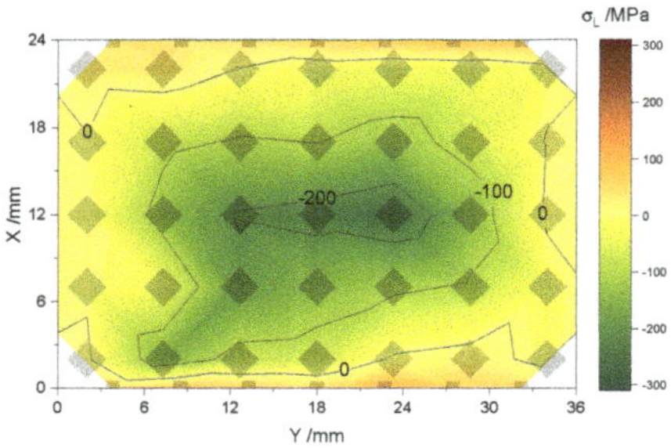

(**a**) ▲ Longitudinal residual stresses (RS) component for CtP border fill strategy

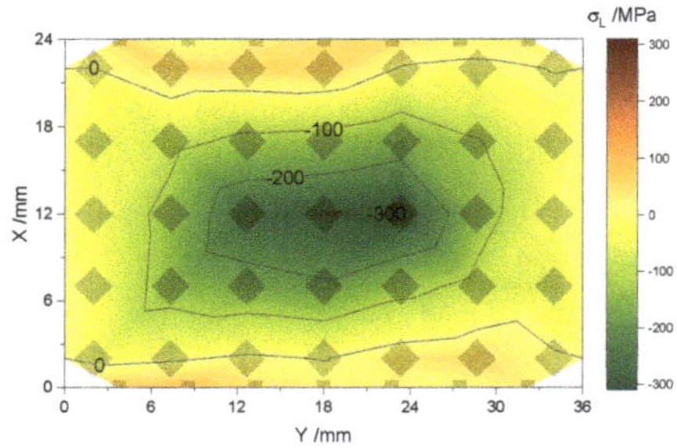

(**b**) ▼ Longitudinal RS component for PtC border fill strategy

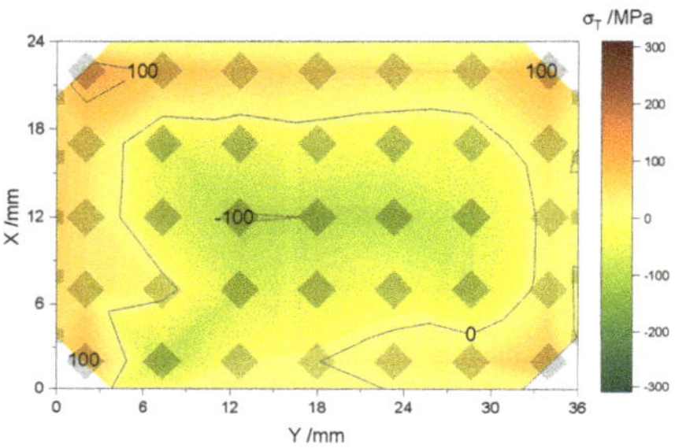

(**c**) ▲ Transversal RS component for CtP border fill strategy

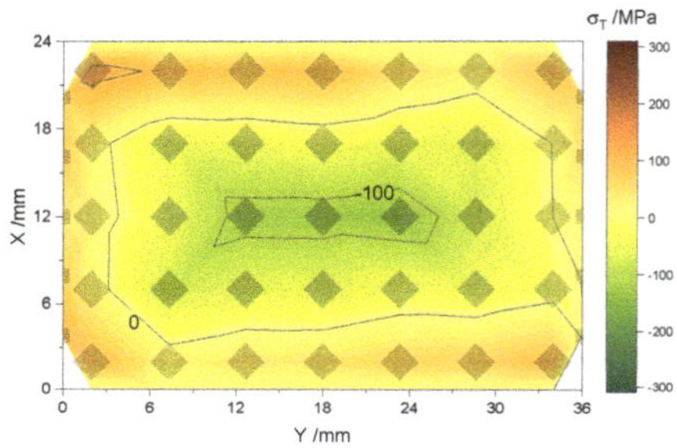

(**d**) ▼ Transversal RS component for PtC border fill strategy

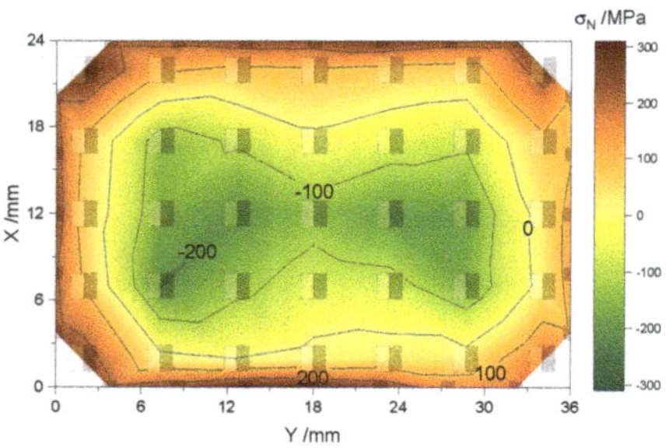

(**e**) ▲ Normal RS component for CtP border fill strategy

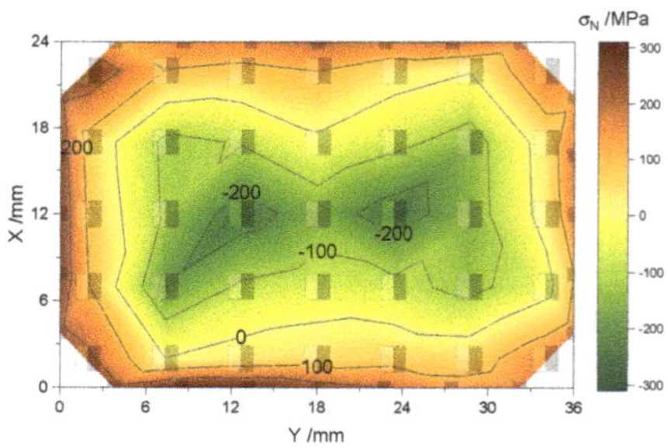

(**f**) ▼ Normal RS component for PtC border fill strategy

Figure 5. Comparison of RS maps of the tw scan strategies including results from lab X-ray diffraction at the surface (The big semi-translucent squares in the bulk represent the almost cubic ND gauge volume (orientated differently for different stress components), whereas the small semi-translucent squares at the edges represent the lab X-ray measurement positions).

The highest magnitude of RS of each specimen (maximum tensile or maximum compressive) were observed in the normal direction. However, the PtC specimen (▼) displayed higher bulk compressive stresses in all three orthogonal directions. The different stress values in the normal direction, as compared to the longitudinal and transversal direction, are also reflected in the lattice spacing of the the d_0 cube. Whereas, the *d*-spacing in the longitudinal and transversal direction are relatively similar to each other, in the normal direction a larger lattice spacing was measured (see

Table 1). The d_0 value was obtained from averaging the three measured directions. This averaged value was used as a stress-free lattice parameter to calculate the RS.

Table 1. Distribution of orthogonal *d*-spacing values of the reference cube.

Orientation	*d*-Spacing	Error
Longitudinal	1.074 49 Å	2.36×10^{-5} Å
Transversal	1.074 48 Å	2.71×10^{-5} Å
Normal	1.074 93 Å	2.41×10^{-5} Å

3.3. Micro Computed Tomography

The reference cube was analysed by μCT to provide an example of local defect distribution in the specimen. Because it was cut from the corner where the laser path started and ended, it represented the area where the highest amount of defects was expected. The μCT results presented in Figure 6 reveal a network of defects at the location where the laser started and ended, as well as between the hatches. Because the same scan vector was used on each hatch and layer, the projection of defects onto one plane (Figure 6c) reveals the lack of fusion between neighbouring hatches. As reported in literature [29,30] alternating the orientation of scan vectors between layers prevents the formation of lack of fusion defects. Because the effect of shrinkage on RS was the subject of this study, scan vectors were not altered between layers to magnify such effect.

The largest defects were situated close to the edge of the sample (Figure 6b,c). A total porosity of 0.28% was observed.

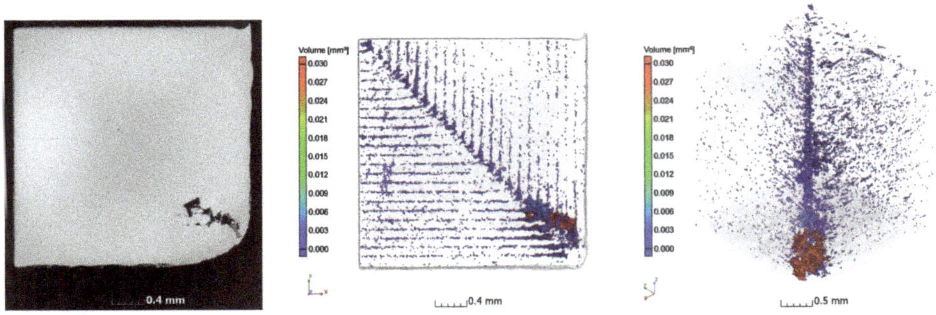

(**a**) slice of the cube perpendicular to the build direction

(**b**) 2D projection of the 3D rendered of the segmented defects within the cube perpendicular the build direction

(**c**) 3D rendering of the segmented defects

Figure 6. μCT reconstructions of the ND reference cube sectioned from a twin PtC specimen (▼).

3.4. Optical Microscopy

Optical microscopy images of the polished and etched specimen's bottom surface are shown in Figure 7. Defects at the turn location of the laser are visible (Figure 7d,e). The bottom surface of both specimens are expected to be less effected by heat accumulation due to the smaller build height at the time the microstructure *froze*, and a better heat conduction into the build plate as compared to the top layers of the specimens. Nonetheless, defects (pores and voids) were observed in both specimens at the positions, where the laser turns by 90°.

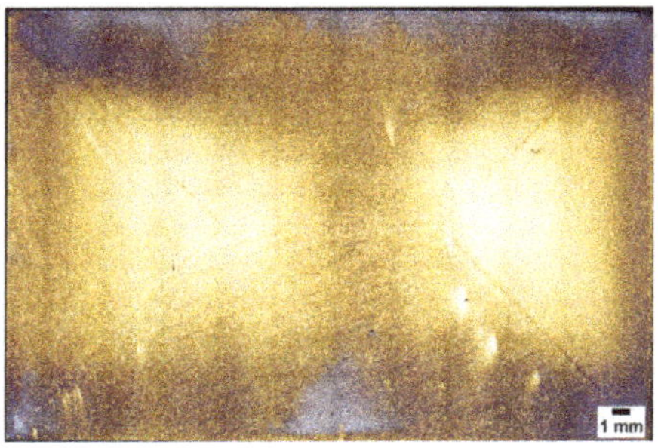

(**a**) full sized image of CtP specimen

(**b**) full sized image of PtC specimen

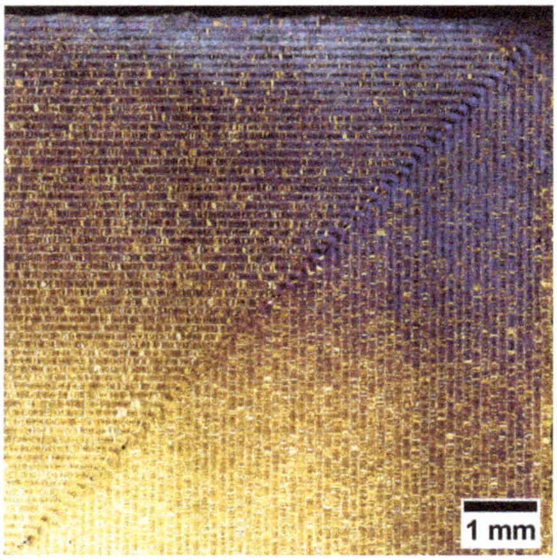

(**c**) magnified upper right section of CtP specimen

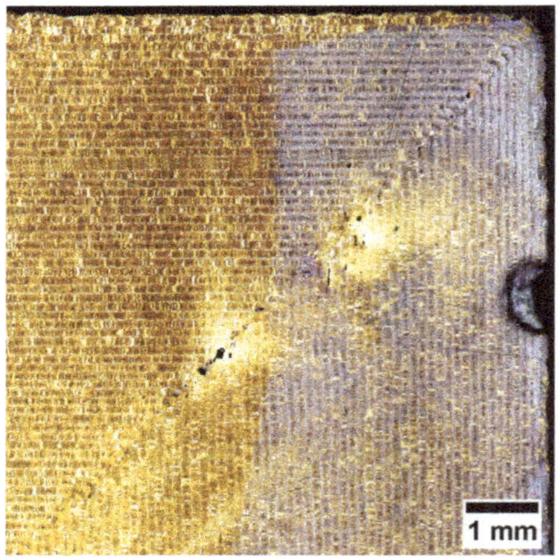

(**d**) magnified upper right section section of PtC specimen

(**e**) magnified section of the centre of CtP specimen

(**f**) magnified section of the centre of PtC specimen

Figure 7. Optical microscopy images of the samples' bottom surface after polishing and etching.

4. Discussion

As mentioned in the introduction, Mercelis and Kruth [3] described two major driving mechanisms for the formation of RS in AM metallic parts made by LPBF: the temperature gradient mechanism (TGM) and the solidification shrinkage mechanism (SSM). Both of the mechanisms induced compressive stresses into the bulk material. The combined data of ND and Lab XRD showed the typical stress distribution pattern for metallic AM parts: compressive stresses in the bulk and tensile stresses close to the surface.

Because of the same geometric dimensions, it can be assumed that the solidification shrinkage effects in the two samples were equal. This implies that any differences in the RS distributions of the two samples should be attributed to the TGM.

According to the SSM, the induced RS distribution by solidification shrinkage depends on the length of the scanned laser tracks. The shrinkage of longer laser tracks introduces higher RS than shorter laser tracks [31]. Therefore, the longer laser tracks parallel to the *y*-axes of the specimens presented in Figure 5a–d) appear to be the main reason for the higher compressive stresses in the longitudinal stress distributions of the two specimens, as compared to the shorter laser tracks parallel to the *x*-axes, which seemed to result in lower compressive stresses for the transversal RS distributions of the two specimens.

The distribution of the normal component of the RS was assumed to be mostly independent of the SSM, since there were almost no restrictions to solidification shrinkage due to the free top surface during the layer-wise production. Therefore, the TGM was assumed to be the main mechanism to shape the RS distribution of the normal component. The RS distribution has the shape of a *butterfly* (see Figure 5e,f). The spikes of this *butterfly* pattern match the location where the laser turned by 90° (see Figure 8).

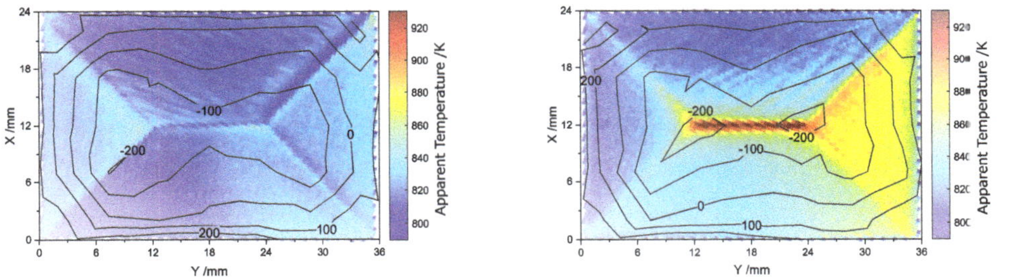

(a) (▲) Normal stress component (Figure 5e) over thermography results (Figure 3c)

(b) (▼) Normal stress component (Figure 5f) over thermography results (Figure 3d)

Figure 8. Normal (i.e., build direction) stress component (contour lines) in MPa (Figure 5e,f) overlayed with thermography results (Figure 3c,d) to highlight the compressive stresses at the laser's turn locations.

It is also observed that the normal component exhibits the highest tensile stresses (see Figure 5 and Table 2). The determined RS values are below the tensile yield strength ranges for LPBF 316L (450 MPa to 590 MPa) as reported by Wang et al. [32].

Table 2. Max. and min. values of the orthogonal stress components in the plane.

Stress Component	Centre to Perimeter (▲)		Perimeter to Centre (▼)	
	Max. [MPa]	Min. [MPa]	Max. [MPa]	Min. [MPa]
σ_L	100.8 ± 16.9	-253.4 ± 15.8	86.4 ± 10.2	-304.2 ± 15.6
σ_T	140.6 ± 10.4	-103.6 ± 15.4	122.6 ± 11.0	-134.0 ± 15.8
σ_N	295.6 ± 10.5	-202.1 ± 14.2	290.5 ± 10.6	-245.8 ± 19.5

Figure 8b shows that higher compressive stresses in the PtC specimen (▼) are localised close to the zone of the highest heat accumulation. This is in contrast to the CtP specimen (▲), which shows lower and more evenly distributed compressive stresses in the plane (see Figure 5). The higher maximum temperatures at the centre of the PtC specimen (▼) in combination with the slower cooling rate seem to result in larger RS in the centre of the plane as compared to the CtP specimen (▲). Lower contributions from the TGM to the compressive stress profile of the CtP specimen (▲) might be a result of the lack of heat accumulation in the centre of its plane (see Figures 3c and 8a) and a faster cooling rate (see Figure 4c).

Line stress profiles (Figure 9) that were derived from the in-plane data presented in Figure 5 at a middle line of the plane at $x = 12$ mm emphasise the conclusion of SSM being the major mechanism to define the *shape* of the RS distribution. However, the TGM appears to influence the *magnitude* of the peak compressive stresses (see Figure 9).

It should be noted that, in Figure 9, only measured values are displayed. Therefore, the boundary condition values of 0 MPa were excluded from the longitudinal stress profile at $y = 0$ and $y = 36$ (see Figure 9a). The measured bulk values close to these surfaces support the assumption of zero stress at these surfaces.

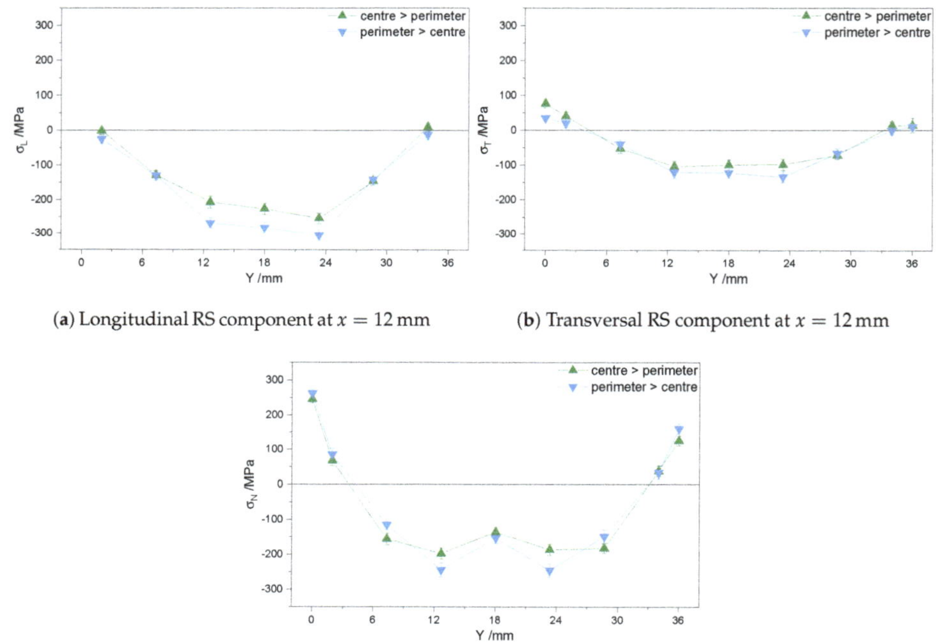

(**a**) Longitudinal RS component at $x = 12$ mm

(**b**) Transversal RS component at $x = 12$ mm

(**c**) Normal RS component (=build direction) at $x = 12$ mm

Figure 9. Line scans in both samples for all three orthogonal directions using the combined data from ND and lab XRD.

Since the TGM does not require melting [33], it can be assumed that the compressive stress components induced by the TGM were formed in subjacent layers within the heat affected zone of the melt pool but below the layers that were remolten. It should be noted that a linear interpolation was applied to combine the surface measurement positions and the neutron data in Figure 9. Therefore, any possible sub-surface tensile peaks that had been reported by Mishurova et al. [11,34] in LPBF Ti-6Al-4V were not be taken into account in this study.

The μCT results (Figure 6) reveal a network of defects at the laser's start and stop position. OM results (Figure 7d,e) show defects at the turn locations of the scan track. These regions have been measured to be under higher tensile RS (in particular, σ_T), which is assumed to arise from the longitudinal and transversal solidification shrinkage, as depicted in Figure 5. Therefore, any defects in these regions might have served as micro-crack initiators. The combination of pores and tensile stress at corners could also explain why the defects observed by μCT (Figure 6b) seem to be larger towards the outer edge: the tensile stress due to solidification shrinkage increases towards the edges of the specimens due to an increase length of the laser tracks. The analysis of the optical microscopy and μCT data indicates that an unknown amount of RS might had been relaxed by micro-cracks at the laser's start and stop location.

5. Conclusions

Two prismatic AISI316L specimens using a border fill scan strategy were produced in order to differentiate the effect of the temperature gradient mechanism from the solidification shrinkage mechanism in AM metallic parts produced by LPBF. The following conclusions could be made:

1. A combination of surface and bulk residual stress results was needed to fully cover the (surface) tensile and (bulk) compressive regions of the in-plane stress distribution.
2. By comparing the two samples, a similar stress distribution was revealed for each of the three orthogonal directions (see Figures 5 and 9), indicating that the solidification shrinkage mechanism is the main mechanism controlling the shape of the RS distribution in these samples. The temperature gradient mechanism seems to influence the magnitude of the compressive stresses without changing the overall pattern of the stress distribution.
3. In-situ thermography results of the sample exposed from the perimeter to the centre revealed a heat accumulation, which corresponds to highly localised compressive stresses due to the temperature gradient mechanism.

6. Outlook

In-situ Thermography recorded a detailed data set of apparent surface temperatures and enables an analysis of cooling rates. These data can be used in future simulations in order to model the RS field at the specimens' mid height. In future studies, the RS distribution in the subsurface region might be resolved using additional techniques, such as hole drilling, slitting, X-ray with layer removal, deep hole drilling, or contour method.

Additionally, twin specimens of the two prisms will allow further experiments to study the effect on maximum values of the RS distribution, if the reference cube is cut from different positions within these twin specimens. In addition, these twins will allow a systematic μCT analysis of different positions within the two specimens, in particular to relate RS and defect distributions.

Author Contributions: Conceptualization, A.U., T.F., G.M. and S.J.A.; methodology, A.U., G.M., M.S., K.S., S.J.A., T.M., I.S.-M., T.F., M.H.; formal analysis, A.U., S.J.A., M.S., K.S.; investigation, A.U., M.S., K.S., S.J.A., T.M., I.S.-M., T.F., M.H.; writing—original draft preparation, A.U., G.M., M.S., K.S., S.J.A.; writing—review and editing, A.U., G.M., M.S., S.J.A., T.M., I.S.-M., M.H., A.E., G.B., T.F.; visualization, A.U., S.J.A.; supervision, A.E., G.B.; project administration, A.E., G.B.; funding acquisition, G.B. All authors have read and agreed to the published version of the manuscript.

Funding: This research received no external funding.

Acknowledgments: This work has been funded by the BAM Focus Area Materials project AGIL "Microstructure Development in Additively Manufactured Metallic Components: from Powder to Mechanical Failure" and ProMoAM "Process monitoring of Additive Manufacturing". We are thankful for the financial support and the fruitful cooperation with all partners. This work is based upon experiments performed at the STRESS-SPEC instrument operated by FRM II at the Heinz Maier-Leibnitz Zentrum (MLZ), Garching, Germany.

Conflicts of Interest: The authors declare no conflict of interest. The funders had no role in the design of the study; in the collection, analyses, or interpretation of data; and in the decision to publish the results.

Abbreviations

The following abbreviations are used in this manuscript:

AM	Additive Manufacturing
FRM II	Forschungs-Neutronenquelle Heinz Maier-Leibnitz (Research Reactor Munich II)
μCT	Micro Computed Tomography
ND	Neutron Diffraction
LPBF	Laser Powder Bed Fusion
OM	Optical Microscopy
RS	Residual Stress
SSM	Solidification Shrinkage Mechanism
TGM	Temperature Gradient Mechanism
VED	Volumetric Energy Density
XRD	(Lab) X-Ray Diffraction

References

1. Ngo, T.D.; Kashani, A.; Imbalzano, G.; Nguyen, K.T.; Hui, D. Additive manufacturing (3D printing): A review of materials, methods, applications and challenges. *Compos. Part B Eng.* **2018**, *143*, 172–196. [CrossRef]
2. Frazier, W.E. Metal Additive Manufacturing: A Review. *J. Mater. Eng. Perform.* **2014**, *23*, 1917–1928. [CrossRef]
3. Mercelis, P.; Kruth, J.P. Residual stresses in selective laser sintering and selective laser melting. *Rapid Prototyp. J.* **2006**, *12*, 254–265. [CrossRef]
4. Li, R.; Shi, Y.; Wang, Z.; Wang, L.; Liu, J.; Jiang, W. Densification behavior of gas and water atomized 316L stainless steel powder during selective laser melting. *Appl. Surf. Sci.* **2010**, *256*, 4350–4356. [CrossRef]
5. Miranda, G.; Faria, S.; Bartolomeu, F.; Pinto, E.; Madeira, S.; Mateus, A.; Carreira, P.; Alves, N.; Silva, F.; Carvalho, O. Predictive models for physical and mechanical properties of 316L stainless steel produced by selective laser melting. *Mater. Sci. Eng. A* **2016**, *657*, 43–56. [CrossRef]
6. Wang, D.; Wu, S.; Yang, Y.; Dou, W.; Deng, S.; Wang, Z.; Li, S. The Effect of a Scanning Strategy on the Residual Stress of 316L Steel Parts Fabricated by Selective Laser Melting (SLM). *Materials* **2018**, *11*, 1821. [CrossRef]
7. Roehling, J.D.; Smith, W.L.; Roehling, T.T.; Vrancken, B.; Guss, G.M.; McKeown, J.T.; Hill, M.R.; Matthews, M.J. Reducing residual stress by selective large-area diode surface heating during laser powder bed fusion additive manufacturing. *Addit. Manuf.* **2019**, *28*, 228–235. [CrossRef]
8. Bacon, G. *Neutron Diffraction*; Clarendon Press: Oxford, UK, 1975. [CrossRef]
9. Withers, P.; Bhadeshia, H. Residual stress. Part 1—Measurement techniques. *Mater. Sci. Technol.* **2001**, *17*, 355–365. [CrossRef]
10. Hutchings, M.; Withers, P.; Holden, T.; Lorentzen, T. *Introduction to the Characterization of Residual Stress by Neutron Diffraction*; CRC Press: Boca Raton, FL, USA, 2005. [CrossRef]
11. Mishurova, T.; Cabeza, S.; Artzt, K.; Haubrich, J.; Klaus, M.; Genzel, C.; Requena, G.; Bruno, G. An Assessment of Subsurface Residual Stress Analysis in SLM Ti-6Al-4V. *Materials* **2017**, *10*, 348. [CrossRef]
12. Nadammal, N.; Cabeza, S.; Mishurova, T.; Thiede, T.; Kromm, A.; Seyfert, C.; Farahbod, L.; Haberland, C.; Schneider, J.A.; Portella, P.D.; et al. Effect of hatch length on the development of microstructure, texture and residual stresses in selective laser melted superalloy Inconel 718. *Mater. Des.* **2017**, *134*, 139–150. [CrossRef]
13. Thiede, T.; Cabeza, S.; Mishurova, T.; Nadammal, N.; Kromm, A.; Bode, J.; Haberland, C.; Bruno, G. Residual Stress in Selective Laser Melted Inconel 718: Influence of the Removal from Base Plate and Deposition Hatch Length. *Mater. Perform. Charact.* **2018**, *7*, 717–735. [CrossRef]
14. Wu, A.S.; Brown, D.W.; Kumar, M.; Gallegos, G.F.; King, W.E. An Experimental Investigation into Additive Manufacturing-Induced Residual Stresses in 316L Stainless Steel. *Metall. Mater. Trans. A* **2014**, *45*, 6260–6270. [CrossRef]
15. Riemer, A.; Leuders, S.; Thoene, M.; Richard, H.; Troester, T.; Niendorf, T. On the fatigue crack growth behavior in 316L stainless steel manufactured by selective laser melting. *Eng. Fract. Mech.* **2014**, *120*, 15–25. [CrossRef]

16. Vrancken, B.; Wauthlé, R.; Kruth, J.P.; Van Humbeeck, J. Study of the influence of material properties on residual stress in selective laser melting. In Proceedings of the Solid Freeform Fabrication Symposium, Austin, TX, USA, 12–14 August 2013; pp. 393–407.
17. Wang, D.; Yang, Y.; Yi, Z.; Su, X. Research on the fabricating quality optimization of the overhanging surface in SLM process. *Int. J. Adv. Manuf. Technol.* **2013**, *65*, 1471–1484. [CrossRef]
18. Mohr, G.; Altenburg, S.J.; Hilgenberg, K. Effects of inter layer time and build height on resulting properties of 316L stainless steel processed by laser powder bed fusion. *Addit. Manuf.* **2020**, *32*, 101080. [CrossRef]
19. Hunnewell, T.S.; Walton, K.L.; Sharma, S.; Ghosh, T.K.; Tompson, R.V.; Viswanath, D.S.; Loyalka, S.K. Total Hemispherical Emissivity of SS 316L with Simulated Very High Temperature Reactor Surface Conditions. *Nucl. Technol.* **2017**, *198*, 293–305. [CrossRef]
20. Mohr, G.; Altenburg, S.J.; Ulbricht, A.; Heinrich, P.; Baum, D.; Maierhofer, C.; Hilgenberg, K. In-Situ Defect Detection in Laser Powder Bed Fusion by Using Thermography and Optical Tomography—Comparison to Computed Tomography. *Metals* **2020**, *10*, 103. [CrossRef]
21. Macherauch, E.; Mueller, P. Das sin2Psi-Verfahren der roentgenographischen Spannungsmessung. *Z. Angew. Phys.* **1961**, *13*, 305–312.
22. Kroener, E. Berechnung der elastischen Konstanten des Vielkristalls aus den Konstanten des Einkristalls. *Z. Phys.* **1958**, *151*, 504–518. [CrossRef]
23. Rangaswamy, P.; Griffith, M.; Prime, M.; Holden, T.; Rogge, R.; Edwards, J.; Sebring, R. Residual stresses in LENS (R) components using neutron diffraction and contour method. *Mater. Sci. Eng. A* **2005**, *399*, 72–83. [CrossRef]
24. Chen, W.; Voisin, T.; Zhang, Y.; Florien, J.B.; Spadaccini, C.M.; McDowell, D.L.; Zhu, T.; Wang, Y.M. Microscale residual stresses in additively manufactured stainless steel. *Nat. Commun.* **2019**, *10*, 4338. [CrossRef] [PubMed]
25. Hofmann, M.; Gan, W.; Rebelo-Kornmeier, J. STRESS-SPEC: Materials science diffractometer. *J. Large-Scale Res. Facil. JLSRF* **2015**, *1*, 6. [CrossRef]
26. Holden, T.M.; Clarke, A.P.; Holt, R.A. Neutron diffraction measurements of intergranular strains in MONEL-400. *Metall. Mater. Trans. A* **1997**, *28*, 2565–2576. [CrossRef]
27. Weck, E.; Leistner, E. *Metallographische Anleitung zum Farbätzen nach dem Tauchverfahren.—3: Nichteisenmetalle, Hartmetalle und Eisenwerkstoffe, Nickel-Basis- und Kobalt-Basis-Legierungen = Metallographic Instructions for Colour Etching by Immersion.—3: Non-Ferrous Metals, Cemented Carbides and Ferrous Metals, Nickel-Base and Cobalt-Base Alloys*; Deutscher Verlag für Schweisstechnik: Duesseldorf, Germany, 1986.
28. Denlinger, E.R.; Gouge, M.; Irwin, J.; Michaleris, P. Thermomechanical model development and in situ experimental validation of the Laser Powder-Bed Fusion process. *Addit. Manuf.* **2017**, *16*, 73–80. [CrossRef]
29. Thijs, L.; Verhaeghe, F.; Craeghs, T.; Van Humbeeck, J.; Kruth, J.P. A study of the microstructural evolution during selective laser melting of Ti–6Al–4V. *Acta Mater.* **2010**, *58*, 3303–3312. [CrossRef]
30. Yadroitsev, I.; Thivillon, L.; Bertrand, P.; Smurov, I. Strategy of manufacturing components with designed internal structure by selective laser melting of metallic powder. *Appl. Surf. Sci.* **2007**, *254*, 980–983. [CrossRef]
31. Bo, Q.; Yu-sheng, S.; Qing-song, W.; Hai-bo, W. The helix scan strategy applied to the selective laser melting. *Int. J. Adv. Manuf. Technol.* **2012**, *63*, 631–640. [CrossRef]
32. Wang, Y.M.; Voisin, T.; McKeown, J.T.; Ye, J.; Calta, N.P.; Li, Z.; Zeng, Z.; Zhang, Y.; Chen, W.; Roehling, T.T.; et al. Additively manufactured hierarchical stainless steels with high strength and ductility. *Nat. Mater.* **2018**, *17*, 63–71. [CrossRef]
33. Kruth, J.; Froyen, L.; Van Vaerenbergh, J.; Mercelis, P.; Rombouts, M.; Lauwers, B. Selective laser melting of iron-based powder. *J. Mater. Process. Technol.* **2004**, *149*, 616–622. [CrossRef]
34. Mishurova, T.; Artzt, K.; Haubrich, J.; Requena, G.; Bruno, G. Exploring the Correlation between Subsurface Residual Stresses and Manufacturing Parameters in Laser Powder Bed Fused Ti-6Al-4V. *Metals* **2019**, *9*, 261. [CrossRef]

metals

MDPI

Article

Residual Stress and Microstructure of a Ti-6Al-4V Wire Arc Additive Manufacturing Hybrid Demonstrator

Tatiana Mishurova [1,*], Benjamin Sydow [2], Tobias Thiede [1], Irina Sizova [2], Alexander Ulbricht [1], Markus Bambach [2,3] and Giovanni Bruno [1,4]

1 Bundesanstalt für Materialforschung und–prüfung (Institute for Materials Research and Testing (BAM), 12205 Berlin, Germany; tobias.thiede@bam.de (T.T.); Alexander.Ulbricht@bam.de (A.U.); giovanni.bruno@bam.de (G.B.)

2 Chair of Mechanical Design and Manufacturing, Brandenburg University of Technology Cottbus-Senftenberg, D-03046 Cottbus, Germany; benjamin.sydow@b-tu.de (B.S.); sizova@b-tu.de (I.S.); bambach@b-tu.de (M.B.)

3 Advanced Manufacturing, Department of Mechanical and Process Engineering, ETH Zurich, 8092 Zurich, Switzerland

4 Institute of Physics and Astronomy, University of Potsdam, 14476 Potsdam, Germany

* Correspondence: tatiana.mishurova@bam.de; Tel.: +49-30-810-445-27

Received: 9 April 2020; Accepted: 23 May 2020; Published: 26 May 2020

Abstract: Wire Arc Additive Manufacturing (WAAM) features high deposition rates and, thus, allows production of large components that are relevant for aerospace applications. However, a lot of aerospace parts are currently produced by forging or machining alone to ensure fast production and to obtain good mechanical properties; the use of these conventional process routes causes high tooling and material costs. A hybrid approach (a combination of forging and WAAM) allows making production more efficient. In this fashion, further structural or functional features can be built in any direction without using additional tools for every part. By using a combination of forging basic geometries with one tool set and adding the functional features by means of WAAM, the tool costs and material waste can be reduced compared to either completely forged or machined parts. One of the factors influencing the structural integrity of additively manufactured parts are (high) residual stresses, generated during the build process. In this study, the triaxial residual stress profiles in a hybrid WAAM part are reported, as determined by neutron diffraction. The analysis is complemented by microstructural investigations, showing a gradient of microstructure (shape and size of grains) along the part height. The highest residual stresses were found in the transition zone (between WAAM and forged part). The total stress range showed to be lower than expected for WAAM components. This could be explained by the thermal history of the component.

Keywords: residual stress; WAAM; Ti-6Al-4V; additive manufacturing; neutron diffraction; hybrid manufacturing

1. Introduction

The titanium alloy Ti-6Al-4V is intensively used in lightweight applications for the aviation and space industry because of its high specific strength [1]. Hot forging is the usual manufacturing route; this allows better material formation and control of the microstructure. Subsequent machining is used to attain the desired tolerances regarding dimensions, shape, and surface condition [2]. One positive effect of hot forging is that the microstructure and the fiber flow can be optimized, which results in excellent mechanical properties. However, forging is limited as far as the achievable geometries is concerned: as an example, due to friction and heat transfer, the material cannot flow into cavities of the forging die with a large depth-to-diameter ratio. Moreover, multistage hot forging operations require expensive heated die sets and lead to parts with large tolerances; this requires further machining to create the final shape. In fact, the "buy-to-fly" ratio (the mass ratio between the originally purchased stock material used to produce a part and the mass of the final finished part) ranges from 12:1 to 25:1 for aircraft titanium components made by traditional manufacturing techniques [3]. This means that 12–25 kg of raw material are required to produce just 1 kg of parts. In this way, more than 90% of the material is machined away. Such a high buy-to-fly ratio is unacceptable from many points of view. For a typical titanium component produced by additive manufacturing (AM) techniques the buy-to-fly ratio drops to 3–12:1 [4] and can even be close to 1:1 [5]. By using AM for production of aerospace components not only costs for expensive titanium alloys can be reduced, but also the machining time can be drastically cut.

AM is an innovative manufacturing technology that allows manufacturing almost arbitrarily complex shapes. In powder-bed metal AM processes, the powder is distributed layer-wise on a substrate or already deposited material and fused by melting the material using an electron (electron beam melting–EB-PBF) or laser beam (laser powder bed fusion–L-PBF). Because of the small (focused) beam, the deposition rates are quite low. PBF AM processes have in general a small layer thickness of a few tens of μm up to 2 mm [6]. This is necessary due to the local bonding of adjacent powder particles and the limited beam penetration depth. Although beam scanning systems can achieve high laser or electron beam scanning speeds, large parts require a high number of layers, which causes high production times. The part size is limited due to the small process chamber (typically 200–300 mm edge size). Also, as the powder-bed has to be established before the AM process begins, no existing parts can be extended by means of PBF. This makes this technology only attractive for small parts. An increasingly popular AM technology that may allow overcoming this issue is Wire Arc Additive Manufacturing (WAAM). WAAM utilizes common and well-known welding technologies. A wire is fed to the desired position and molten by an electric arc. The melt pool solidifies and forms a weld bead. One advantage of WAAM compared to other AM technologies is the high deposition rate. This is especially important for large scale products such as landing gear parts or wing ribs, which are difficult to forge due to their size or have a large material waste when machined. WAAM achieves deposition rates as high as several kg/h [7] Also, a wide range of materials is available as wires, which offers multiple possible applications. The microstructure and mechanical performance of WAAM manufactured Ti-6Al-4V parts will not reach the level of forged material without additional (often expensive) post-processing techniques such as hot isostatic pressing (HIP) or thermomechanical processing [8]. One strategy is to combine the benefits of the two technologies (forging and WAAM) to produce so-called hybrid parts with the desired mechanical properties (even at local scale). This approach is of particular interest for the production of the components, where the forged sections are heavily loaded during service and the added AM features should be capable to withstand only moderate loads. Moreover, in aerospace industry many parts possess a symmetry axis. To produce such components by means of conventional processing a number of different forging die sets should be used. Instead, the hybrid approach can be more efficient, since the required structural or functional features can be added to the forged part without using additional tools for every part design. In so doing, both the cost of forging tools and material waste can be reduced. Figure 1 represents a possible hybrid manufacturing route, where the basic structure can be produced by hot forming and the missed structural features can be added using AM.

Basically, two process routes are possible: (i) AM and subsequent forming or (ii) forming and subsequent AM. The first hybrid processing route was extensively studied by authors of the present work in regard to Ti-6Al-4V [9–11]. It was observed that Ti-6Al-4V pre-forms for forging made either by L-PBF or by WAAM show a good hot workability. Moreover, voids could be closed applying compressive stresses at elevated temperature, so that the final microstructure and mechanical properties could be improved. Extensive work on the second hybrid processing route can also be found in the literature.

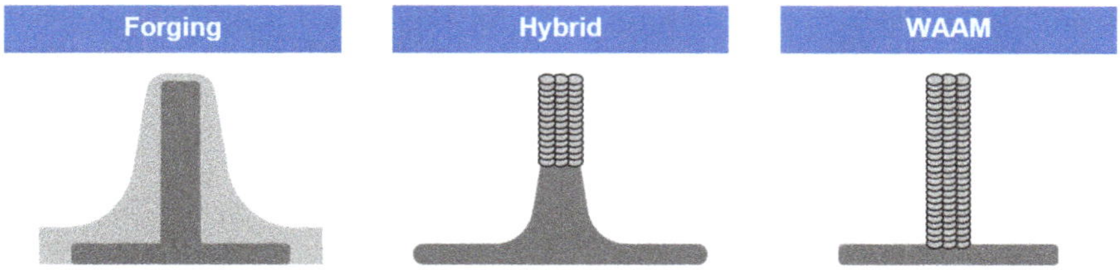

Figure 1. Illustration of the possible hybrid manufacturing route combining metal hot forming and WAAM.

Bambach et al. [11] examined the same demonstrator shape and process route (forging with subsequent WAAM) used in this study and proved that its mechanical properties exceeded the minimum requirements for forged parts regarding yield strength (YS = 837 MPa > 830 MPa) and ultimate tensile strength (UTS = 934 MPa > 900 MPa). Also, only minor anisotropic behavior was observed. In another work, Bambach et al. [12] showed that laser cladding technology could be used to produce flexibly applicable local patches to locally increase the stiffness or the thickness of sheet metal components. Papke et al. [13] analyzed tensile bonding strength of hybrid parts made of Ti-6Al-4V produced by combination of laser beam melting and warm bending. It was reported, that sheet material and AM material possess different hardness values, and the sheet thickness strongly influences the bonding strength. Moreover, it was confirmed, that the contact zone between sheet and AM is the most significant for the strength of hybrid components [14]. Hirtler et al. [15] investigated the production of modifying stiffening ribs made of AlSi12 on conventionally produced pre-forms (EN-AW 6082). The authors proved the feasibility of the process combination between WAAM and hot forming. In general, most of the existing work on hybrid manufactured of Ti-6Al-4V concentrates on the increase of the strength of the sheet metal components using AM.

To produce a sound Ti-6Al-4V hybrid part the microstructure should be thoroughly controlled. During WAAM of Ti-6Al-4V β-grains (BCC) grow epitaxially, similar to PBF AM techniques [16]. Such microstructures are hard to avoid because at the low concentration Al and V have a high solubility in the alloy and, thus, do not partition ahead of the solidification front [17]. The length of prior β grains along the solidification direction can reach a few cm or even cover the whole height of the sample, possessing a strong <001> fiber texture [18]. After cooling below the β-transus temperature (approximately 995 ± 25 °C) β grains typically transform to fine α laths (HCP) retaining β lamellar structure (Widmanstätten structure). Here the thickness of α-platelets and the sizes of β-grains are the fundamental parameters that affect the mechanical performance of Ti-6Al-4V alloy. Very high cooling rates (>410 °C/s) promote martensitic transformation. Slower cooling rates (<20 °C/s) promote formation of Widmanstätten microstructures [19]. Furthermore, the cooling rate determines the thickness and the presence of α-phase on prior β grain boundaries, which strongly affect the mechanical performance, because they induce anisotropy [20]; such anisotropy is unwanted in many applications. In general, the Widmanstätten-type microstructure is characterized by relatively low tensile ductility, good creep resistance, moderate fatigue properties and crack growth resistance [21].

Together with the microstructure, residual stress (RS) is one of the major obstacles in the development of the AM techniques [22]. It may lead to cracking and geometrical distortion of the parts, thereby limiting the design freedom of the component [23,24]. In the case of WAAM, RS is also an important factor influencing the structural integrity [20]. Colegrove et al. have reported that in as-deposited state the RS values in steel WAAM part can reach the yield stress of the material [25]. Also, the presence of RS highly affects the mechanical behavior of the part. Zhang et al. have reported [26] that RS facilitates crack propagation from a WAAM Ti-6Al-4V part to its substrate. One of the strategies for the reduction of RS and the refinement of the microstructure in WAAM is high pressure rolling after every deposited layer (or a few of them) [25,27]. This approach, however, imposes some limitations on the part design and degrades the main advantage of the WAAM technique: the high deposition rates. In general, according to ASTM F2924 [28], stress-relief heat treatment is mandatory for Ti-6Al-4V AM parts. The connection between microstructure and RS for welded and AM materials has been reported in several studies [29–32]. Due to high cooling rate needed for martensitic transformation ($\beta \rightarrow \alpha'$) in Ti-6Al-4V, the martensitic microstructure is often linked with high tensile RS [30]. It has been shown that the presence of β phase in PBF Ti-6Al-4V, introduced by intrinsic heat treatment, leads to the RS relaxation [31]. Thus, for the general application of WAAM, it is important to understand the microstructure and the RS in as-manufactured WAAM components. Such properties depend also on the location in the hybrid part. An investigation of the microstructure and the RS at the interface between forged and AM features is here of particular importance. In AM parts, high tensile RS is usually reached at the interface between the substrate and the deposited material [33]. This could be critical for the utilization of hybrid parts.

The main aim of the current study is to investigate the microstructure and the RS distribution in a hybrid part produced by the combination of conventional hot forming and WAAM. A T-section geometry manufactured by means of WAAM on a hot forging pre-form is investigated. The paper is structured as follows: Section 2 presents the manufacturing of hybrid Ti-6Al-4V parts using WAAM and hot forging. Also, the procedures of microstructural analysis and of determination of RS by means of neutron diffraction are presented. Section 3 gives an overview of results of metallographic examinations and RS. Finally, the results are discussed, and conclusions are presented.

2. Materials and Methods

2.1. Sample Manufacturing

The considered hybrid process route encompassed forging and subsequent WAAM of Ti-6Al-4V. A T-shaped pre-form was forged, as shown in Figure 2. It contained a 10 mm wide and ~90 mm long rib, which was milled flat at a total height of 42 mm. The WAAM process was used to increase the height of the rib by 66 mm.

Hot forging was performed in the $\alpha + \beta$-temperature range using a 2500 t crank press at OTTO FUCHS KG (Meinerzhagen, Germany). The forged part was machined on the top side to obtain a flat surface to ensure a stable WAAM process.

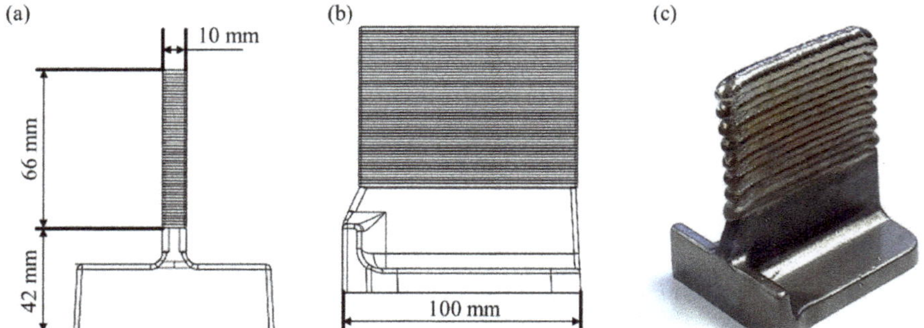

Figure 2. Drawing of the final pre-form with dimensions. Front view (**a**), side view (**b**), final appearance of the hybrid near-net shape demonstrator (**c**).

WAAM was performed using a Fanuc six-axis robot (FANUC Europe Corporation S. A., Echternach, Luxembourg) and a TPS 500i welding power source (Fronius®, Wels, Austria) utilizing the Cold Metal Transfer (CMT) variant of the Gas Metal Arc Welding (GMAW) process. The welding was carried out in an argon filled, sealed chamber to avoid oxidation. The chamber was equipped with an O_2 sensor. Pre-heating of the sample was not applied. The used WAAM set-up is displayed in Figure 3a. A Ti-6Al-4V wire with a diameter of 1.0 mm was used. One hybrid demonstrator was produced for investigations. The shielding gas flow rate at the torch was 15 L/min, also the build chamber was flooded with 50 L/min before the deposition process. During deposition the chamber gas flow rate was reduced to 20 L/min to compensate the small gas leakage. The argon gas purity was ≥99.99%. The chemical composition for as-built Ti-6Al-4V WAAM is shown in Table 1.

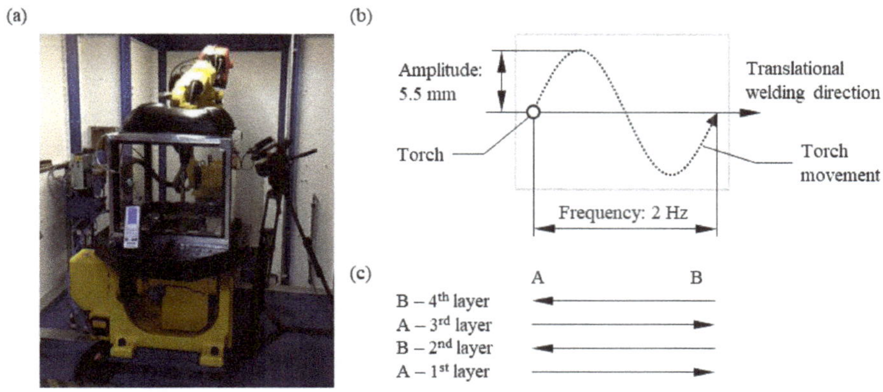

Figure 3. For the WAAM process (**a**) used WAAM machine, (**b**) applied sinus wave movement and (**c**) swapping of the starting point each layer.

Table 1. Chemical composition of Ti-6Al-4V as-built by WAAM (max. weight %).

Ti	Al	V	Fe	C	Other
Bal.	6.45	3.87	0.19	0.07	0.17

The deposition pattern is shown in Figure 3b,c. The starting point (A/B) changed for every layer using a bidirectional tool path on a single bead. A sinus wave was superposed onto the translational motion of the torch to obtain the desired wall thickness. The layer thickness was approximately 4.4 mm. The number of deposited layers was 15.

The welding parameters are shown in Table 2. The first three layers were welded with a higher electrical current to achieve a good fusion with the forged substrate and to increase the average temperature (this ensures a stable WAAM process). The current was than reduced to 100 A for the following layers.

Table 2. WAAM manufacturing parameters.

Parameter	Value
Wire diameter	1.0 mm
Amplitude	5.5 mm
Frequency	2 Hz
Welding speed	15 cm/min
Cooling time between deposition of each layer	130 s
Average electrical voltage	13.5 V
Average electrical current	1st layer: 135 A 2nd layer: 115 A 3rd layer: 105 A ≥4th layers: 100 A

2.2. Microstructural Characterization

In order to investigate the microstructure, the manufactured hybrid part (including the forged region) was cut along the build direction. The sample was mounted with the cut cross-sections on top and ground flat with successively finer grades (from 320 to 1200 μm) of silicon carbide (SiC) papers. The specimen was then polished with 0.05 μm silica solution (OP-S Suspension, Struers GmbH, Willich, Germany) with the addition of H_2O_2, HNO_3, and HF. Sample was etched with Kroll's agent solution to reveal the microstructure. The sections were analyzed using an optical microscope Carl Zeiss Axiotech by Carl Zeiss Microscopy (Jena, Germany).

2.3. Residual Stress Analysis

The neutron diffraction experiment was conducted on the instrument E3 at BER II reactor (Helmholtz Zentrum Berlin, Germany) [34]. A monochromatic neutron beam of wavelength λ = 1.476 Å was used. The three orthogonal strain components, assumed to be principal directions based on the sample geometry, were measured (longitudinal, transversal, normal, see Figure 4). The longitudinal and normal directions coincided with the deposition and the build direction of WAAM, respectively. The acquisition time was set to 75 min for each stress component. Additionally, to improve the diffraction signal, a constant ω oscillation of ±10° around the scattering vector (this is the bisectrix between the incident and the diffracted beam directions and corresponds to the direction in which the strain is measured) was performed. For the longitudinal stress component, a primary slit with opening of 4 mm × 4 mm was used. For the normal and transversal component, a primary slit with vertical opening of 18 mm and horizontal opening of 2 mm was used. This allowed at the same time increasing the signal and keeping high spatial resolution along the build direction of the sample. A secondary collimator with a focus of 2 mm was used for all measurements. The strain was measured along two lines: along the deposition direction at the middle height of the WAAM part (L-line) and along the build direction (N-line). The measured points and coordinate system are schematically presented in Figure 4a.

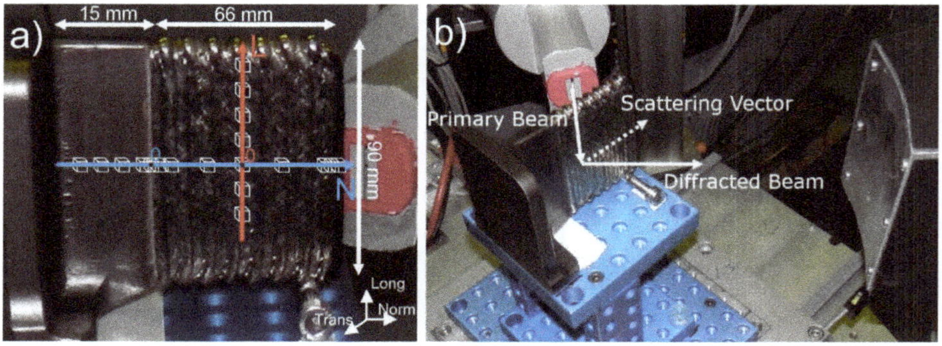

Figure 4. Photo of sample during neutron diffraction experiment on E3 with the schematics of (**a**) the measured points and the coordinate system, (**b**) the set-up. (Note that the sample is aligned for the measurement on the normal strain component).

The measurements were performed at the diffraction angle $2\theta = 76°$, which allowed the simultaneous detections of three diffraction peaks for α-Ti: $11\bar{2}2$, $20\bar{2}1$ and 0004. However, only the $11\bar{2}2$-α reflection appeared at every measured point and sample direction, therefore the RS was calculated for this crystallographic family. The lattice spacing d was calculated according to Bragg's law:

$$d_{11\bar{2}2} = \frac{\lambda}{2 \cdot \sin \theta_{11\bar{2}2}} \tag{1}$$

The lattice spacing (averaged within the gauge volume) was determined by fitting the diffraction peak with a Gaussian function using the software StressTexCalculator 1.53 (TU-Clausthal, Germany). In order to calculate lattice strains ε a strain-free reference d_0 has to be used, since:

$$\varepsilon = \frac{d_{11\bar{2}2} - d_0}{d_0} = \frac{\sin\theta_0}{\sin\theta_{11\bar{2}2}} - 1 \tag{2}$$

One of the main problems in diffraction-based RS analysis is the estimation of such a reference [35,36]. There are different approaches for the determination of d_0: measurements using a reference powder or stress-relieved coupons; or calculations using a global average or stress balance/boundary conditions. In the case of AM materials the variation of the microstructure inside the part and the different thermal history of the part from the raw material (power, wire) makes the determination of d_0 even more challenging [35]. In our case, a global d_0 (or θ_0) was taken as the average of all measured points, as proposed in [37]. This value was $2\theta_0$ = 72.856° with an average error of ±0.007° (coming from the fit of all diffraction peaks). As an alternative, the stress balance condition was applied to the L-line scan for the normal component (Figure 4): $\int \sigma_{Norm} dL = 0$. For this estimation the stress profile was extrapolated to the surface and interpolated between the points with a B-spline function. Symmetry of the stress profile was assumed, i.e., $\sigma_N(+L) = \sigma_N(-L)$. Using this approach, we obtained $2\theta_0$ = 72.859°. This value lies within the error range of the $2\theta_0$ obtained using the global average approach and shifts the resulting stress values only by +10 MPa. Therefore, only stress profiles calculated satisfying the above-mentioned stress balance conditions will be reported.

RS were calculated according to the tensorial Hooke's Law: $\sigma = C\varepsilon$, with C as the stiffness tensor. With the assumption that the principal geometric directions are also principal stress directions, Hooke's Law in the case of a quasi-isotropic solid reads:

$$\sigma_{L,T,N} = \frac{E^{11\bar{2}2}}{\left(1+\nu^{11\bar{2}2}\right)\left(1-2\nu^{11\bar{2}2}\right)}\left[\left(1-\nu^{11\bar{2}2}\right)\varepsilon_{L,T,N} + \nu^{11\bar{2}2}\left(\varepsilon_{T,N,L} + \varepsilon_{N,L,T}\right)\right] \tag{3}$$

where $\sigma_{L,T,N}$ and $\varepsilon_{L,T,N}$ are stresses and strains along the longitudinal, transversal and normal direction, and $E^{11\bar{2}2}$ = 112.7 GPa and $\nu^{11\bar{2}2}$= 0.321 are the diffraction elastic constants for α-Ti $11\bar{2}2$ reflection calculated by the Kröner's model [38].

Hydrostatic stress σ_H and von Mises stresses σ_{vM} were calculated according to:

$$\sigma_H = \frac{\sigma_T + \sigma_L + \sigma_N}{3} \tag{4}$$

$$\sigma_{vM} = \sqrt{\frac{1}{2}\left[(\sigma_L - \sigma_T)^2 + (\sigma_T - \sigma_N)^2 + (\sigma_N - \sigma_L)^2\right]} \tag{5}$$

3. Results and Discussion

Figures 5 and 6 show the macro- and microstructure of the hybrid part, respectively. To interpret the microstructural features, we must recall that the first layers were produced with higher current (see Table 2) to compensate the heat sink effect of the substrate. Then, the heat input (i.e., current) was gradually decreased, and after the fourth layer it was set to a constant value. The macrostructure of the WAAM Ti-6Al-4V is characterized by the epitaxial growth of large columnar prior β-grains, which stretch through several deposited layers (Figure 5). The average dimensions of such grains are about 1.3 mm perpendicular and 8 mm parallel to the build direction.

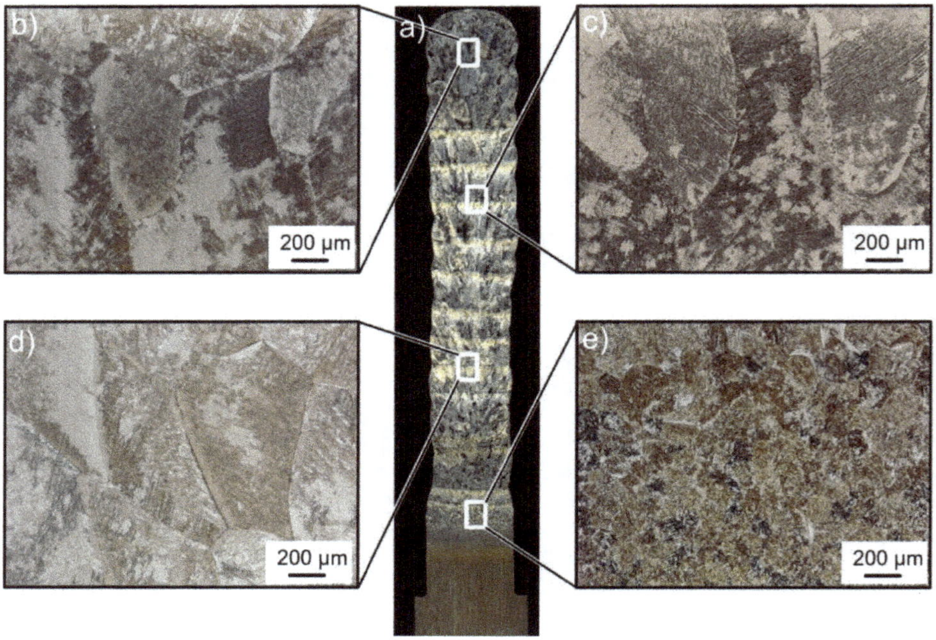

Figure 5. (**a**) Micrograph showing prior β-Ti grains in the middle section of the sample with enlarged view for (**b**) top of WAAM part, (**c**,**d**) middle height of WAAM part, (**e**) transition region between WAAM and forged parts.

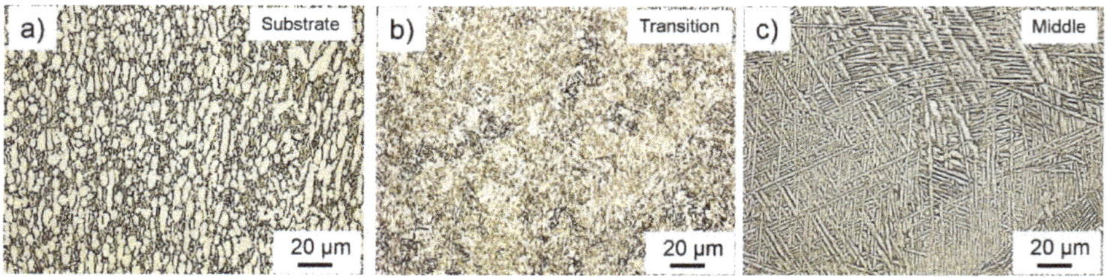

Figure 6. Typical micrographs obtained (**a**) in the substrate, (**b**) in the transition zone close to the substrate, (**c**) in the center of the WAAM part.

The forged Ti-6Al-4V T-section shows, instead, a bi-modal α+β microstructure (Figure 6a), typically obtained in Ti-6Al-4V after conventional thermomechanical processing [21]. The first layers of the WAAM material are characterized by a higher cooling rate compared to the following layers, due to heat conduction to the substrate. Therefore, in the transition zone (e.g., in the first layers, up to approximately 7 mm distance from the seam line), a fine martensitic α′ microstructure is found within small globular prior β-grains with average dimensions of approximately 50 μm (Figure 6b).

At sample mid-height, the micrographs show an α_W-Widmanstätten microstructure with a thickness of the α-platelets of approximately 2 μm within the columnar prior β-grains (Figure 6c). Such microstructures are often reported in the literature [3]. Moreover, the α-phase also appears as a thin layer on the prior β-grain boundaries with a thickness comparable to the thickness of the α-platelets. This microstructure was generated by the repeated rapid heating and cooling that occur during the WAAM process. It is important to stress that the coarse prior β grains may critically affect the mechanical properties (since β is more ductile than α).

A flat panel detector was used in the neutron diffraction experiment. The 2D distributions of the diffraction signal give valuable information about the microstructure (Figure 7): especially in the longitudinal direction, diffraction spots (instead of the typical Debye rings observed for polycrystalline materials) were observed. This is called the "coarse grain effect" (Figure 7a). The grain size of α-Ti laths is quite small, with an average thickness of about 2 μm and an average length of about 25 μm (see Figure 6). However, the origin of α-Ti laths create a distinct crystallographic texture. All the grains of α-Ti inside the prior β-grain are only variants of the same crystallographic orientation, according to Burgers orientation relation [39]. Thus, from the crystallographic point of view, such laths form coarse grains, which scatter similarly to single crystals in the longitudinal direction. In spite of this effect, the projections of the few spots inside a 2D detector image could be analyzed as a Bragg peak, and the lattice parameter could be reliably determined.

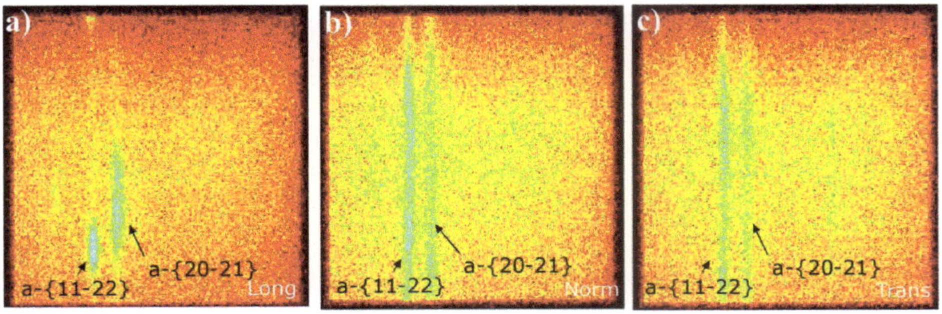

Figure 7. Image of 2D detector obtained during neutron diffraction experiment on the WAAM part (at 46 mm) for (**a**) longitudinal direction, (**b**) normal direction, (**c**) transversal direction. Note: the images are normalized to the maximum intensity, the pixel size is 1.17 mm × 1.17 mm.

The neutron diffraction patterns (i.e., integrated along the Debye ring) collected along the L-line for three orthogonal directions are presented in Figure 8. The patterns look similar at every position, thereby showing no microstructural variations along the deposition direction. The peak intensities (of both $11\bar{2}2$ and $20\bar{2}1$ reflections) do not vary for different orientations, highlighting the absence of a strong crystallographic texture.

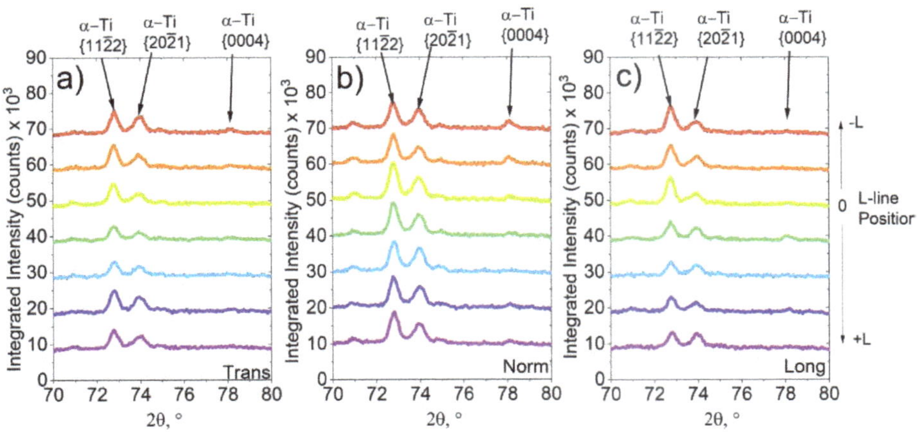

Figure 8. Diffraction patterns along the L-line for (**a**) transversal component, (**b**) normal component, (**c**) longitudinal component.

The intensity ratios follow the theoretical predictions for the neutron diffraction pattern of HCP-Ti powder (i.e., assuming random crystallographic texture, calculated by PowderCell [40]): the intensity of the $11\bar{2}2$ reflection is the highest out of the three reflections observed, followed by $20\bar{2}1$ and 0004 reflection.

In contrast to the diffraction patterns acquired along the deposition direction (L-line, Figure 8), a decrease of the intensity towards the substrate for all reflections of the transversal strain component can be observed along the build direction (Figure 9a). As shown in Figure 2, the thickness of the sample is the same as the thickness of the substrate (in the analyzed region up to 15 mm of substrate), so this cannot be caused by higher neutron absorption. In other words, the path of the neutron beam stays the same for the WAAM and the forged part. Therefore, this intensity variation is an evidence of the microstructural changes reported in Figure 6 and, in particular, of the crystallographic texture. The ratio between the $11\bar{2}2$ and $20\bar{2}1$ peak intensities changes with the height for the longitudinal component (Figure 9c). The 0004 peak increases its intensity for the points in the transition zone and inside the substrate for the normal component (Figure 9b). Thus, an increase in intensity from the prismatic crystallographic plane (1000) for the transversal component could be expected. Since we mostly observed mixed reflections ($11\bar{2}2$ and $20\bar{2}1$), it is hard (and anyway out of the scope of this study) to quantify the effect of crystallographic texture from the current experiments.

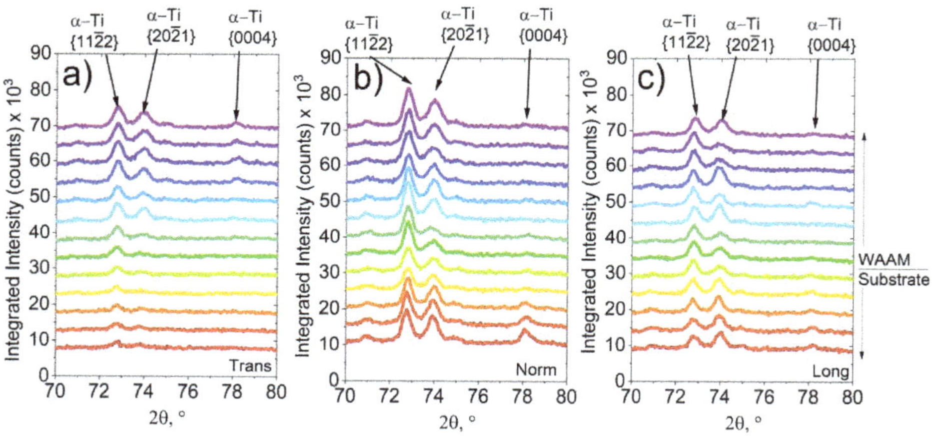

Figure 9. Diffractograms along the N-line for (**a**) transversal component, (**b**) normal component, (**c**) longitudinal component.

The normalized integrated intensity and the full width at half maximum (FWHM) of diffraction peaks along the N-line are presented in Figure 10. Although the coarse grain effect was observed for the longitudinal component (Figure 7a), the FWHM and the integrated intensity of the peak did not vary within the sample (Figure 10b). As for the transition zone and for the substrate, the intensity drops for each measured strain component. In the substrate (Figure 6a), the intensity increases for the normal component. This can be caused by microstructural and crystallographic texture changes (see Figures 5 and 6). It should be noticed that most changes happen in the transition region (from +10 mm to −10 mm) and stabilize inside the WAAM section and the substrate. An increase of the FWHM is often connected to plastic deformation of the material, however, in the present case the FWHM remains constant for the whole height of the sample. Only the transversal component of some points in the substrate shows an increase of the FWHM; this is an artifact due to the low intensity of the peaks (i.e., higher fitting error), see Figure 9a.

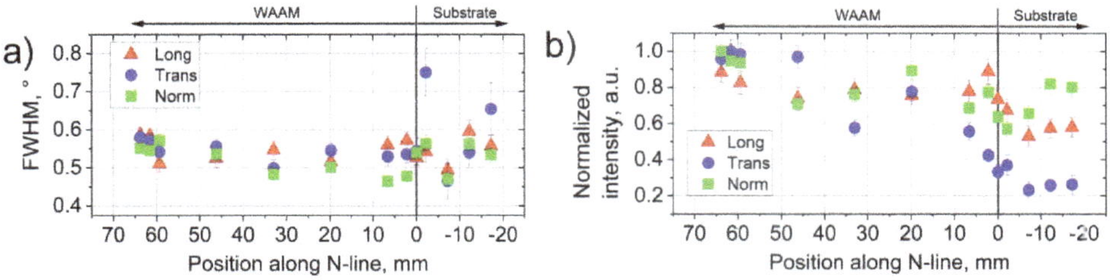

Figure 10. (**a**) FWHM and (**b**) normalized integrated intensity of the $11\bar{2}2$ -Ti diffraction peak along the build direction.

The hydrostatic σ_H and von Mises stress σ_{vM} profiles along L- and N-lines are presented in Figure 11. The stress range for both is around 100 MPa. The hydrostatic stresses cannot be released by post-processing heat treatment and, hence, could be critical for structural integrity of the part under load. In the present case, σ_H reaches its maximum value of around 75 MPa, this is only 10% of the yield stress of WAAM Ti-6Al-4V (around 750 MPa according to [41]). While this value is not critical for the part integrity, it should be taken into account when assessing the life of the component (e.g., under fatigue).

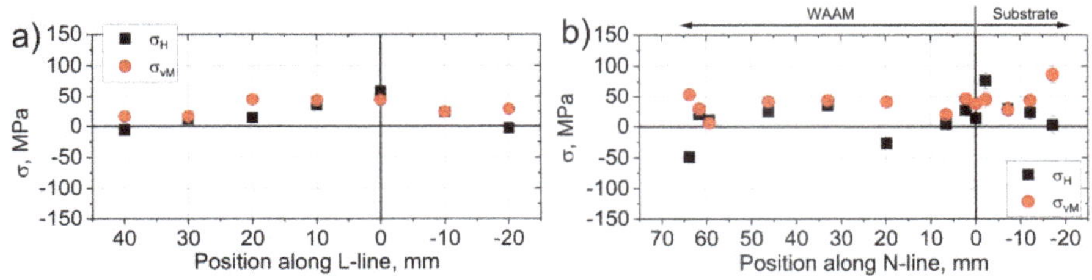

Figure 11. Hydrostatic and von Mises stress along (**a**) L-line, (**b**) N-line. Error bars are contained in the markers.

The longitudinal RS profile along the sample L-line shows a maximum (about 100 MPa) in the middle of the sample (Figure 12a). The profile is symmetric and is similar to that simulated by Ding et al. [42] for steel WAAM parts. The stress profiles along N-line are more complicated (Figure 12b). In the literature the longitudinal component of the RS (coinciding with the deposition direction) usually shows the highest values and the highest gradients. Several experimental RS determination methods (such as the contour method [43] and neutron diffraction [44]) and modeling [42] have revealed the typical longitudinal RS profile generated along the height of WAAM parts. There is some consensus that high tensile stress appears in the region near the substrate and constantly decreases to compressive towards the top of the sample (see schematic in Figure 12c). This is also typical for stress profiles across heterogeneous junctions (see e.g., [45]). The RS rapidly decreases from the transition zone (WAAM/substrate) to the bottom of the substrate. In fact, the whole part should satisfy the stress balance conditions. In our case, the RS decreases to around −50 MPa at 20 mm and then increases to slightly tensile values and decreases again at the very top of the sample (Figure 12b). This follows typical RS profiles (see Figure 12c), with the exception of the point at 20 mm. As mentioned above, during production the energy input was decreased for the first four layers (around 18 mm, see Table 2); therefore, this region of the sample had different thermal input compared to the rest of the WAAM part. This may also be the reason why in this region low RS is found. The sensitivity of RS to variations of the production parameters has been reported for many AM materials [31,36].

Along the build direction all components of the RS lie in the range between −100 MPa and 100 MPa (Figure 12b). This stress range is lower than that reported for WAAM Ti-6Al-4V (reaching around 600–700 MPa [33,37,43,44], with a maximum tensile stress value of around 500 MPa). High RS could be induced by large cooling rates, which are favored by large contact areas (WAAM fabrication, Figure 1) and small substrate thickness, as they were used in the cited works. In the present study, the part of the pre-form used as substrate (a thin upright wall, as in hybrid fabrication, Figure 1), allowed a larger heat accumulation. In this case the geometry of the hybrid part played an important role in RS formation. Furthermore, the used deposition strategy with a sinusoidal path induced a higher energy per unit length (Figure 3b). Thus, the average temperature level was higher during production. This caused a lower temperature gradient, lower flow stresses, and, therefore, smaller RS. Furthermore, the repeated sinusoidal passes induced a so-called intrinsic heat treatment (IHT), analogous to certain scan strategies for L-PBF (see e.g., [31,36]), thereby offering a mechanism for RS relaxation.

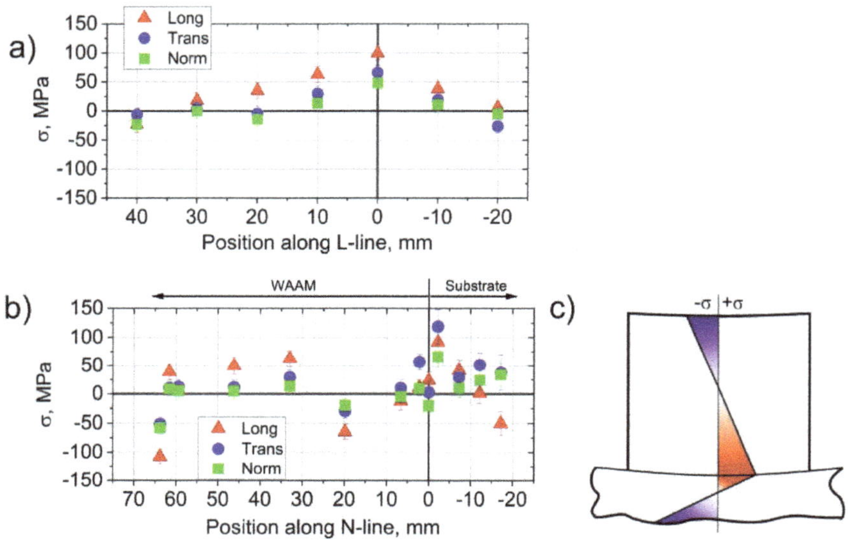

Figure 12. RS profile (**a**) along the L-line and (**b**) along the N-line, (**c**) schematic of the typical stress profile along the WAAM sample height.

4. Summary

We investigated a so-called hybrid Ti-6Al-4V part, made of a WAAM wall and a forged T-shape substrate. Such a component showed the following microstructure along its height: bi-modal ($\alpha+\beta$) in the forged part, martensitic in the transition zone, and Widmanstätten laths inside the WAAM part (as observed by optical microscopy). Neutron diffraction patterns revealed the presence of preferential crystallographic orientation, changing along the sample height (build direction) but not within deposited layers. Residual stress analysis in the bulk of such hybrid part (by means of neutron diffraction) showed a strong stress gradient in the transition region for every stress component. The hybrid forging+WAAM production induced lower residual stress (maximum values around 100 MPa) compared to purely WAAM components reported in the literature (with maximum stress around 500 MPa). We explained such a low stress by the higher heat accumulation and lower cooling rate during hybrid production (the thin substrate hinders heat accumulation), as well as by the sinusoidal shape of the torch movement. An intrinsic heat treatment (successive sinusoidal passes) also assists stress relaxation. The residual stress profiles along the build direction are also expected to be affected by the heat input variation within the part during production. Simulation work is ongoing to understand the formation of residual stress and microstructure in such hybrid WAAM parts.

Author Contributions: Conceptualization, T.T., B.S.; neutron diffraction experiment, T.T., A.U., T.M.; microstructural investigation, I.S.; data curation, I.S., T.M.; writing—original draft preparation, T.M., G.B.; writing—review and editing, all authors.; supervision, G.B., M.B. All authors have read and agreed to the published version of the manuscript.

Funding: The authors gratefully acknowledge the financial support provided by Federal Ministry for Economic Affairs and Energy (BMWi) for the LUFO SAMT64 Project "Forging and additive manufacturing as a process combination for the resource-efficient production of aerospace structural components made of TiAl6V4 on flexible production scales" (20W1719D).

Acknowledgments: The authors would like to thank Robert Wimpory for the support during the beam time. The authors are also grateful to Frank Meiners from OTTO FUCHS KG for providing the material used in this study.

Conflicts of Interest: The authors declare no conflict of interest.

References

1. Boyer, R.R. An overview on the use of titanium in the aerospace industry. *Mater. Sci. Eng. A* **1996**, *213*, 103–114. [CrossRef]
2. Beal, J.D.; Boyer, R.; Sanders, D. Forming of Titanium and Titanium Alloys. In *Metalworking: Sheet Forming*; Semiatin, S.L., Ed.; ASM International: Geauga, OH, USA, 2006; Volume 14B.
3. Huang, R.; Riddle, M.; Graziano, D.; Warren, J.; Das, S.; Nimbalkar, S.; Cresko, J.; Masanet, E. Energy and emissions saving potential of additive manufacturing: The case of lightweight aircraft components. *J. Clean. Prod.* **2016**, *135*, 1559–1570. [CrossRef]
4. Allen, J. An Investigation into the Comparative Costs of Additive Manufacture vs. Machine from Solid for Aero Engine Parts. In *Cost Effective Manufacture via Net-Shape Processing, Proceedings of the Meeting Proceedings RTO-MP-AVT-139, 15–17 May 2006, Amsterdam, The Netherlands*; RTO: Neuilly-sur-Seine, France, 2006; p. 171-1.
5. Dutta, B.; Froes, F.H. Chapter 1—The Additive Manufacturing of Titanium Alloys. In *Additive Manufacturing of Titanium Alloys*; Dutta, B., Froes, F.H., Eds.; Butterworth-Heinemann: Oxford, UK, 2016; pp. 1–10. [CrossRef]
6. DebRoy, T.; Wei, H.L.; Zuback, J.S.; Mukherjee, T.; Elmer, J.W.; Milewski, J.O.; Beese, A.M.; Wilson-Heid, A.; De, A.; Zhang, W. Additive manufacturing of metallic components—Process, structure and properties. *Prog. Mater. Sci.* **2018**, *92*, 112–224. [CrossRef]
7. Williams, S.W.; Martina, F.; Addison, A.C.; Ding, J.; Pardal, G.; Colegrove, P. Wire + Arc Additive Manufacturing. *Mater. Sci. Technol.* **2016**, *32*, 641–647. [CrossRef]
8. Martina, F.; Williams, S.; Colegrove, P. Improved microstructure and increased mechanical properties of additive manufacture produced TI-6AL-4V by interpass cold rolling. In Proceedings of the 24th International Solid Freeform Fabrication Symposium, Austin, TX, USA, 12–14 August 2014.
9. Sizova, I.; Hirtler, M.; Günther, M.; Bambach, M. Wire-arc additive manufacturing of pre-forms for forging of a Ti–6Al–4V turbine blade. In Proceedings of the 22nd International Conference on Material Forming, Vitoria-Gasteiz, Spain, 8–10 May 2019; p. 150017.
10. Bambach, M.; Sizova, I.; Emdadi, A. Development of a processing route for Ti-6Al-4V forgings based on preforms made by selective laser melting. *J. Manuf. Process.* **2019**, *37*, 150–158. [CrossRef]
11. Bambach, M.; Sizova, I.; Sydow, B.; Hemes, S.; Meiners, F. Hybrid manufacturing of components from Ti-6Al-4V by metal forming and wire-arc additive manufacturing. *J. Mater. Process. Technol.* **2020**, *282*, 116689. [CrossRef]
12. Bambach, M.; Sviridov, A.; Weisheit, A.; Schleifenbaum, J. Case Studies on Local Reinforcement of Sheet Metal Components by Laser Additive Manufacturing. *Metals* **2017**, *7*, 113. [CrossRef]
13. Papke, T.; Huber, F.; Geyer, G.; Schmidt, M.; Merklein, M. Characterisation of the Tensile Bonding Strength of Ti-6Al-4V Hybrid Parts Made by Sheet Metal Forming and Laser Beam Melting. In *Advances in Production Research*; Schmitt, R.G.S., Ed.; Springer: Cham, Switzerland, 2019.
14. Butzhammer, L.; Dubjella, P.; Hubera, F.; Schaub, A.; Aumüller, M.; Baum, A.; Petrunenko, O.; Merklein, M.; Schmidt, M. Experimental investigation of a process chain combining sheet metal bending and laser beam melting of Ti-6Al-4V. In Proceedings of the World of Photonics Congress: Lasers in Manufacturing, Munich, Germany, 26–29 June 2017.

15. Hirtler, M.; Jedynak, A.; Sydow, B.; Sviridov, A.; Bambach, M. Investigation of microstructure and hardness of a rib geometry produced by metal forming and wire-arc additive manufacturing. In Proceedings of the 5th MATEC Web of Conferences, Bremen, Germany, 18–21 September 2018; EDP Sciences: Les Ulis, France, 2018; p. 02005.
16. Antonysamy, A.A.; Meyer, J.; Prangnell, P.B. Effect of build geometry on the β-grain structure and texture in additive manufacture of Ti6Al4V by selective electron beam melting. *Mater. Charact.* **2013**, *84*, 153–168. [CrossRef]
17. Bermingham, M.J.; McDonald, S.D.; Dargusch, M.S.; StJohn, D.H. Grain-refinement mechanisms in titanium alloys. *J. Mater. Res.* **2011**, *23*, 97–104. [CrossRef]
18. Wang, F.; Williams, S.; Colegrove, P.; Antonysamy, A.A. Microstructure and Mechanical Properties of Wire and Arc Additive Manufactured Ti-6Al-4V. *Metall. Mater. Trans. A* **2012**, *44*, 968–977. [CrossRef]
19. Ahmed, T.; Rack, H.J. Phase transformations during cooling in a+b titanium alloys. *Mater. Sci. Eng. A* **1998**, *243*, 206–211. [CrossRef]
20. Wu, B.; Pan, Z.; Ding, D.; Cuiuri, D.; Li, H.; Xu, J.; Norrish, J. A review of the wire arc additive manufacturing of metals: Properties, defects and quality improvement. *J. Manuf. Process.* **2018**, *35*, 127–139. [CrossRef]
21. Leyens, C.; Peters, M. *Titanium and Titanium Alloys: Fundamentals and Applications*; Wiley-VCH Verlag GmbH & Co. KGaA: Hoboken, NJ, USA, 2003.
22. Liu, S.; Shin, Y.C. Additive manufacturing of Ti6Al4V alloy: A review. *Mater. Des.* **2019**, *164*, 107552. [CrossRef]
23. Yadroitsev, I.; Yadroitsava, I. Evaluation of residual stress in stainless steel 316L and Ti6Al4V samples produced by selective laser melting. *Virtual Phys. Prototyp.* **2015**, *10*, 67–76. [CrossRef]
24. Patterson, A.E.; Messimer, S.L.; Farrington, P.A. Overhanging Features and the SLM/DMLS Residual Stresses Problem: Review and Future Research Need. *Technologies* **2017**, *5*, 15. [CrossRef]
25. Colegrove, P.A.; Coules, H.E.; Fairman, J.; Martina, F.; Kashoob, T.; Mamash, H.; Cozzolino, L.D. Microstructure and residual stress improvement in wire and arc additively manufactured parts through high-pressure rolling. *J. Mater. Process. Technol.* **2013**, *213*, 1782–1791. [CrossRef]
26. Zhang, J.; Wang, X.; Paddea, S.; Zhang, X. Fatigue crack propagation behaviour in wire+arc additive manufactured Ti-6Al-4V: Effects of microstructure and residual stress. *Mater. Des.* **2016**, *90*, 551–561. [CrossRef]
27. Donoghue, J.; Antonysamy, A.A.; Martina, F.; Colegrove, P.A.; Williams, S.W.; Prangnell, P.B. The effectiveness of combining rolling deformation with Wire–Arc Additive Manufacture on β-grain refinement and texture modification in Ti–6Al–4V. *Mater. Charact.* **2016**, *114*, 103–114. [CrossRef]
28. ASTM. *ASTM F2924—Standard Specification for Additive Manufacturing Titanium-6 Aluminum-4 Vanadium with Powder Bed Fusion*; ASTM: West Conshohocken, PA, USA, 2014.
29. Mehdi, B.; Badji, R.; Ji, V.; Allili, B.; Bradai, D.; Deschaux-Beaume, F.; Soulié, F. Microstructure and residual stresses in Ti-6Al-4V alloy pulsed and unpulsed TIG welds. *J. Mater. Process. Technol.* **2016**, *231*, 441–448. [CrossRef]
30. Rae, W.; Lomas, Z.; Jackson, M.; Rahimi, S. Measurements of residual stress and microstructural evolution in electron beam welded Ti-6Al-4V using multiple techniques. *Mater. Charact.* **2017**, *132*, 10–19. [CrossRef]
31. Mishurova, T.; Artzt, K.; Haubrich, J.; Requena, G.; Bruno, G. New aspects about the search for the most relevant parameters optimizing SLM materials. *Addit. Manuf.* **2019**, *25*, 325–334. [CrossRef]
32. Serrano-Munoz, I.; Mishurova, T.; Thiede, T.; Sprengel, M.; Kromm, A.; Nadammal, N.; Nolze, G.; Saliwan-Neumann, R.; Evans, A.; Bruno, G. The residual stress in as-built Laser Powder Bed Fusion IN718 alloy as a consequence of the scanning strategy induced microstructure. *Sci. Rep.* **2020**. under revision.
33. Martina, F.; Roy, M.; Colegrove, P.; Williams, S.W. Residual stress reduction in high pressure interpass rolled wire+arc additive manufacturing Ti-6Al-4V components. In Proceedings of the 25th International Solid Freeform Fabrication Symposium, Austin, TX, USA, 4–6 August 2014.
34. Boin, M.; Wimpory, R.C. E3: Residual Stress Neutron Diffractometer at BER II. *J. Large Scale Res. Facil. JLSRF* **2016**, *2*, 100. [CrossRef]
35. Mishurova, T.; Serrano-Munoz, I.; Thiede, T.; Ulbricht, A.; Sprengel, M.; Evans, A.; Kromm, M.; Madia, M.; Bruno, G. A critical discussion on the diffraction-based experimental determination of Residual Stress in AM parts. *Astm. Sel. Tech. Pap. (Stp)* **2020**, in press.

36. Thiede, T.; Cabeza, S.; Mishurova, T.; Nadammal, N.; Kromm, A.; Bode, J.; Haberland, C.; Bruno, G. Residual Stress in Selective Laser Melted Inconel 718: Influence of the Removal from Base Plate and Deposition Hatch Length. *Mater. Perform. Charact.* **2018**, *4*, 717–735. [CrossRef]
37. Szost, B.A.; Terzi, S.; Martina, F.; Boisselier, D.; Prytuliak, A.; Pirling, T.; Hofmann, M.; Jarvis, D.J. A comparative study of additive manufacturing techniques: Residual stress and microstructural analysis of CLAD and WAAM printed Ti–6Al–4V components. *Mater. Des.* **2016**, *89*, 559–567. [CrossRef]
38. Kröner, E. Berechnung der elastischen Konstanten des Vielkristalls aus den Konstanten des Einkristalls. *Z. Phys.* **1958**, *151*, 504–518. [CrossRef]
39. Burgers, W.G. On the process of transition of the cubic-body-centered modification into the hexagonal-close-packed modification of zirconium. *Physica* **1934**, *1*, 561–586. [CrossRef]
40. Kraus, W.; Nolze, G. POWDER CELL—A program for the representation and manipulation of crystal structures and calculation of the resulting X-ray powder patterns. *J. Appl. Crystallogr.* **1996**, *29*, 301–303. [CrossRef]
41. Rodrigues, T.A.; Duarte, V.; Miranda, R.M.; Santos, T.G.; Oliveira, J.P. Current Status and Perspectives on Wire and Arc Additive Manufacturing (WAAM). *Materials* **2019**, *12*, 1121. [CrossRef]
42. Ding, J.; Colegrove, P.; Mehnen, J.; Ganguly, S.; Almeida, P.S.; Wang, F.; Williams, S. Thermo-mechanical analysis of Wire and Arc Additive Layer Manufacturing process on large multi-layer parts. *Comput. Mater. Sci.* **2011**, *50*, 3315–3322. [CrossRef]
43. Hönnige, J.R.; Colegrove, P.A.; Ahmad, B.; Fitzpatrick, M.E.; Ganguly, S.; Lee, T.L.; Williams, S.W. Residual stress and texture control in Ti-6Al-4V wire + arc additively manufactured intersections by stress relief and rolling. *Mater. Des.* **2018**, *150*, 193–205. [CrossRef]
44. Martina, F.; Roy, M.J.; Szost, B.A.; Terzi, S.; Colegrove, P.A.; Williams, S.W.; Withers, P.J.; Meyer, J.; Hofmann, M. Residual stress of as-deposited and rolled wire+arc additive manufacturing Ti–6Al–4V components. *Mater. Sci. Technol.* **2016**, *32*, 1439–1448. [CrossRef]
45. Vila, M.; Prieto, C.; Zahr, J.; Pérez-Castellanos, J.L.; Bruno, G.; Jiménez-Ruiz, M.; Miranzo, P.; Osendi, M.I. Residual stresses in ceramic-to-metal joints: Diffraction measurements and finite element method analysis. *Philos. Mag.* **2007**, *87*, 5551–5563. [CrossRef]

metals

MDPI

Article

Effect of Stress-Relieving Heat Treatment on the High Strain Rate Dynamic Compressive Properties of Additively Manufactured Ti6Al4V (ELI)

Amos Muiruri [1,*], Maina Maringa [1], Willie du Preez [1] and Leonard Masu [2]

1. Department of Mechanical and Mechatronics Engineering, Central University of Technology, Free State, 20 President Brand St, Bloemfontein Central, Bloemfontein 9301, South Africa; mmaringa@cut.ac.za (M.M.); wdupreez@cut.ac.za (W.d.P.)
2. Department of Mechanical Engineering, Vaal University of Technology, Andries Potgieter Blvd, Vanderbijlpark 1900, South Africa; leonard@vut.ac.za

* Correspondence: amos.mwangi.muiruri@gmail.com; Tel.: +27-63-014-1847

Received: 16 April 2020; Accepted: 11 May 2020; Published: 18 May 2020

Abstract: A study was undertaken on the compressive high strain rate properties and deformation behaviour of Direct Metal Laser-Sintered (DMLS) Ti6Al4V (ELI) parts in two separate forms: as-built (AB) and stress relieved (SR). The high strain rate compression tests were carried out using a Split Hopkinson Pressure Bar test system at ambient temperature. The average plastic strain rates attained by the system were 400 s^{-1} and 700 s^{-1}. Comparative analyses of the performance (flow stresses and fracture strains) of AB and SR specimens were carried out based on the results obtained at these two plastic strain rates. Microstructural analyses were performed to study the failure mechanisms of the deformed specimens and fracture surfaces. Vickers microhardness test values were obtained before and after high strain rate compression testing. The results obtained in both cases showed the strain rate sensitivity of the stress-relieved samples to be higher in comparison to those of as-built ones, at the same value of true strain.

Keywords: direct metal laser sintering; Ti6Al4V (ELI); dynamic properties; Split Hopkinson Pressure Bar; flow stress; fracture strain

1. Introduction

Direct metal laser sintering (DMLS) is a technology that has been available commercially since 1995. This technology was developed by EOS GmbH of Münich, Germany [1]. In DMLS, material is added one cross-sectional layer at a time to create a three-dimensional object [1]. The process, therefore, makes solid 3D objects of almost any shape from a computer-aided design (CAD) model. The CAD model, which constitutes the part geometry, is created, and, once optimised, is then "sliced numerically" into the layers of thickness at which the additive manufacturing (AM) machine will build it. Next, the file is transferred to the DMLS machine's software, allowing file-based building to begin. Layers of powder are laid down successively and fused selectively by a laser, with each fused layer corresponding to a particular two-dimensional slice of the CAD file. Through repetition of this process, and by fusing layers onto each other, these successive layers form a shape conforming to the 3D CAD model [2].

Varieties of metal and alloy powders can be used, as long as they conform to the DMLS process in terms of atomised particle size and shape. Some potential materials that have been developed for DMLS include, but are not limited to, aluminium alloys, steels and titanium alloys [3]. However, Ti6Al4V has been the main alloy used in DMLS processes. This is mainly due to its application in the medical and aerospace industries [1–4]. The alloy offers good characteristics, such as specific strength and ductility, low thermal conductivity and corrosion resistance, making it a preferred alloy for use in the aerospace industry. In airframes, it is used for general structural components such as undercarriages, fuselages, wings and fasteners [3]. Due to its relatively moderate maximum allowable temperature being between 347 °C and 417 °C [5], it is used for the manufacturing of fan blades, fan cases and compressor blades in the intake sections of aircraft turbo-engines [4]. Other applications include biomedical devices and implants, and high-performance automotive engine parts for racing cars [4]. The opportunities for use of AM technologies such as DMLS in the aerospace sector has led to a reduction of the required quantities of raw materials used to produce in-service components, known as the "buy-to-fly" ratio [6].

The use of DMLS-produced Ti6Al4V parts in the aforementioned applications is expected to increase in the near future [7]. Therefore, it will be imperative to determine how such manufactured parts respond under dynamic conditions of loading. Residual stresses and high surface roughness developed during the building process are the major challenges facing the DMLS process [8]. While the residual stresses result from the rapid cooling of the process, the surface roughness of the manufactured parts is due to partially melted or un-melted powder from the surrounding powder bed sticking to the surface. Porosity and generation of microstructures that are different from those of wrought or cast materials are other challenges associated with AM processes. The production of parts that are free of from pores requires precise control of a number of process parameters [9]. Therefore, the material properties of the DMLS parts are often very different from those of cast or wrought material with the same composition. For instance, the yield strengths of cast and wrought Ti6Al4V are 885 MPa and 945 MPa, respectively, whereas a value of 1075 MPa has been reported for DMLS-produced parts [10–12]. The higher yield strength value in the DMLS process is attributed to the resulting α'-martensitic microstructure as a result of the rapid cooling rate associated with the process.

The behaviour of conventionally manufactured Ti6Al4V under dynamic loading has been the subject of various studies [13–15]. Generally, these studies demonstrated the positive strain rate sensitivity of the Ti6Al4V parts made by conventional manufacturing processes. In the recent past, some studies have attempted to demonstrate the dynamic deformation behaviour of AM Ti6Al4V [16,17]. In their research, Peng et al. [16] paid special attention to both the effects of strain rate as well as temperature on the tensile and compressive plastic flow and fracture characteristics of Ti6Al4V alloy produced via 3D laser deposition technology. The study observed that the flow stress under compressive and tensile loading increased with strain rate. However, the flow characteristics of the material decreased with an increase in temperature under both loading conditions. The authors concluded that compressive fracture occurred as a result of adiabatic shear bands (ASBs) that formed and expanded during deformation for all the test high strain rates. Elsewhere, Mohammadhosseini et al. [17] investigated the dynamic compressive properties of Ti6Al4V alloy produced by electron beam melting (EBM). The study concluded that the strain at fracture was lower in dynamic compared to static compression deformation. Microstructure investigation revealed the presence of ASBs in high strain rate samples. ASBs are narrow zones in which considerable local increases of temperature can be experienced due to the degeneration of part of the mechanical work of high strain rate deformation. The ASBs are usually a precursor to the final ductile or brittle fracture as a result of voids and small cracks growing and coalescing within them [18,19].

In a manner similar to convectional manufacturing processes, parts produced by various AM processes may not necessarily show similar mechanical properties and behaviour, due to the differences of processing parameters, as well the variability in individual processes. A comparative study indicated that the tensile strength of DMLS Ti6Al4V is approximately 30% higher than that of EBM [9]. This is due to the fact that the EBM process results in an $\alpha + \beta$ microstructure, whereas the DMLS process results in an α' phase, which has a comparatively higher yield strength [11]. The layer-by-layer building process of metal powder in the DMLS process with non-optimal parameters influences the porosity, whereas rapid cooling influences the formation of the resulting microstructure. Together, these two factors determine the mechanical properties of the part produced.

The DMLS-produced parts are functional and can be used for different applications experiencing wide ranges of strain rate loadings, such as biomedical implants (low strain rate loading), and in high-performance parts, such as in automotive engine, aircraft engine and structural components (high strain rate loading). Table 1 shows the reported values of strain rate reached under various environmental conditions.

Table 1. Actual levels of strain rate in the human body, and structural parts under various conditions.

Strain Rates (s^{-1})	Conditions	Strained Parts	Ref.
0.004	Walking	Tibia and cortical bones	[20]
0.05	Sprinting and downhill running	Human bones	[21]
1–50	Traumatic fracture events (accidents)	Human bones	[22]
1–500	Automotive crash	Automobile body parts	[23]
10–1000	Aircraft undercarriage	Landing gears	[24]

Aircraft turbo-engine frontal components are prone to impact by foreign objects, such as what is commonly known as a "bird strike" [25]. Bird strikes may occur during take-off, climb, approach and landing. A typical case that recently occurred in South Africa is that of Mango Airlines Flight JE 147, which had to turn around after one of the aircraft's turbo-engines was damaged by a bird strike a few minutes after take-off from O.R. Tambo International Airport, Johannesburg [26]. Therefore, it is critical to design structures and components to withstand such eventuality. Data on the response of materials to high strain rates is necessary to analyse problems related to foreign object damage (FOD), and also in modelling and numerical simulation. Fan casings, compressor casings and turbine casings are designed for containment of debris as a result of blade release. In such an event, a loose/broken blade has to be contained within the casing to avoid further damages to other components, such as controls, which may lead to a loss of control of the aircraft, in turn leading to tragic losses or even an unsafe landing.

The aim of the work reported on in this paper was to develop an understanding of the mechanical behaviour of DMLS-fabricated Ti6Al4V (ELI) parts under high strain rate compressive loading using a Split Hopkinson Pressure Bar (SHPB). The two forms of as-built (AB) and stress-relieved (SR) DMLS specimens were used in this testing. Stress relieving was done to reduce the residual stresses developed during fabrication through the DMLS process. The variations of microhardness, stress–strain behaviour and microstructural characteristics of the two forms of alloy under high strain rate compressive loading were studied in this work, and are reported here.

2. Materials and Methods

2.1. Materials

Table 2 shows the chemical composition of the Ti6Al4V (ELI) powder used for the production of specimens for high strain rate tests. The powder complied with the ASTM F3001-14 standard and was supplied by TLS Technik GmbH & Co. Spezialpulver KG (Bitterfeld–Wolfen, Germany).

Table 2. Chemical composition of TLS Technik GmbH Ti6Al4V (ELI) alloy powder (wt. %).

Materials	Al	V	Fe	C	O	N	H	Ti
Wt. %	6.34	3.944	0.25	0.006	0.082	0.006	0.001	Bal.

The Ti6Al4V (ELI) experimental high strain rate compression specimens were built in a DMLS EOSINT M280 system (EOS GmbH, Munich, Germany) with a 200 W ytterbium fibre laser that has a laser diameter of 80 μm. The samples were fabricated with a laser power setting of 175 W, a hatch distance (the distance between two adjacent scan vectors) of 100 μm and a layer thickness of 30 μm. The Ti6Al4V (ELI) powder particle size was an average of ≤40 μm diameter. A back-and-forth raster scanning pattern with shift angle of 67° for each pass was used in the production of the specimens. Argon gas was used as a protective inert atmosphere in the manufacturing process. Thirty-two cylindrical specimens, with a diameter of 6 mm, were fabricated, and half of them were stress relieved in an argon gas atmosphere at 650 °C, with a soaking period of three hours, and then cooled down in this protective atmosphere. These specimens are herein referred to as "stress-relieved (SR) specimens". The rest of the specimens are herein referred to as "as-built (AB) specimens". The AB and SR specimens were cut off the base plate by electrical discharge machining (EDM) wire cutting. The cylindrical compression test specimens were made in two sets: one with a height of 10 mm and diameter of 6 mm, and the other with a height of 6 mm and diameter of 6 mm. Sixteen specimens (8 AB and 8 SR specimens) were built for each of the two heights. Two specimens in each category (AB and SR) were used for the calibration of the test equipment. Two different heights of specimens were used during SHPB compression testing in order to attain two different strain rates using the same equipment setting (striker velocity). The specifications of the SHPB test require a height-to-diameter ratio of at most 2, in order to facilitate equilibrium of the dynamic forces during testing [27]. The ends of the specimens were faced off on a lathe machine to ensure that they were flat and parallel within a tolerance of 0.02 mm. The parallel ends of specimens are normally required to ensure best possible contact between the specimen and the pressure bars of the SHPB apparatus during impact loading.

2.2. Dynamic Compression Test

The classical SHPB is used to characterise high strain rate behaviour of materials in compression, tension and shear. The device is capable of attaining strain rates ranging from 10^2 s^{-1} to 10^4 s^{-1} [28]. Such range of strain rates is sufficient to simulate high-speed crashes and high strain rate impacts. The typical compression SHPB test set-up is as shown in Figure 1. The incident and the transmitter bars were made of high-strength 4340 steel. The incident and transmitter bars were 3000 mm long, with a diameter of 20 mm. The striker bar was 710 mm long, and was fired into the incident bar at velocity of approximately 11 m/s.

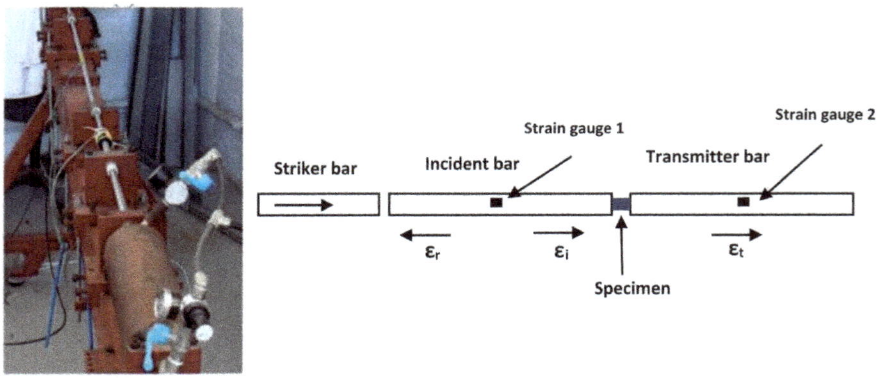

Figure 1. The SHPB compression test setup, reproduced from [29], with permission from Elsevier, 2019.

Strains in the incident and the transmitter bars were measured with strain gauges in a full Wheatstone bridge configuration. The bridge was constructed from four Micro-Measurement (Raleigh, North Carolina, USA), ED-DY-125AC10C 1000 strain gauges placed at the centre of both the incident and transmitter bars. An excitation voltage of 1.50 V was provided through two Agilent E 2630A power supplies. To reduce the friction and maintain uniaxial compression in the sample, molybdenum disulphide grease was used to lubricate the interface between the bars and the specimen. The changes in resistance on the strain gauge as the striker bar struck the incident bar were converted to proportional changes in voltage, and then amplified, digitally sampled and stored as numerical data. The strain gauge readings in volt were then converted into strain using the Equation (1), where the strain gauge data (resistance, gauge factor and voltage) were known [28].

$$Strain = \frac{Voltage\ reading\ x\ Gauge\ resistance}{Calibrated\ voltagexGauge\ factor\ (Gauge\ resistance + Calibrated\ resistance)} \quad (1)$$

Figure 2 shows a typical output signal from the strain gauges on the bars of the SHPB test rig in a typical high strain rate compression test, obtained for DMLS Ti6Al4V (ELI).

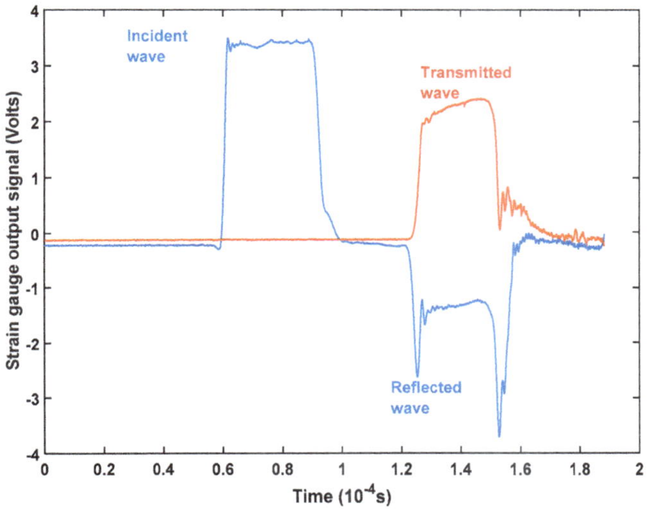

Figure 2. The voltage output from the strain gauges after filtering out noise in a typical compression test.

It is clear that the time at which the waves start picking up in magnitude is not the same for the transmitted and the reflected wave. The estimated picking up times in Figure 2 for the reflected and the transmitted wave are 0.0001214 s and 0.0001229 s, respectively. The reason for this difference in time between these two waves is that the incident wave is immediately reflected at the incident bar/specimen interface, while the transmitted wave occurs upon the wave travelling through the specimen. The reflected signal ε_r(t) was captured from the strain wave of the incident bar, whereas the transmitted strain signal ε_t(t) was captured from the strain wave of the transmitter bar for the same time range as the reflected strain wave. These signals were converted into strain using Equation (1), before input into an existing MatLab programme that converts such signal in stress (σ_s), strain (ε_s) and strain rate $\left(\dot{\varepsilon_s}\right)$ using Equations (2)–(4):

$$\sigma_s = EA_0\varepsilon_t/A_s \quad (2)$$

$$\varepsilon_s = 2C_0/l_s \int_0^t \varepsilon_r dt \quad (3)$$

$$\dot{\varepsilon_s} = 2\varepsilon_r C_0/l_s = \mathrm{v}_{s1} - v_{s2}/l_s \quad (4)$$

where E is the Young's modulus of the incident and transmitter bars; A_0 and A_s are the cross-sectional areas of the bars and the specimen, respectively; l_s is the length of the specimen; and C_0 is the longitudinal wave velocity in the incident and transmitter bars.

Therefore, during the experiment, testing at each strain rate range was repeated seven times, and in all instances the stress and strain data captured were averaged using MatLab codes in order to enhance the credibility of data used. This was to reduce uncertainties that could affect the recorded pulse, leading to spurious results.

2.3. Preparation of Fractured and Cut Surfaces (in the Unfractured Specimens)

In order to study and characterise the deformed surfaces of those specimens that did not fracture after compressive high strain rate testing, they were first sectioned along their respective longitudinal loading axes. The cut was done using EDM wire cutting. Before grinding and polishing of the cut surfaces, the specimens were mounted in conductive Bakelite/Epoxy using a Struers Citopress-1 mounting machine (Cleveland, Ohio, USA). Grinding and polishing of the specimens then followed, and were performed in accordance with the suggested protocol by Struers for titanium alloys. The polished parts were then cleaned individually under tap water, and thereafter dried using a strong stream of compressed air. Metallographic samples were then etched using a solution of Kroll's reagent, a dilute aqueous solution containing HF and HNO_3 that is commonly used for commercial titanium alloys.

The surfaces of those specimens that fractured upon testing were first cleaned with an ultrasonic cleaner for a period of three minutes, using ethanol as the cleaning solvent. The surfaces were then rinsed under running water, before being dried using a strong stream of compressed air. An optical microscope (OM) and a scanning electron microscope (SEM) were used to examine the fracture surfaces and the cut surfaces of the aforementioned loaded DMLS Ti6Al4V (ELI) specimens.

Microhardness testing of the DMLS Ti6Al4V (ELI) wire cut specimens that were exposed to compression high strain rates without fracturing was performed using a Future Tech Vickers hardness tester. A 200 g load, with a dwell time of 10 s, was used, and a minimum of 30 indentations were made on the polished cut surfaces of each sample. The indentations were made approximately 0.5 mm from the edges of the cut surfaces, in both the transverse and longitudinal directions, and a set of indentations were made along the middle in the loading direction, as illustrated in Figure 3. To study the hardness in the regions of critical deformation (adiabatic shear zones), a smaller load of 10 g, with a dwelling time of 10 s, was used. A smaller load was used because the regions were too narrow to accommodate a load of 200 g.

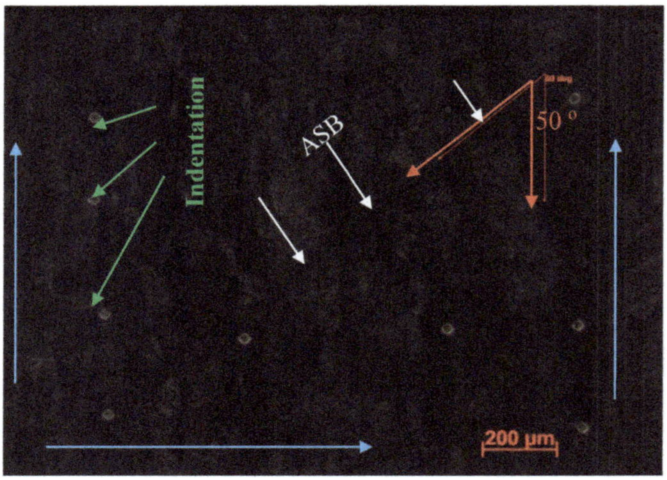

Figure 3. Illustration of typical indentations made on the cut surfaces of the specimens. The blue and red arrows show the direction of indentation and inclination of adiabatic shear bands (ASB), respectively.

3. Results and Discussion

3.1. Microstructure

The optical microstructural observations of the longitudinal sections of the two forms of the alloy shown in Figure 4, revealed prior β-grains elongated approximately parallel to the build direction, before and after stress-relieving heat treatment. An almost equiaxed grain morphology of β-grains on the transverse sections were observed for both the AB- and SR-sectioned samples. The diameters of these grains were approximately 100 μm, which corresponds with the hatch spacing, a DMLS process parameter mentioned in Section 2.1. The interior of the elongated and equiaxed prior β-grains shown in Figure 4 consist of fine acicular-type structures referred to as "α′-martensite". Previous X-ray diffraction (XRD) and transmission electron microscopy (TEM) analyses of AB and SR specimens by Yadroitsev et al. [30] confirmed them to consist of fine acicular-type structures with no traces of β-grain precipitation, even upon stress-relieving heat treatment for the latter.

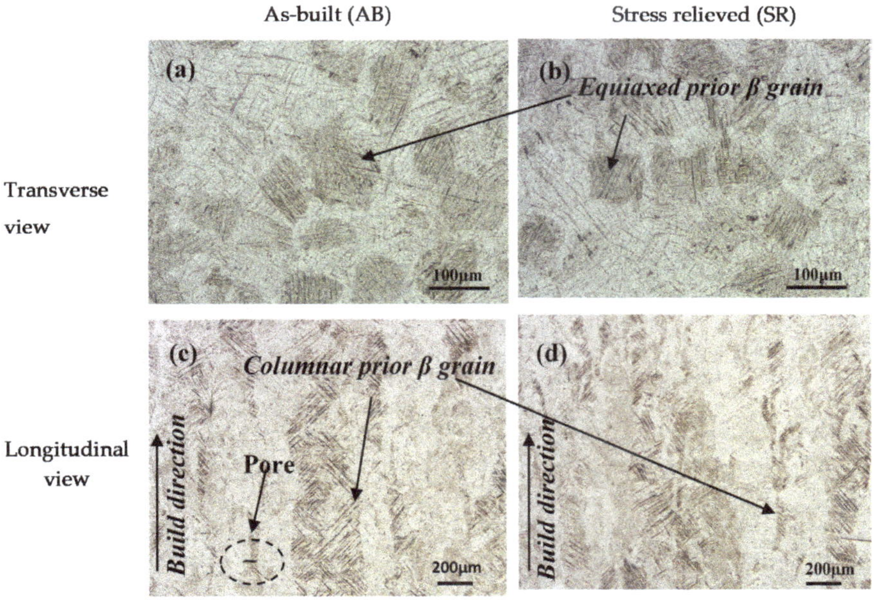

Figure 4. Typical optical micrographs of the as-built (AB) (**a**,**c**) and stress-relieved (SR) (**b**,**d**) transverse and longitudinal sections of the specimens in relation to the build direction, respectively.

3.2. Flow Stress and Strain Rate Dependency

The 6 mm and 10 mm long compression specimens for the two forms of alloy, AB and SR, yielded two different high strain rates. Figure 5 shows the strain rate against time history for the two dimensions of the specimen used.

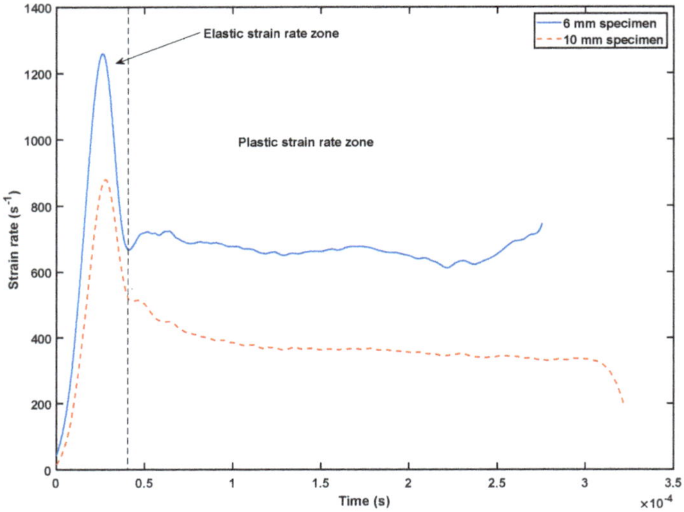

Figure 5. Typical imposed strain rate vs. time history for the 6 mm and 10 mm AB specimens.

This figure shows that the length of the compression specimen has an effect on the strain rate for the same striker velocity, which further confirms Equation (4) in Section 2.2. From the figure, it is seen that the strain rate varied significantly with time, particularly during elastic deformation of the specimens, and remained relatively constant during plastic deformation. The 6 mm and 10 mm specimen for the two forms of alloy yielded average plastic strain rates of approximately 700 s^{-1} and 400 s^{-1}, respectively, for the same applied impact velocity of approximately 11 m/s.

Based on the strain signal obtained, the typical computed strain, strain rate and stress graphs for Ti6Al4V (ELI) as a function of time are shown in Figure 6. Over most of the test period, the specimens experienced linearly increasing strains as the period of testing increased, as is clear from Figure 6. It is also evident from the figure that the compressive stresses induced in the specimens increased more or less linearly with time, up to a certain value (elastic limit). Beyond this point, the change in compressive stress is seen in Figure 6 to become nonlinear (plastic deformation). The compressive strain rate is seen in Figure 6 to rise rapidly, and then drop sharply, before levelling off in the zone of plastic deformation, as previously shown in Figure 5. Like most stress–time curves of metals and alloys during deformation, the curves in this figure exhibit three distinct regions: the initial region of elastic deformation; the second region of uniform plastic deformation, which terminates in the ultimate strength of the material; and the third region of unloading, referred to as a region of non-uniform plastic deformation.

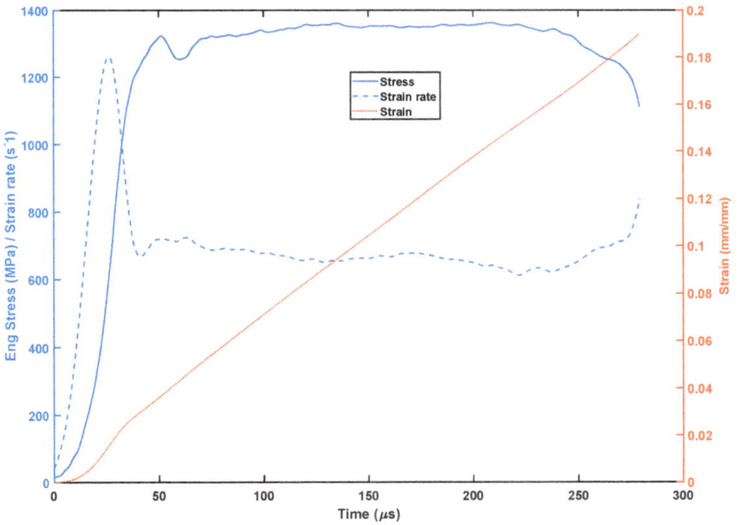

Figure 6. Typical stress, strain rate and the strain against time graph for a 6 mm long SR Direct Metal Laser-Sintered (DMLS) Ti6Al4V specimen.

The curves shown in Figure 7 also exhibit the three regions of elastic and plastic deformation discussed above. After dissipation of the effect of the applied impact, the stresses induced in the specimens dropped. Most reported cases of fracture occurred in this region, after which the stress dropped to zero, as shown in Figure 7.

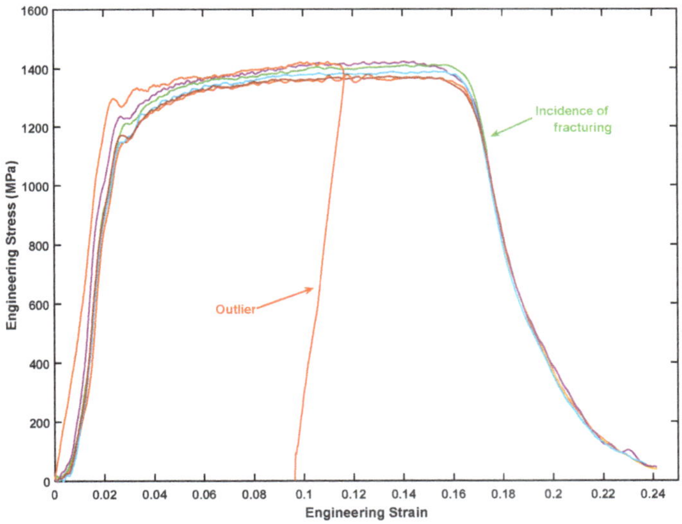

Figure 7. Stress–strain curves for the six SR specimens of 6 mm length, tested at a strain rate of approximately 400 s^{-1}.

At least seven specimens were tested in each category (AB and SR) and at each strain rate (400 and 700 s^{-1}). However, due to instability of the SHPB equipment data acquisition system, some tests failed to record results, even though the specimens were loaded. For instance, in Figure 7, although all seven specimens were loaded, only six of them recorded results. All specimens tested experienced plastic deformation, as shown in Figure 7. Moreover, as shown in Table 3, 14% and 29% of the AB and SR specimens, respectively, fractured during testing conducted at a plastic strain rate of approximately 400 s^{-1}. It was noted that 42% and 71% of AB and SR specimens, respectively, failed during testing conducted at a plastic strain rate of approximately 700 s^{-1}. It is apparent, therefore, that more specimens fractured during testing that was conducted at the higher strain rate. The compressive fracture strength of as-built (AB) Ti6Al4V (ELI) is expected to decrease upon stress-relieving heat treatment, while ductility is expected to improve resulting in the higher numbers of fractured SR specimens, as seen in Table 3. Generally, this is due to the reduction of residual stress of the AB microstructure as a result of heat treatment.

Table 3. Conditions of the specimens after dynamic compression testing.

Specimens	As-built (AB)		Stress Relieved (SR)	
Plastic strain rates (s^{-1})	400	700	400	700
No. of fractured specimens	1	3	2	5
No. of unfractured specimens	6	4	5	2
Total	7	7	7	7

The average values of compressive stresses and strains for the seven specimens, less outliers, tested at each strain rate were computed using MatLab codes, up to the point where the first incidence of fracture was reported in each case, at average plastic strain rates of 400 s^{-1} and 700 s^{-1} for both the AB and SR specimens. The curves for engineering stress (σ_{eng}) and strain (ε_{eng}) were converted into curves for true stress–strain using the following relationships.

$$\sigma_{true}(t) = \sigma_{eng}(t)\big(1 - \varepsilon_{eng}(t)\big) \tag{5}$$

$$\varepsilon_{true}(t) = -In\big(1 - \varepsilon_{eng}(t)\big) \tag{6}$$

The high strain rate average compression stress–strain curves for the AB and SR samples, such as the set of curves shown in Figure 7, are presented in Figure 8.

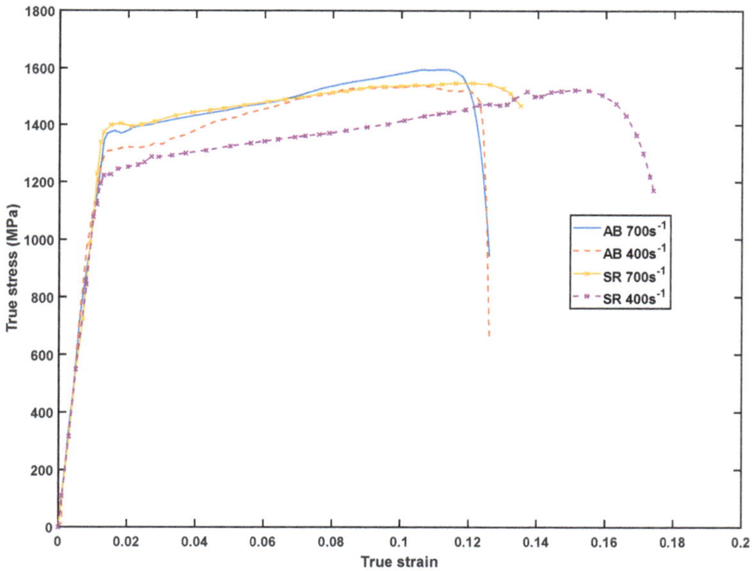

Figure 8. The resultant average compressive true stress–strain curves for the AB and SR samples at two different strain rates.

The sudden impact on the specimens during the SHPB test loaded the material out of the elastic limit, thus leading to instantaneous strain hardening (pile up of dislocations). The result of this was higher values of stress than the normal yield stress of 1080 MPa for the alloy required to initiate plastic deformation, as seen in Figure 8. The yield point varied for the two forms of the alloy tested at the two strain rates. At the higher strain rate of 700 s^{-1}, the yield stresses of the AB and SR specimens are seen in the two figures to be 1384 MPa and 1402 MPa, respectively. Both forms of alloy yielded at lower values of stresses of 1304 MPa and 1225 MPa for the AB and SR specimens, respectively, at a strain rate of 400 s^{-1}.

Further increase of the applied strain initiated plastic deformation for both forms of the alloy. For all the specimens that were tested, the true flow stresses are seen in Figure 8 to have increased after yielding up to a maximum flow stress, which is known to be due to strain hardening, and then decreased thereafter. Most of the plastic work done during this period is dissipated as heat. However, as materials are rarely pure and without flaws, inhomogeneities in the material form areas of stress concentration, and therefore centres of concentration for plastic deformation [31]. The heat generated in such areas is likely to be more than the heat dissipated for high strain rates, even more so for materials with a low thermal conductivity, such as Ti6Al4V. This would then give rise to higher temperatures in the alloy at such areas, which in turn may lead to thermal softening. Thermal softening stops further strain hardening by enhancing the motion of dislocations, leading to flow localisation. The decrease in flow stresses with increasing strain is due to the effect of unstable deformation caused by such flow localisation, which then leads to the development of voids and microcracks. This unstable deformation occurs at the maximum true stress (σ_{max}), and the strain at this stress is defined as the strain of unstable deformation. As is evident from Figure 8, the value of maximum true stress obtained in the present work was higher for the AB samples, at 1596 MPa, compared to a value of 1550 MPa for the SR samples at the same strain rate of 700 s^{-1}. At the lower strain rate of 400 s^{-1}, the values of maximum true stress for the AB and SR still show the same trend, although with smaller differences, namely, 1539 MPa and 1525 MPa, respectively.

The curves in Figure 8 were plotted up to the points at which fracture occurred. The strain at such points is therefore a fracture strain. The fracture strains were higher at the lower strain rate of 400 s^{-1}. At the lower strain rate, the AB specimens fractured at an average strain of 0.126, compared to that of the SR specimens at 0.174. At the higher strain rate of 700 s^{-1}, the strains at fracture were 0.124 and 0.135 for the AB and SR specimens, respectively.

At fixed strain and temperature, the effect of strain rate on the flow stress (σ) can be described by the power-law expression [32]:

$$\sigma = C\dot{\varepsilon}^m \tag{7}$$

where C is the material constant exponent, m is the strain rate sensitivity and $\dot{\varepsilon}$ the strain rate. For material deforming at two levels of strain rate, Equation (7) can further be expressed as [32]

$$\sigma_2/\sigma_1 = (\dot{\varepsilon}_2/\dot{\varepsilon}_1)^m \tag{8}$$

The computed values of the relative strain rate sensitivity m for the AB and SR samples from the average true stress–strain curves at the same strain are as shown in Table 4.

Table 4. Experimental values of relative strain rate sensitivity.

Strain	0.02	0.04	0.08	0.12
Form of samples	Strain rate sensitivity (m)			
AB	0.074	0.064	0.001	−0.04
SR	0.183	0.182	0.15	0.09

The relative strain rate sensitivity is seen from Table 4 to be higher for the stress-relieved samples in comparison to the as-built ones. This could suggest that the stress-relieved DMLS Ti6Al4V (ELI) material becomes more work hardened under high strain rate loading conditions in comparison to the as-built parts under the same loading.

It is common practice to expose the DMLS parts to stress relieving heat treatment in order to relieve the residual stresses that develop during the build process. Therefore, a comparison of the results obtained in the present study for the SR samples and those of Ti6Al4V (ELI) commonly used in industry and manufactured via the conventional methods is presented in Table 5.

Table 5. A comparison of the flow properties of SR DMLS Ti6Al4V (ELI) samples in the present work with those reported in the literature for wrought Ti6Al4V (ELI).

Ti6Al4V	Microstructure	Strain Rate (s^{-1})	YS (MPa)	UCS (MPa)	ε_f	Ref.
DMLS-SR	Martensitic	400	1225	1525	0.174	Present study
DMLS-SR	Martensitic	700	1402	1550	0.135	Present study
Wrought	A + β lamellar	1100	1130	1180	0.180	[33]
Wrought	α + β lamellar	4100	1376	1400	0.300	[33]
Wrought	α + β Equaxied	2630	1210	1346	0.320	[34]
Wrought	A + β Equaxied	6400	1385	1410	0.270	[34]

YS: Yield Strength, UCS: Ultimate Compressive Strength: ε_f-fracture strain.

The materials in the table are of different microstructures: martensitic, lamellar and equiaxed. The yielding stresses and ultimate compressive strength for stress-relieved DMLS Ti6Al4V (ELI) parts are seen in this table to be higher than those of the wrought material, even though the latter were tested at higher strain rates than in the SR specimens in the present study. The fracture strains on the other hand are lower for the DMLS parts compared with those of the wrought materials tested at higher strain rates. An increase in the strain rate of materials generally causes an increase in flow stress and a decrease in fracture strain due to greater multiplication of dislocations. The observation made with reference to these two parameters in this table, in comparing their values for SR and wrought specimens, suggests that the microstructure plays a significant role in the mechanical properties of the materials at high strain rate. Thus, the higher flow stresses and lower fracture strains for SR, DMLS Ti6Al4V (ELI) samples with fine martensitic microstructure, compared to lamellar and equaxied structures of wrought Ti6Al4V. This suggests a need for further heat treatment of DMLS Ti6Al4V (ELI) parts to optimize the strength and ductility for industrial applications.

3.3. Hardness Test Result

The obtained values of the Vickers microhardness test for the unloaded samples and the two sets of specimens (AB and SR) that were exposed to compressive high strain rate testing at the average plastic strain rates of 400 s^{-1} and 700 s^{-1} showed variations over the range of indentations done. The average values (bars) and standard deviation (error bars) of Vickers microhardness for the unloaded and loaded AB and SR samples are presented in Figure 9.

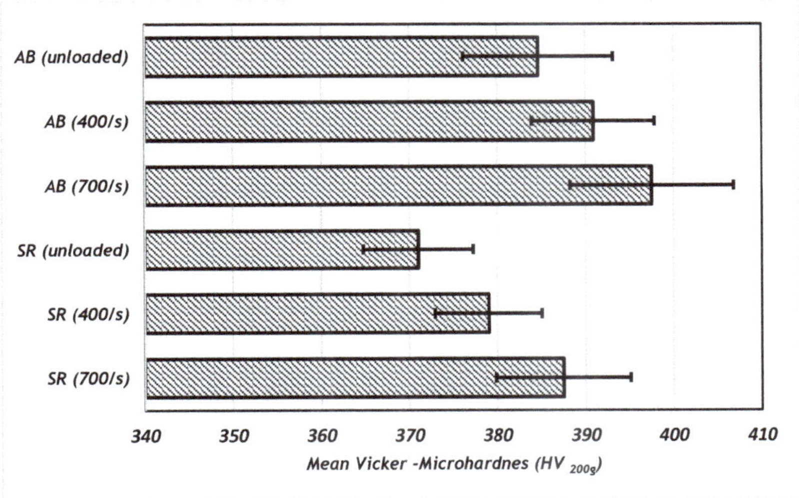

Figure 9. The values of mean Vickers microhardness for DMLS Ti6Al4V (ELI) for different load conditions in HV_{200g}.

The stress-relieving heat treatment resulted in a decrease in micro-hardness of the as-built parts as seen in this figure. A hypothesis of the reduction in a large amount of tensile residual stress created in DMLS process due to rapid solidification can be proposed to have resulted in the observed higher values of micro hardness of the AB specimens. The values of Vickers microhardness for both the AB and SR specimens were higher on the average for the higher compressive strain rate. It is evident from Figure 10 that the change in compressive high strain rate from 400 s^{-1} to 700 s^{-1} does not lead to significant changes in the mean values of microhardness for both the AB (6.63 HV_{200g}) and SR (8.50 HV_{200g}) specimens. However, the change is greater for the cases of the unloaded and loaded samples at the higher strain rate, giving a mean difference of 12.87 HV_{200g} and 16.47 HV_{200g} for AB and SR specimens, respectively. It is noteworthy that, in view of preceding statements on the relative hardness of AB and SR specimens and the possible causes of the difference of values for the two forms of alloy, the change is higher for the SR than for the AB specimens.

The increase in the values of microhardness with increasing strain rate seen here can be attributed to barriers to the movement of dislocations created by strain rate deformation that increases with the increasing strain rate. This phenomenon can further be investigated for specimens loaded under a wider range of strain rates. As the hardness of material is known to be proportional to the yield strength, determination of the Johnson and Cook (JC) model parameters is useful in studying this phenomenon [33].

Careful consideration of the values of microhardness using a small load (HV_{10g}) for the area near, within and far from the ASBs for the specimens loaded at a strain rate of 700 s^{-1} indicated some significant variations. The microhardness was greater near the ASBs, and gradually faded away towards the regions far from them, as shown in Figure 10.

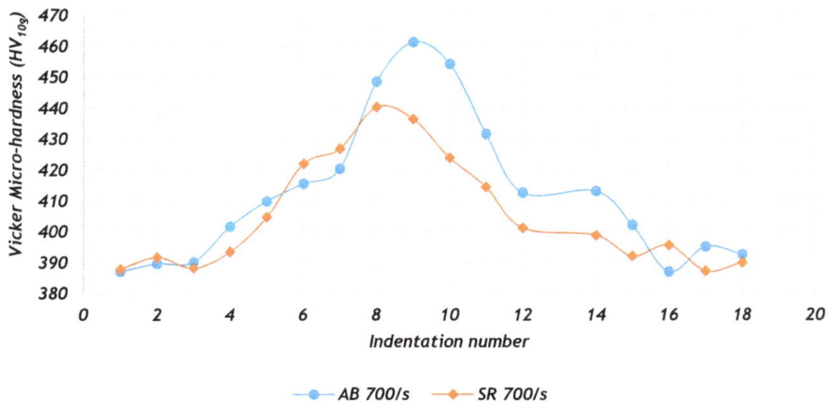

Figure 10. Variation in the microhardness of a typical section of the ASB at a small load of 10 g.

Two key aspects are proposed here to explain this phenomenon: First, the strain hardening of the microstructure within and in the regions neighbouring the ASBs, which are conjectured as being more plastically deformed sections of the shear-strained samples due to shear–strain failure. This increases the dislocation density in such regions, resulting in interaction of dislocation strain fields, which hinders motion of dislocations. The second proposal is related to the higher temperature gradient occurring between the regions of ASBs and the rest of the material. Corresponding higher cooling rates upon removal of load, and therefore strain in such regions, cause the ASBs to develop hard and brittle structures, which result in lower resistance to fracture, but higher values of hardness.

3.4. Observation and Analysis of Sectioned and Fractured Surfaces

The fractured features of the specimens that were impacted at strain rates of 400 s^{-1} and 700 s^{-1} were studied using both optical and scanning electron microscopy. The analysis showed that localised shear dominates the fracture behaviour of the two forms of the Ti6Al4V (ELI) alloy. Narrow transformed ASBs running across the specimens, were observed using optical microscopy in all cases of the sectioned specimens. A typical optical micrograph of a fractured specimen and a schematic description of its features are shown in Figure 11.

In a secondary electron image (SEI) mode of a SEM, deformed and transformed ASBs zones were observed on the longitudinal surfaces of the two forms of specimen that did not fracture. The micrographs, for the specimens that were deformed under either one of the two strain rate regimes of 400 s^{-1} and 700 s^{-1} are shown in Figures 12 and 13. The deformed ASBs are considered as zones of intense concentration of shear strain with no definite boundaries, whereas dynamic recrystallisation or phase transformation does occur in transformed ASBs, and they also have clear boundaries [31]. In Figure 12a, the martensitic structures are seen to be under shear strain, with no definite boundaries observed between the matrix and ASBs. Shear zones, as well as the transformed ASBs, with definite boundaries forming at the centre of the deformed zones, are observable in Figure 12b. Figure 12c shows the transformed ASB under high magnification. The narrow zones pointed to as transformed ASBs are known to be areas of dynamic recrystallisation [18].

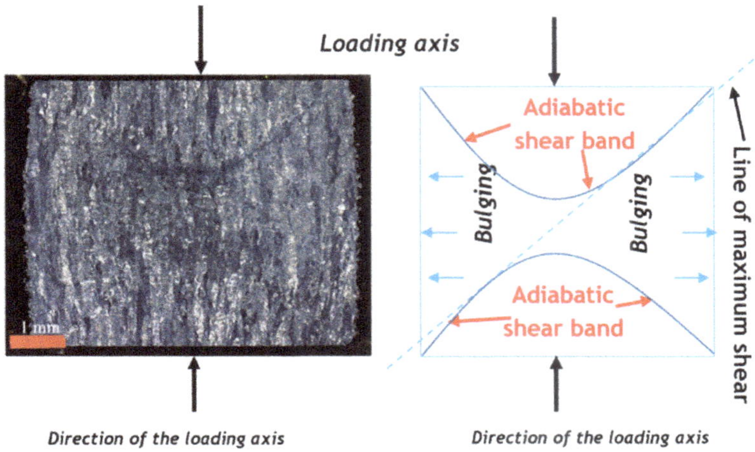

Figure 11. Typical optical micrograph and schematic description of the failure surfaces of a loaded specimen.

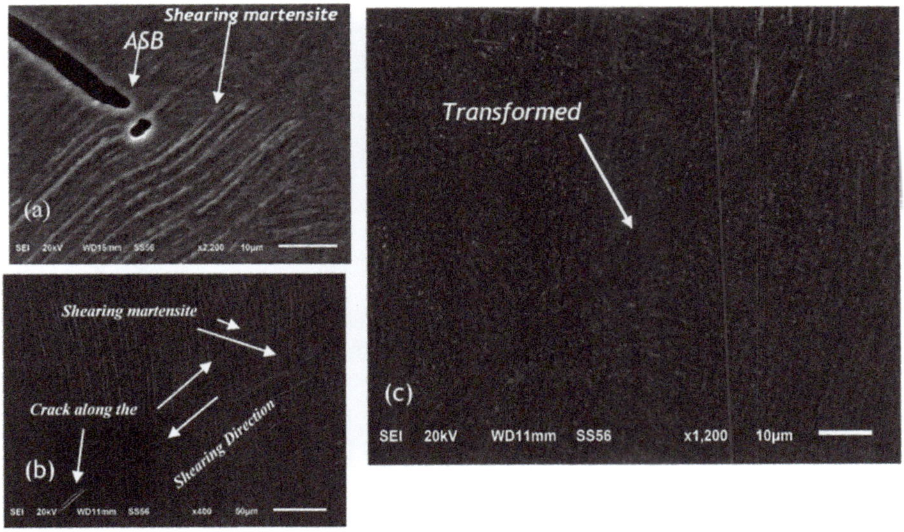

Figure 12. SEM-SEI of deformed (shearing martensite) (**a**,**b**) and transformed adiabatic shear bands (**b**,**c**) in DMLS Ti6Al4V (ELI).

In Figure 13, voids and cracks are apparent, with clear signs of coalescence within the formed ASBs. It is evident from the micrographs in Figure 13 that the locations of ASBs can be considered as sites for further damage and future failure when the voids and cracks coalesce along them. Figure 13a,b shows a typical array of coalesced voids in well-developed ASBs. Initially, the voids can typically be considered spherical or elliptical, as shown in Figure 13b. However, upon their growth, and therefore their diameters reaching the thickness of the shear bands, the voids merged, and extended along the shear bands, forming into elongated cavities. The width of these resulting cavities may therefore be considered to be the same as the width of the transformed shear band. The widths of these transformed shear bands are seen to be narrower in the case of AB specimens (Figure 13a,c) in comparison to those of SR specimens (Figure 13b,d) at the same test strain rate. The measured average width along the transformed shear bands for the AB and SR specimens at the strain rate of approximately 400 s^{-1}, are seen in the figure to be 2 μm and 5 μm, respectively. Those at a strain rate of approximately 700 s^{-1} are slightly higher, at 3 μm and 7 μm for AB and SR specimens, respectively.

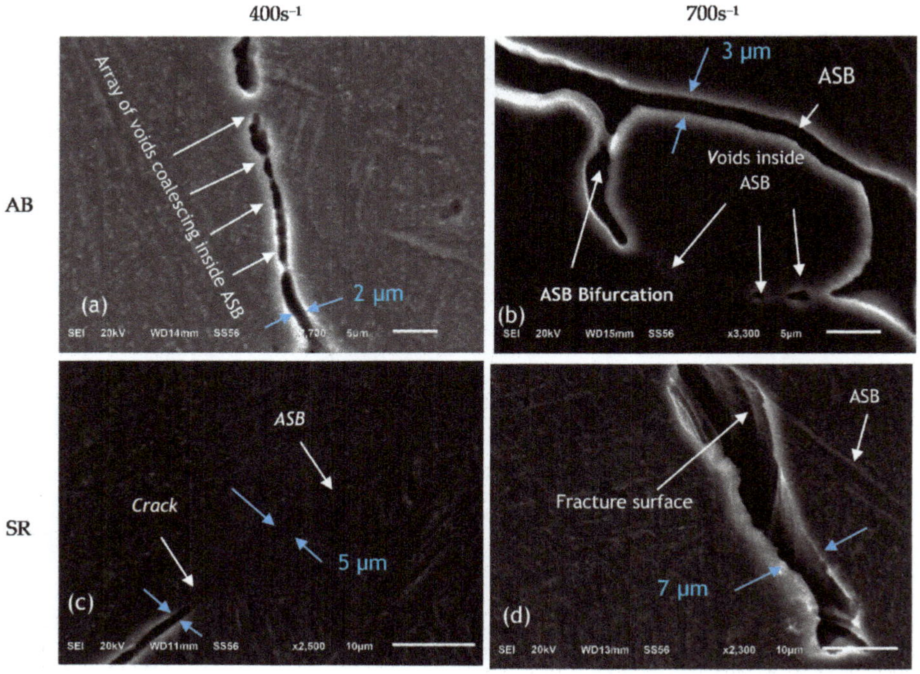

Figure 13. SEM-SEI of deformed adiabatic shear bands for AB samples (**a**,**b**), while panels (**c**,**d**) show the SR samples (**a**,**c**) tested at a strain rate of 400 s^{-1} and (**b**,**d**) at 700 s^{-1}.

The work of Wang [35] demonstrated that, at higher strain rates, the peak stress and the relative average plastic shear strain are higher, transformation occurs earlier and the shear stress relative to this transformation is higher. Therefore, a wider transformed ASB may be expected. Moreover, at high strain rates, high peak temperatures in the ASBs, and steeper profiles of temperature in ASBs, are expected. The variation of the width of the transformed ASBs at the same strain rate for the AB and SR specimens may be considered a result of the difference in hardness of the two forms of the alloy. The research of Dodd and Bai [18] showed that the width of the shear bands is primarily determined by the material hardness, and it diminishes with increasing hardness of the material. This ties in with the observations made in Section 3.2 that stress relieving at a temperature of 650 °C and a soaking period of 3 h led to a decrease in the values of Vickers microhardness, from an average value of 384 ± 8.49 HV_{200} for AB specimens, to 371 ± 6.23 HV_{200} for SR specimens.

Studies of fractographs of DMLS AB and SR Ti6Al4V (ELI) that were loaded at compressive high strain rates highlighted the fact that the fracture surfaces consisted of both shiny and fibrous zones, as shown in Figure 14. It is evident in the images shown in Figure 14 that the tension and the shear deformation regions coexist, and form an x-shape on the fracture surface, defining the shear bands. Deformation under compressive loading is initially homogenous, with no induced radial or circumferential stresses. As seen in Figures 11–13, plastic instability, which occurs at given magnitudes of load, leads to deformation that is localised in narrow zones: the ASBs. A further increase in compressive strain causes a transverse bulging of the specimens and circumferential/hoop stresses appear at the equatorial plane of a cylindrical specimen. This creates a tensile loading state, characterised by fibrous tearing zones with elongated dimples on the sides of the loaded specimens. The shiny, smooth zone is a result of shearing action between the two opposing halves of the loaded specimen.

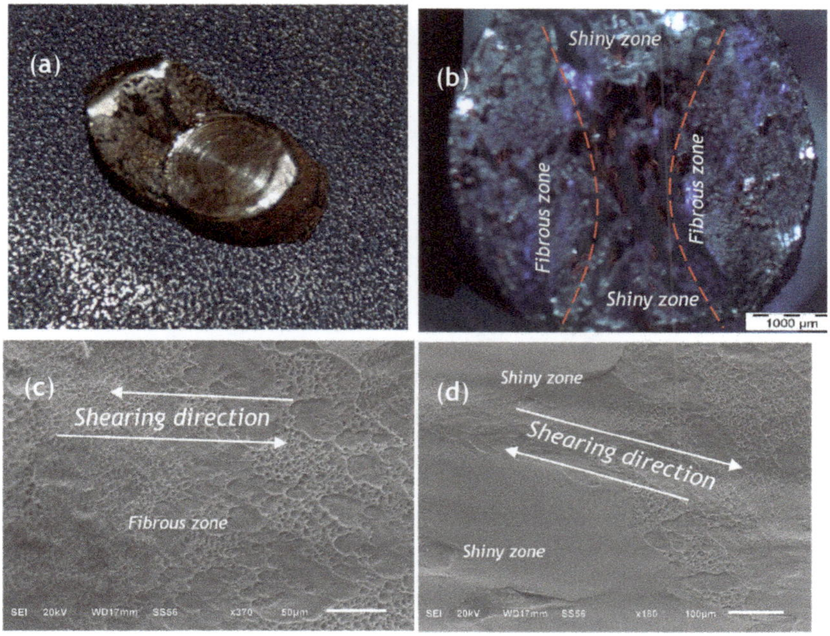

Figure 14. Typical photograph (**a**), optical micrograph (**b**) and secondary electron images (**c**,**d**) of the fracture surfaces of DMLS Ti6Al4V (ELI), showing fibrous and shiny zones.

4. Conclusions

The experimental high strain rate behaviour of as-built (AB) and stress-relieved (SR) DMLS Ti6Al4V (ELI) presented here lead to the following conclusions.

(a) The compressive fracture strength of the AB DMLS Ti6Al4V (ELI) decreased as a result of stress-relieving heat treatment, while ductility increased for both forms of the alloy.
(b) The flow stress under the compressive loading increased, while the fracture strain decreased with a higher strain rate for both forms of the alloy.
(c) The strain rate sensitivity of the SR specimens was higher in comparison to that of AB specimens at the same true strain, implying that SR DMLS Ti6Al4V (ELI) specimens are strengthened much faster under dynamic loading conditions than AB specimens under the same loading.
(d) Stress-relieving heat treatment resulted in the reduction of the microhardness of the specimens. This is thought to be due to the reduction in tensile residual stresses created in the DMLS process.
(e) The values of microhardness for both the AB and SR specimens were higher for the higher compressive strain rates. However, microhardness was greater near the ASBs, and declined gradually towards the regions far from the ASBs.
(f) SEM image analysis indicated that the compressive fracturing of both forms of the alloy was a result of the development of ASBs for the two test strain rates.
(g) At the same strain rate, the transformed ASBs for AB samples were narrower than those for the SR specimens.

Future work will be aimed at developing an ideal microstructure of DMLS-produced Ti6Al4V (ELI), with optimal strength and ductility, thereby enhancing high strain rate properties, for application in the aerospace sector, where such mechanical properties are vital for the reliability of parts.

Author Contributions: Conceptualisation, A.M., M.M., and W.d.P. and L.M.; methodology, A.M.; formal analysis, A.M.; writing—original draft preparation, A.M.; writing—review and editing, A.M., M.M., and W.d.P. and L.M.; supervision, M.M., W.d.P., and L.M.; funding acquisition, W.d.P. All authors have read and agreed to the published version of the manuscript.

Funding: This research was funded by the South African Department of Science and Technology (DST) through the Council for Scientific and Industrial Research (CSIR), for the Collaborative Program in Additive Manufacturing, Contract No.: CSIR-NLC-CPAM-18-MOA-CUT-01.

Acknowledgments: All Ti6Al4V (ELI) specimens were produced by the Centre for Rapid Prototyping and Manufacturing (CRPM) of the Central University of Technology, Free State (CUT), for which the authors express their gratitude. The Department of Mechanical Engineering of the University of Cape Town (UCT), through the Blast Impact & Survivability Research Unit (BISRU), is acknowledged for providing the facilities that were used for testing.

Conflicts of Interest: The authors declare no conflict of interest.

References

1. Ivanova, O.; Williams, C.; Campbell, T. Additive manufacturing (AM) and nanotechnology: Promises and challenges. *Rapid Prototyp. J.* **2013**, *19*, 353–364. [CrossRef]
2. Ralf, C.; Claus, A.; Coube, O.; Keith, M. *Introduction to Additive Manufacturing Technology: A Guide for Designer and Engineers*, 2nd ed.; European Powder Metallurgy Association: Shrewsbury, UK, 2015.
3. DebRoy, T.; Wei, L.; Zuback, S.; Mukherjee, T.; Elmer, W.; Milewski, O.; Zhang, W. Additive manufacturing of metallic components–process, structure and properties. *Prog. Mater. Sci.* **2018**, *2*, 112–224. [CrossRef]
4. Ikuhiro, I.; Tsutomu, T.; Yoshihisa, S.; Nozomu, A. *Application and Features of Titanium for Aerospace Industry*; Technical Review—Nippon Steel Technical Report No 106; Nippon Steel Corporation: Osaka, Japan, 2014.
5. Properties: Titanium Alloys—Ti6Al4V Grade 5—AzoM.com. Available online: https://www.azom.com/properties.aspx?ArticleID=1547 (accessed on 28 April 2020).
6. Reducing the Buy-to-Fly Ratio with Metal 3D Printing. Available online: http://www.materialise.com/en/manufacturing/whitepaper-buy-to-fly-ratio-cutting-costs-metal-3d-printing (accessed on 19 April 2018).
7. Uriondo, A.; Esperson-Miguez, M.; Perinpanyangam, S. The present and future of additive manufacturing in aerospace sector: A review of important aspects. *Proc. Inst. Mech. Eng. Part G J. Aerosp. Eng.* **2015**, *229*, 1–16. [CrossRef]
8. Patterson, E.; Sherri, L.; Messimer, L.; Farrington, P. Overhanging features and the SLM/DMLS residual stresses problem. Review and future research need. *Addit. Manuf. Tech.* **2017**, *5*, 15. [CrossRef]
9. Rafi, H.K.; Karthik, N.V.; Gong, H.L.; Starr, S.E. Microstructures and mechanical properties of Ti6Al4V parts fabricated by selective laser melting and electron beam melting. *Mater. Eng. Perf.* **2013**, *22*, 3372–3383. [CrossRef]
10. Todd, M.M.; Long, M.J. Mechanical behaviour of additive manufactured, powder-bed laser-fused materials. *Mater. Sci. Eng. A* **2015**, *651*, 198–213. [CrossRef]
11. Moletsane, G.; Krakhmalev, P.; Du Plessis, A.; Yadroitsava, I.; Yadroitsev, I.; Kazantseva, I. Tensile properties and microstructure of direct metal laser-sintered Ti-6Al-4V (ELI) alloy. *SAJIE* **2016**, *27*, 110–121. [CrossRef]
12. Du Plessis, A.; Ina Yadroistava, I.; Roux, S.; Yadroitsev, I.; Fieres, J.; Reinhart, C.; Rossouw, P. Prediction of mechanical performance of Ti6Al4V cast alloy based on microCT-based load simulation. *J. Alloys Compd.* **2017**, *724*, 267–274. [CrossRef]
13. Yatnalkar, R. Experimental Investigation of Plastic Deformation of Ti-6Al-4V under Various Loading Conditions. Master's Thesis, The Ohio State University, Columbus, OH, USA, 2010.
14. Wulf, G.L. High strain rate compression of titanium and some titanium alloys. *IJMSCI* **1976**, *21*, 713–718. [CrossRef]
15. Lee, W.-S.; Lin, M.-T. The effects of strain rate and temperature on the compressive deformation behaviour of Ti6Al4V alloy. *J. Mater. Process. Technol.* **1997**, *71*, 235–246. [CrossRef]

16. Li, P.H.; Guo, W.G.; Huang, W.D.; Su, Y.; Lin, X.; Yuan, K.B. Thermomechanical response of 3D laser-deposited Ti–6Al–4V alloy over a wide range of strain rates and temperatures. *Mater. Sci. Eng.* **2015**, *647*, 34–42. [CrossRef]
17. Mohammadhosseini, A.; Masood, S.; Fraser, D.; Jahedi, M. Dynamic compressive behaviour of Ti-6Al-4V alloy processed by electron beam melting under high strain rate loading. *Adv. Manuf.* **2015**, *3*, 232–243. [CrossRef]
18. Bradley, D.; Yilong, B. *Introduction to Adiabatic Shear Localization*, Revised ed.; Imperial College Press: London, UK, 2015; pp. 29–32.
19. Lee, W.-S.; Lin, C.-F. Plastic deformation and fracture behaviour of Ti–6Al–4V alloy loaded with high strain rate under various temperatures. *Mater. Sci. Eng.* **1998**, *241*, 48–56. [CrossRef]
20. Lanyon, L.E.; Hampson, W.G.; Goodship, A.E.; Shah, J.S. Bone deformation recorded in vivo from strain gauges attached to the human tibial shaft. *Acta Orthop. Scand.* **1975**, *46*, 256–268. [CrossRef] [PubMed]
21. Burr, D.B.; Milgrom, C.; Fyhrie, D.; Forwood, M.; Nyska, M.; Finestone, A.; Hoshaw, S. In vivo measurement of human tibial strains during vigorous activity. *Bone* **1996**, *18*, 405–410. [CrossRef]
22. Hansen, U.; Zioupos, P.; Simpson, R.; Currey, D.; Hynd, D. The effect of strain rate on the mechanical properties of human cortical bone. *J. Biomed. Eng.* **2008**, *130*, 11–18. [CrossRef]
23. Wood, P.K.; Schley, C.A. *Strain Rate Testing of Metallic Materials and Their Modelling for Use in CAE Based Automotive Crash Simulation Tools (Recommendations and Procedures)*; Ismithers: Shrewsbury, UK, 2009.
24. Wiesner, C.; McGillivray, H. *Loading Rate Effects on Tensile Properties and Fracture Toughness of Steel*; TAGSI Seminar; Imperial College: London, UK, 1999; Available online: https://www.twi-global.com/technical-knowledge/published-papers/loading-rate-effects-on-tensile-properties-and-fracture-toughness-of-steel (accessed on 3 April 2018).
25. Wang, X.; Feng, Z.; Wang, F.; Yue, Z. Dynamic response analysis of bird strike on aircraft windshield based on damage-modified nonlinear viscoelastic constitutive relation. *Chin. J. Aeronaut.* **2007**, *20*, 511–517.
26. Mango Planes Engine Damaged in Bird Strike at OR Tambo. Available online: http://www.traveller24.com/News/Flights/watch-mango-planes-engine-damaged-in-bird-strike-after-or-tambo-take-off-20171119 (accessed on 19 May 2018).
27. Noori, S.A. Dimension effect on dynamic stress equilibrium in SHPB tests. *IJMP* **2014**, *5*, 15–26.
28. Gray, G.T., III. Classic split-Hopkinson pressure bar testing. In *ASM Handbook: Mechanical Testing and Evaluation*; ASM International: Materials Park, OH, USA, 2008; Volume 8, pp. 462–476.
29. Kariem, M.A.; Santiago, R.C.; Govender, R.; Shu, D.W.; Ruan, D.; Nurick, G.; Alves, M.; Lu, G.; Langdon, G.S. Round-Robin test of split Hopkinson pressure bar. *Int. J. Impact Eng.* **2019**, *126*, 62–75. [CrossRef]
30. Yadroitsev, I.; Krakhmalev, P.; Yadroitsava, I.; Du Plessis, A. Qualification of Ti6Al4V ELI alloy produced by laser powder bed fusion for biomedical applications. *JOM* **2007**, *70*, 372–377. [CrossRef]
31. Xue, Q.; Meyers, M.; Nesterenko, V. Self–organization of shear bands in Titanium and Ti6Al4V alloy. *Acta Mater.* **2002**, *50*, 575–596. [CrossRef]
32. William, F.H. *Mechanical Behaviour of Materials*, 2nd ed.; Cambridge University Press: New York, NY, USA, 2010; p. 92.
33. Sudhanshu, S.; Anuradha, N.M.; Chavan, V.M. Deformation response of Titanium alloy under static and dynamic loading. *Procedia Eng.* **2007**, *173*, 1894–1900.
34. Kim, J.Y.; Shim, I.O.; Kim, H.K.; Hong, S.S.; Hong, S.H. Dynamic deformation and high velocity impact behaviors of TI-6Al-4V Alloys. *Mater. Sci. Forum.* **2007**, *539*, 2269–2274. [CrossRef]
35. Wang, X.B. Effects of temperature and strain rate on the evolution of thickness of transformed adiabatic shear band. *Solid State Phenom.* **2008**, *138*, 385–392. [CrossRef]

metals

MDPI

Article

In-Situ Defect Detection in Laser Powder Bed Fusion by Using Thermography and Optical Tomography—Comparison to Computed Tomography

Gunther Mohr [1,2,*], Simon J. Altenburg [2], Alexander Ulbricht [2], Philipp Heinrich [3], Daniel Baum [3], Christiane Maierhofer [2] and Kai Hilgenberg [1,2]

1 Institute of Machine Tools and Factory Management, Chair of Processes and Technologies for Highly Loaded Welds, Technische Universität Berlin, Straße des 17. Juni 135, 10623 Berlin, Germany; kai.hilgenberg@bam.de

2 Federal Institute for Materials Research and Testing (BAM; Bundesanstalt für Materialforschung und prüfung), Unter den Eichen 87, 12205 Berlin, Germany; simon.altenburg@bam.de (S.J.A.); alexander.ulbricht@bam.de (A.U.); christiane.maierhofer@bam.de (C.M.)

3 Konrad-Zuse-Zentrum für Informationstechnik Berlin (ZIB), Mathematics for Life and Materials Sciences Takustraße 7, 14195 Berlin, Germany; heinrich@zib.de (P.H.); baum@zib.de (D.B.)

* Correspondence: gunther.mohr@bam.de; Tel.: +49-30-8104-4865

Received: 24 November 2019; Accepted: 20 December 2019; Published: 9 January 2020

Abstract: Among additive manufacturing (AM) technologies, the laser powder bed fusion (L-PBF) is one of the most important technologies to produce metallic components. The layer-wise build-up of components and the complex process conditions increase the probability of the occurrence of defects. However, due to the iterative nature of its manufacturing process and in contrast to conventional manufacturing technologies such as casting, L-PBF offers unique opportunities for in-situ monitoring. In this study, two cameras were successfully tested simultaneously as a machine manufacturer independent process monitoring setup: a high-frequency infrared camera and a camera for long time exposure, working in the visible and infrared spectrum and equipped with a near infrared filter. An AISI 316L stainless steel specimen with integrated artificial defects has been monitored during the build. The acquired camera data was compared to data obtained by computed tomography. A promising and easy to use examination method for data analysis was developed and correlations between measured signals and defects were identified. Moreover, sources of possible data misinterpretation were specified. Lastly, attempts for automatic data analysis by data integration are presented.

Keywords: laser powder bed fusion (L-PBF); selective laser melting (SLM); additive manufacturing (AM); process monitoring; infrared thermography; optical tomography; computed tomography (CT); data fusion; lack-of-fusion

1. Introduction

Additive manufacturing (AM) technologies started as manufacturing instruments for prototyping applications in the late 1980s [1,2]. The further development eventually has enhanced its capability for real part production of metallic components using several applicable alloys, such as different titanium, aluminum, steel and nickel base alloys [3–5]. Among AM technologies, the laser powder bed fusion (L-PBF) is one of the most important technologies to produce metallic components [2,6]. The increasing interest of several industries, particularly aerospace and medical, and also governmental investments boosted the efforts of academia as well as research and development departments to improve the capability of these technologies [2]. The process inherent freedom of design and tool-free production are the key advantages of the layer-based process, but it comes at the expense of time and costs, when considering mass production. However, for small and medium lot sizes, L-PBF is able to demonstrate its advantages in lightweight structures, complex designs and the integration of functions [7–9]. In addition to manufacturing costs, part quality is of utmost importance for many critical applications and quality assurance is one of the big challenges in AM. The layer-wise nature of L-PBF allows on the one hand to monitor the process but on the other hand it is a potential source of defect initiation. Tapia and Elwany [10] denoted quality and repeatability as the "Achilles Heel of AM" and asserted that monitoring and control of AM is one of the key topics in numerous publications. Grasso and Colosimo [11] emphasized the relevance of process monitoring by its high priority in recent roadmaps and keynote studies that are cited in their extensive review paper on that topic.

The manufacturing process of L-PBF comprises essentially three steps, which are repeated iteratively until the part is finished: powder recoating; laser melting or exposing; lowering of the base plate. A fine metallic powder is homogeneously spread over a metallic base plate with a defined layer thickness, which typically ranges between 25 μm and 50 μm. During the exposing phase, the powder is selectively melted by laser radiation according to the 2D geometrical information, that is gained by so-called slicing of a 3D volumetric model. Due to small melt pools, the molten powder solidifies rapidly, and material cohesion is created. Thereafter, the base plate is lowered by the layer thickness and the next iterative loop starts. The process runs in inert gas atmosphere in a closed build chamber. The physics behind the L-PBF process can be compared to a micro-welding process. The produced parts can consist of several kilometers of welding lines. Therefore, process deviations and resulting defects can be expected. Some typical defects for L-PBF are shown in Figure 1. Grasso and Colosimo [11] compiled L-PBF defect categories: porosity; cracking and delamination; balling; geometric defects; surface defects; microstructural inhomogeneities and impurities.

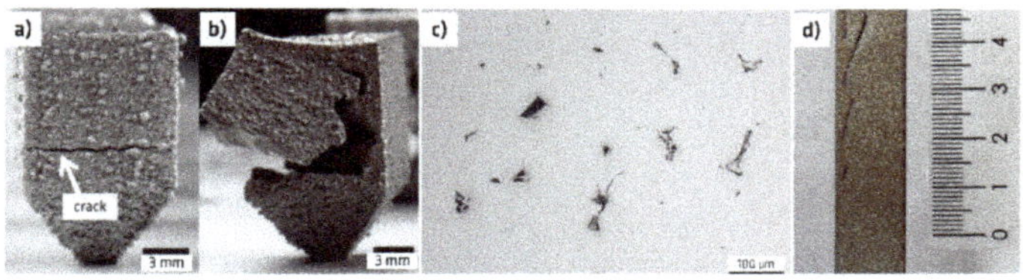

Figure 1. Photographs of characteristic defects in L-PBF (laser powder bed fusion): (**a**) and (**b**) delamination and cracks [9] (**c**) voids; (**d**) surface defects.

L-PBF is known as a process with diverse influencing factors, which might cause defects and affect part quality [12–15]. Whether they are related to the process itself, to the particular machine concept, to the part geometry or to the feedstock material [11,16] cannot be differentiated with high certainty in every case. Therefore, several different in-situ monitoring approaches have been developed to observe distinct process parameters and objects such as laser power [17,18], powder recoating and powder bed surface [19], powder bed compaction [20], plume and spatter behavior [21,22], particle gas emissions [23] or part distortion [24]. In addition, and mostly applied [2], monitoring the spatial and temporal temperature conditions within or close to the laser-material interaction zone has been studied by many authors; most of these studies focusing on temperature measurements in L-PBF are based on contactless measurement techniques such as diodes or cameras [1,25–28].

Diode- or camera-based measurement techniques for thermal radiation can be categorized into on-axis and off-axis methods [29]: On-axis systems use the optical path of the processing laser; these systems allow emission detection following the path of the scanner mirrors. In contrast, off-axis systems observe the processing plane or parts of it from another position, often slightly tilted to the build plane.

Zenzinger et al. [26] were able to capture the thermal emissions of the melting process by using a CMOS camera equipped with a suitable band pass filter. Deviations in the signals derived from optical tomography (OT) were attributed to welding disturbances, which were correlated to micro computed tomography (μCT) data in [30] and data from digital radiography [31]. They achieved a pixel size of approximately 0.1 mm × 0.1 mm. However, a detailed high-resolution signal to defect correlation is not yet possible [31]. Due to their longtime exposure concept, their system differs from infrared camera systems with high temporal dynamics.

Bartlett et al. [27] used a long-wave infrared (LWIR) camera to monitor the L-PBF production of aluminum specimens, inspecting the process from outside the build chamber through a germanium window. They used a very low frame rate of 7 Hz. No information about spatial resolution was given but could be roughly estimated from information of the displayed camera captures as between 150 μm to 200 μm per pixel. The angle of observation remained unclear. With their approach they were able to detect 82% of lack-of-fusion pores by signal deviations and 50% of defects smaller than 50 μm.

Mitchell et al. [32] installed an off-axis two-color pyrometer containing two CMOS cameras directly within the build chamber of a L-PBF machine. They achieved a resolution of 21 μm per pixel on a very small field of view of 1.4 mm × 1.7 mm during the manufacturing of an AISI 316L specimen. The effective sensitive spectral range was around 750 nm and 900 nm, measuring at 6 kHz to 7 kHz. Due to the two-color approach, they were able to calculate absolute temperatures. They focused on the melt pool and its temperatures and derived process irregularities from deviations in melt pool peak temperatures and melt pool shapes. They were able to refer pyrometric data deviations to cavities of a cross section length of 120 μm. The detection of smaller features was hindered by the sampling frequency, which was still too low for the interruption-free data acquisition at scanning velocities of 1400 mm/s. In addition, some diode- or camera-based monitoring techniques for thermal radiation are already integrated into commercial L-PBF systems and can be purchased as machine add-on with different trade names and different working principles, e.g., Melt-Pool-Monitoring MPM (SLM Solutions Group AG, Lübeck, Germany), EOSTATE MeltPool (EOS GmbH, Krailling, Germany) and QM MeltPool 3D (Concept Laser GmbH, Lichtenfels, Germany) [11]. A lack of access to the measured data or missing details about the equipment and measurement principle often limits their use for in-depth research.

Plenty of academic research on thermal detection monitoring systems for L-PBF has been published in the last decade [11]. Different approaches were introduced above. No available system combines all single advantages of appropriate inspection wavelengths, resolutions and frame rates at once. In fact, the question is still open: what is necessary and what is sufficient for in-situ monitoring. Still further development is required also to enhance process understanding and quality assurance.

Hence, this study introduces a new contactless off-axis thermal detection monitoring setup, which is independent of a particular machine concept. The synchronous use of an infrared camera and a VIS NIR camera combines the advantages of high framerate and high spatial resolution, respectively. The capability of the system was tested in a process by building a specimen with forced changes of energy inputs and artificially designed defects. The measured responses in thermal radiation were analyzed and compared to defect mappings from volumetric non-destructive inspection by micro X-ray computed tomography (μCT).

2. Materials and Methods

2.1. L-PBF Processsing Conditions

2.1.1. Machine and Material

AISI 316L austenitic stainless steel powder was used on a commercial L-PBF system SLM280 HL (SLM Solutions Group AG, Lübeck, Germany), equipped with a single 400 W continuous wave (cw) ytterbium fiber laser emitting at a wavelength of 1070 nm with a spot size of approximately 80 μm in focal position. The gas atomized powder was specified by its supplier having an apparent density of 4.58 g/cm^{-3} and a mean diameter of 34.69 μm. The particle size distribution was specified by the supplier by means of laser diffraction: D_{10} = 18.22 μm, D_{50} = 30.50 μm and D_{90} = 55.87 μm The laser melting process was performed in an argon gas atmosphere with an oxygen content below 0.1%. The observed specimen was manufactured on a 100 mm × 100 mm stainless steel baseplate mounted on a 280 mm × 280 mm plate, which was not heated prior to or during the build process by the machine's preheating system. A layer thickness of 50 µm was chosen for the process. The machine was equipped with the commercial Melt Pool Monitoring (MPM) system, using two diodes for data sensing. The working principle and the technical data of this on-axis system can be read in [18]. MPM data was acquired during the process.

2.1.2. Description of the Specimen and Processing Parameters

Figure 2 illustrates the geometry of the specimen. It can be described by two cubes (B and C) of 5 mm edge length on top of a base cuboid (A) of the dimensions 10 mm × 10 mm × 5 mm, as cubes of this edge length are often used for L-PBF parameter studies. It also kept the field of view small for the optical monitoring equipment. Next to the cubes B and C inclined ramps (D) with an inclination angle of 40° and 45°, respectively, were placed. However, these ramps D were not subject of investigation in the course of this study. The specimen was removed from the build base plate by saw-cutting, the approximate position of the cutting line is depicted by the white dotted line. The base cuboid A contained artificial defects: In layers 61–73, an area of 0.6 mm × 3 mm was not exposed to the laser scan, followed by a non-exposed area of 0.37 mm × 3 mm in layer 74 at the same *x*-*y*-position. This region was called cavity in the course of this study. In layer 88, an area of 1 mm × 2.04 mm was not exposed at one corner. Both defects were open defects in the meaning that they were not enclosed within the part but are located at the edges.

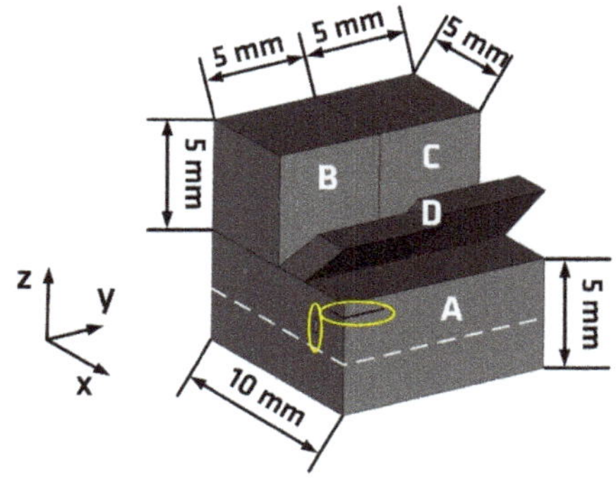

(a)

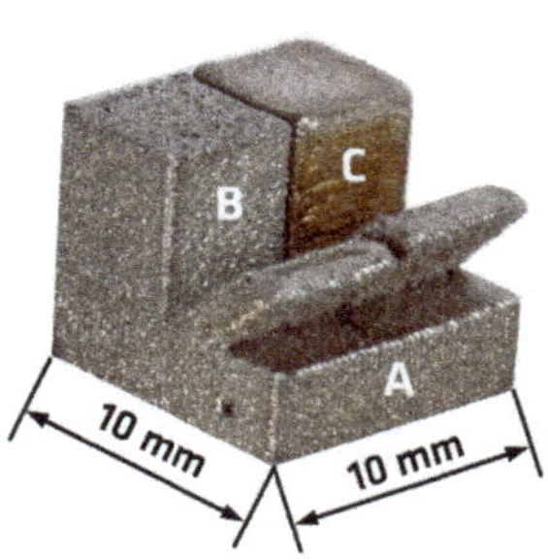

(b)

Figure 2. (**a**) Schematic of the specimen; volumes A, B and C were processed using different volume energy densities; volume D consisted of two ramps inclined by 40° and 45°. The white dotted line displays the cutting plane for part removal. The yellow circles enclose artificial defects. (**b**) Macroscopic photography of the specimen after cutting from the build base plate. The chapter 3, "Results and discussion", focuses on volumes A, B and C.

A stripe scanning pattern with a maximum stripe length of 10 mm was used as scanning strategy for the entire part with scanning vectors parallel to the specimen's edges. A rotation of the scanning vectors by 90° from layer to layer was applied. The hatch distance was set to 0.12 mm. No separate contour scanning was conducted, but a border offset of 0.08 mm was used, resulting in a separation of the single volume sections in *x*-*y*-direction on top of the base cuboid A. Three distinct scan parameters were used to produce the different volume sections to enforce the development of typical L-PBF defects during the process. While volume A was exposed by a parameter set which is normally used for high quality parts, volumes B and C were exposed by parameters resulting in a much lower or a much higher volumetric energy density (VED), as defined in [9]. The specific parameter sets are listed in Table 1. From previous investigations, the applied parameter sets were known for creating lack-of-fusion defects (in case of low VED, volume B) or keyhole porosity (in case of high VED, volume C).

Table 1. L-PBF (laser powder bed fusion) parameters used for the different sections of the specimen.

Parameters of Sections	A	B	C
Parameter Set	Standard volumetric energy density (VED)	Low VED	High VED
VED in J/mm^3	65.5	35.7	152.7
Laser Power P_L in W	275	150	275
Scanning Velocity v_s in mm/s	700	700	300

2.2. *In-Situ Monitoring by Thermograpy and Optical Tomography*

2.2.1. Optical Setup

For the in-situ monitoring of the L-PBF process, two different cameras were used simultaneously: A mid-wave infrared (MWIR) camera of the type ImageIR8300 (InfraTec GmbH Infrarotsensorik und Messtechnik, Dresden, Germany) sensitive in the spectral range from 2 μm to 5.7 μm was chosen for thermographic measurements. For optical tomography (bulb exposure of each layer exposition [30]), a CMOS camera of type M4020 (Teledyne Digital Imaging Inc., Billerica, MA, USA), which works in the visible and near infrared spectral range (VIS NIR), was used. It was equipped with a bandpass filter having a central wavelength of 880 nm and a spectral width of 50 nm. The cameras met the requirements of having a high enough spatial resolution to monitor on a melt track range of size (hatch distance was 120 μm) at different wavelengths. The second requirement of enough temporal dynamics to acquire information of cooling behavior is met by the MWIR camera at the displayed configuration. Figure 3 schematically depicts the optical setup. Both cameras were placed outside of the build camber on top of the machine and observe the same region of the build plate via a beam splitter through a sapphire window in the ceiling of the build chamber. Additionally, two gold mirrors were mounted at the ceiling inside the chamber using standard mounting plates which were screwed at the chamber ceiling. They offset the optical path to allow for a less inclined (close to parallel to the build plane) observation plane of the field of view (FOV). In addition, they enabled a comfortable selection of the FOV by manual adjustment of the mirrors. More technical data of the cameras are listed in Table 2. The maximum size of the FOV of the set-up in its presented configuration was limited by approx. 64 mm × 51.2 mm for the MWIR camera.

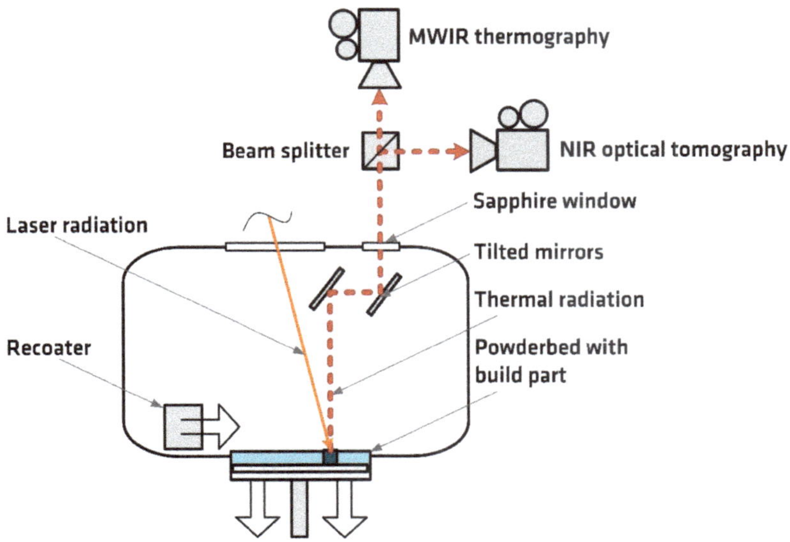

Figure 3. Schematic of the off-axis camera setup.

Table 2. Technical data of the used cameras.

Camera System	Thermography	Optical Tomography
Camera	Infratec ImageIR8300	DALSA Genie Nano-M4020
Detector	cooled InSb-focal-plane array	CMOS
Detector Size	640 × 512 pixels	4112 × 3012 pixels
Used Detector Elements	192 × 176 pixels	4112 × 3012 pixels
Lens Focal Length	100 mm	50 mm
Resulting Optical Resolution	100 μm/pixel	50 μm/pixel
Framerate and Exposure Time	900 Hz, 90 μs	bulb exposure of layer expositions
Effective Sensitive Spectral Range	2 μm–5.7 μm	855 nm–905 nm
Black Body Calibration Range	623 K–973 K	no calibration

2.2.2. Data Acquisition

For MWIR thermography, a thermal video of each layer exposition was recorded using a framerate of 900 Hz and an optical resolution of approximately 100 μm per detector element on the build surface. The acquisition of each layer was started by using a software trigger, which launched the recording when the signal of a single pixel in the image was above a certain threshold value, with pretrigger function. For the OT, the VIS NIR camera was operated in bulb mode, i.e., it was integrating intensity during a high level at the trigger input. Thus, an uninterrupted high-level signal during each layer exposition was needed. To this end, an analog output of the L-PBF machine that is proportional to the laser output power was processed by low pass filtering to remove short low-level phases during the time of jump vectors in between hatch vectors. A subsequent use of a Schmitt-Trigger ensured a high level of well-defined voltage. This signal processing, especially the lowpass filtering, led to a short delay between the first laser activation of a layer exposition and the start of the camera acquisition. Thus, a small dummy part was built outside the FOV of the cameras, that was always exposed first in each layer to counteract this delay. As a result, the camera chip was exposed from shortly before the start of the exposition of the part in the FOV until shortly after exposition for each layer.

The thermographic videos of the build process presented a very large amount of data containing a lot of information, e.g., the size of this study's thermographic data was 12 GB for a sample cube of 10 mm edge length, despite data compression. Extracting the most useful information for a certain monitoring objective is a very challenging task. In this study, only first qualitative and manually obtained results regarding the detection of defects are presented.

2.3. Ex-Situ Inspection by Computed Tomography, Metallography and Data Integration

The specimen was inspected by a custom-made industrial 3D micro CT scanner. This scanner was equipped with a 225 kV micro focus X-ray source (X-ray WorX GmbH, Garbsen, Germany) and a flat panel detector with 2048 × 2048 pixels. More details about this micro CT scanner can be found in [33]. The achieved voxel size of 7.12 μm enabled the detection of defects with a diameter larger than 14 μm. The commercial software VG Studio Max 3.1 (Volume Graphics GmbH, Heidelberg, Germany) was used to analyze the μCT data. The whole specimen was scanned. The volume of the volumes A, B and C was determined within VG Studio 3.1. The volumetric porosity in each 3D dataset of the three volumes was calculated using VG studio's porosity analysis module.

In addition to non-destructive inspection by μCT, destructive metallography was used to analyze the specimen. Therefore, polished cross sections of the specimen were etched using Beraha II etching detergent for 10 s to 30 s. Measurements of melt pool depth were conducted at light-microscopy micrographs for the three distinct parameter sets used in volumes A, B, and C. Light-microscopy was conducted using a Polyvar Met (C. Reichert Optische Werke AG, Wien Austria) which was equipped with a camera DFC290 (Leica Microsystems GmbH, Wetzlar, Germany).

The image registration software elastix 4.9 (University Medical Center Utrecht, Utrecht and contributors, The Netherlands) by Klein et al. [34] was utilized to overlay the extracted datasets of thermography and optical tomography onto the μCT data. Even though it is most frequently used in medical applications, it was well suited for the purpose of this study. It utilized affine transformation with the results getting visualized in Amira ZIB Edition 2019 (Thermo Fisher Scientific, FEI SAS, Illkirch-Graffenstaden, France).

3. Results and Discussion

3.1. Influence of Processing Parameters Analysed by μCT and Metallography

The ex-situ inspection of the specimen confirmed the expected presence of distinctive internal defects due to the variation of the VED in the volumes A, B and C. In addition to defect development, the melt pool depth and shape was influenced by the differences in VED, which is exemplified by the white lines in Figure 4 right depicting the melt pool boundaries of the top layer. Grain size determination was not in the focus of this study. Quantitative measurements from metallography (mean of ten melt pool depths of the topmost layers) and μCT (porosities with diameter larger than 14 μm) are summarized in Table 3.

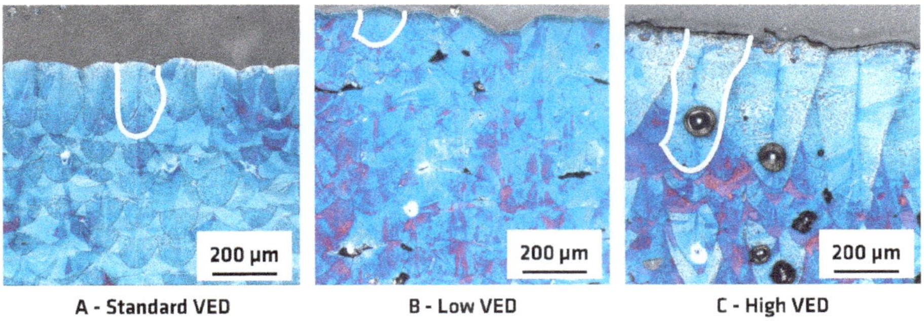

Figure 4. Light-microscopy of etched cross sections. Magnifications of the particular volumes **A**, **B** and **C**; the white line indicates the melt pool borders of the topmost layer.

Table 3. Quantitative measurements of melt pool depth and porosity percentage.

Melt Pool Depth and Porosity in Different Sections	A	B	C
Parameter Set	Standard VED	Low VED	High VED
Melt Pool Depth in μm	213 ± 19	117 ± 14	471 ± 54
Pore Volume in % of Part Volume Measured by μCT	<0.1	2.7 *	7.4

* Underestimation of the real value due to partly entrapped powder in the voids.

Qualitatively, the three parts could be described as: free of major defects (A), containing lack-of-fusion voids (B), containing keyhole porosity (C), as depicted in Figures 4 and 5. Due to the huge penetration depth of the laser in volume C and the resulting melt pool depth, which is considerably deeper than the layer thickness (factor 9), some keyhole pores could also be found in cuboid A underneath the cube C. The pores and voids of the cubes B and C, lack-of-fusion voids and keyhole pores, respectively, showed typical morphological differences: Keyhole pores, as entrapped vapor pores, are characteristic for deep penetration welding when having unstable welding conditions. They are predominately of spherical shape. Lack-of-fusion voids develop due to insufficient fusing of the molten powder material between the adjacent melt tracks and layers. They have predominately irregular and sharp shapes. More information about these well-known defects in L-PBF can be found in the literature, e.g., [11,16,35].

The scanning strategy determined the distribution of the pores. Keyhole pores were found predominately in-line with the scanning vector, whereas the lack-of-fusion voids were mainly located between adjacent scanning lines. Due to the chosen scanning pattern with stripes rotated by 90° after every layer, a clear pore or void pattern was observed in μCT data. In addition to these repeated defect patterns, bigger defect clusters were also observed within the cube B. Both are visualized in Figure 5. Missing distinctive reference points due to solidification shrinkage of the part complicated a precise layer-wise comparison of the data sets from process monitoring and μCT.

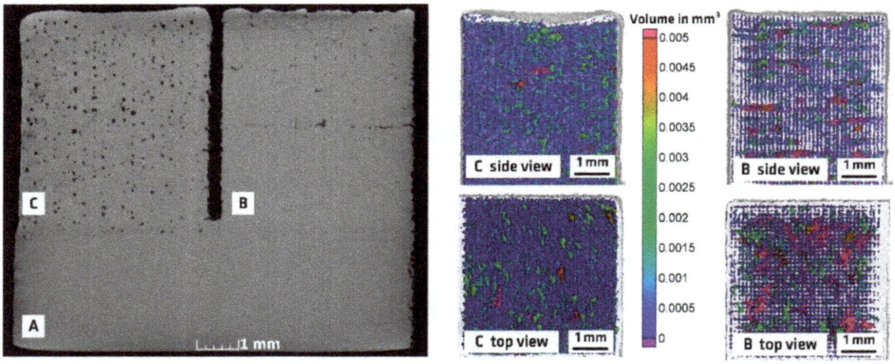

Figure 5. μCT images with achieved voxel size of 7.12 μm. (**a**) μCT cross section of the specimen. (**b**) Visualization of porosity according to their volume.

3.2. *In-Situ Monitoring*

3.2.1. Data analysis Concepts

Thermography–Apparent Temperatures

Since the MWIR camera was calibrated for black body radiation by its manufacturer, the temperatures that were acquired in this work were only apparent temperatures and well below the actual temperatures in the process. There were a number of reasons for this: Firstly, the emissivity of the melt pool, the built part surface and the powder are well below unity, the case for which the camera calibration is valid. For example, the emissivity value depends on surface roughness, wavelength, oxidation state and state of matter. For the used material AISI 316L in the unoxidized solid state, values of the total hemispherical emissivity between roughly 0.25 and 0.35 were expected [36]. In addition, to the material dependent emissivity, differences in heat conduction and heat capacity as well as differences in the processability could lead to different results for other materials. Secondly, the optical elements in the optical path, i.e., the beam splitter, the gold mirrors and the sapphire window attenuated the emitted thermal radiation, which was not taken into account by the vendor calibration. Additional thermal radiation emitted from the optical elements themselves in the optical path may, however, be neglected in the current setup, since their temperature was close to room temperature and well below the lower calibration limit of the camera of 623 K. Thirdly, the limited spatial resolution of 100 μm/pixel of the setup led to an additional error in the temperature calibration. The extremely high spatial temperature gradients on the surface caused each detector element to measure an average intensity of the corresponding area of the build surface. Since the relationship between intensity and temperature is strongly non-linear, the acquired temperature close to and within the melt pool was not an average temperature of this area but it was rather overestimated. Contrarily, if one is interested in the maximum temperature in the melt pool, this value will be underestimated since the spot size of the maximum temperature was expected to be much smaller than 200 μm × 200 μm, which would be the needed minimum size to avoid spatial undersampling. However, these resolution-dependent effects diminished with cooling down after solidification, since the spatial thermal gradients diminished as well.

As the temperature values given by the camera calibration were not valid, the analysis of the raw data intensity values of the camera would have been an opportunity instead. However, in that case, the strong nonlinearity of the intensity-temperature dependence would have complicated the data analysis and other methods to reduce the signal dynamics would have been required, such as analyzing the logarithm of the signal. Instead, in the course of this study, the apparent temperatures as given by the camera were analyzed. Therefore, the measured temperature values and measured temperature differences were only apparent.

Figure 6 gives an overview of typical thermographic data. Figure 6a shows a single thermogram of the process, illustrating that there was a certain amount of spatter ejected from the melt pool traveling through the image, which might complicate certain image processing steps. Figure 6b shows the time development of the apparent temperature at two different single pixel positions in the sample, as depicted in Figure 6a,c. As expected for the chosen build parameters, the apparent temperature rose multiple times, thus each volume element was reheated several times during each layer exposition.

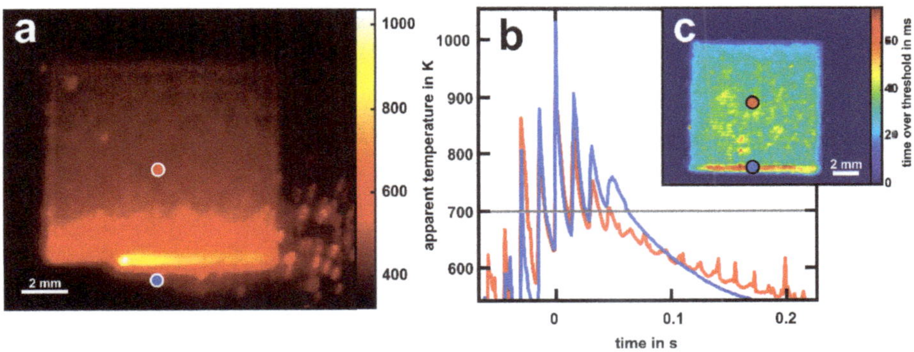

Figure 6. Thermographic data. (**a**) Thermogram of layer 45. In this layer, the exposition was performed in horizontal stripes from top to bottom in the image. In the shown frame, the laser moves from right to left; (**b**) time evolution of the apparent temperature in the positions (single pixels) marked in a. For better comparison, the time axes were shifted for each position to match the maximum apparent temperature at 0 s; (**c**) map of the time over threshold (TOT) for a threshold value of 700 K for layer 45 (see text).

Thermography–Time over Threshold (TOT)

Here, as a feature extracted from the thermographic data, the focus was set on the time a surface element of the specimen had an apparent temperature above a certain threshold value during the build of the respective layer, called time over threshold (TOT). This value could easily be extracted from the data. TOT showed increased (reduced) values if the cooling rate was decreased (increased) in the relevant temperature regime, depending on the chosen threshold value. Such deviations would have been expected if significant process deviations from the ideal case leading to defect formation were present. However, different kinds of defects or their formation mechanisms as well as the part geometry itself might influence the TOT in different ways. For example, the ejection of loosely bounded material that cools down relatively slowly leads to a locally strongly increased TOT directly in the layer of occurrence. In the subsequent layer, however, a small increase in TOT at this location gets possibly registered, due to heat accumulation above the defect. In contrast, the incomplete melting of powder leads to lack-of-fusion pores which might only lead to an increase of TOT in the subsequent layer and not in the layer of origin. Thus, for a thorough understanding of the process and its influence on the TOT, thorough studies are needed. The first results are presented here.

In Figure 6c, a map of the TOT of an apparent temperature of 700 K is shown for the layer formation of Figure 6, which did not contain any defects, as found by reference measurements. The increased

TOT at the lower side of the image was caused by a reduced cooling rate in this area, probably due to a heat accumulation due to the absence of solid material on the further lower hand side (only powder with reduced thermal conductivity is present here). Contrarily, as observable in Figure 6b, the final cooldown at the end of the exposition (blue curve) was faster, since the laser stopped the continuous heating of the part after the last scan track was exposed.

In fact, Lough et al. [37] also performed this kind of time over threshold analysis in the L-PBF process. However, they did not show a correlation between actual defects in the samples and this quantity. They chose the intensity of the solidification temperature plateau as a threshold value and analyzed the intensity values directly. In the present study, the framerate of the thermographic video was insufficient for the use of such a high temperature as threshold value. The maximum apparent temperature of the melt pool moved several pixels between subsequent frames and a strong stroboscope effect became visible in the TOT map due to temporal undersampling. Thus, a threshold value of 700 K was chosen for the analysis, since it was close to the highest possible value without being affected by stroboscopic effects. The presented correlation between extracted features and real defects in the final part was rather phenomenological and the spatial overlap between lack-of-fusion pores and TOT values was best for higher values, which were not affected by stroboscopic effects.

Optical Tomography

Contrary to thermography, the optical tomography camera recorded only a single image per layer exposition. Optical tomography recorded a signal which was proportional to the radiation intensity, emitted from the area on the specimen that was imaged onto the respective detector element (pixel), integrated over the duration of the entire layer exposition. Under normal process conditions, the highest temperatures within the melt pool would have contributed predominantly to the signal due to the nonlinearity of the temperature-intensity dependence despite the very rapid cooling. Therefore, the hatching pattern of the layer exposition can be seen in the optical tomography images, as depicted in Figure 7. However, since the measured signal was only proportional to the time integral of the intensity, it could not be distinguished between areas on the build surface that had an unusually high temperature and those that had an unusually low cooling rate at moderate temperatures. Nevertheless, both occurrences might have been indications for process deviations that might have induced the formation of defects. Thus, for defect detection, threshold analysis operating on the raw signal data were applied.

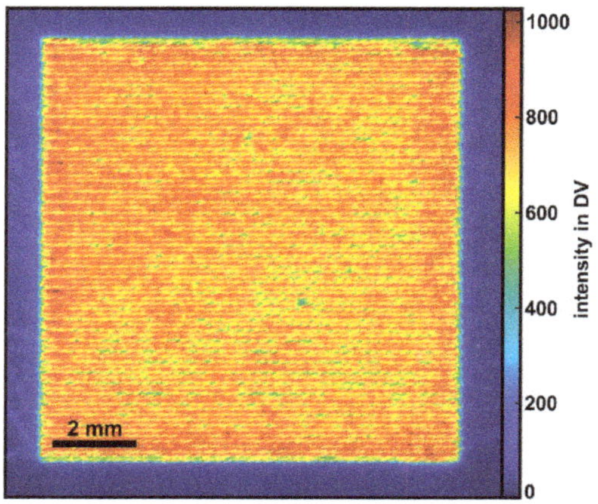

Figure 7. Typical optical tomography (OT) image of the base volume A (layer 99), built with standard parameters. The intensity distribution is quite homogeneous, and the hatching pattern can be observed.

3.2.2. Influence of Processing Parameters on In-Situ Monitoring Signatures

Standard Process Parameters

In the base volume A built with standard parameters, only small spatial deviations were observed in the TOT maps and OT images. Figure 8 shows TOT maps of two different layers in volume A, i.e., the first exposed layer (layer 1, Figure 8a) and the last horizontally exposed layer (layer 99, Figure 8b). Three regions of interest (ROIs) are marked for further analysis: A region at the upper hand side of the part, which was exposed at first (red), the center region (gray) and a region at the lower hand side, which was exposed at last (blue). The mean TOT values developed in these regions with increasing layer numbers as depicted in Figure 6c. Two main observations could be made: Firstly, the average value of the TOT was lowest in layer 1, which could be explained by the decreasing cooling rate for increasing built height. In layer 1, the heat could be conducted directly into the substrate plate, while in layer 99 it had to be conducted through a reduced cross section, i.e., the geometry of the already built part, which was thermally insulated laterally by the surrounding unmolten powder. In addition, with repeated heating of the growing part by melting of further layers, the temperature of the part rose initially. This effect could be seen for all three ROIs in Figure 8c. Secondly, in layer 99, the TOT in the lower ROI (blue) was considerably increased compared to the other ROIs, which was not observable in layer 1. As explained in Section 3.2.1 this feature was caused by the lowered heat conduction into the built part, due to the lack of solid material at the further lower hand side. This hypothesis was supported by the absence of this TOT increase in the first layers and the increasing difference between the TOT values of the upper ROI (red) and the middle ROI (grey) in subfigure c. Here it could also be seen that the TOT in the upper region was always lower than in the other ROIs, which was consistent with higher expected cooling rates at the beginning of a layer exposition due to missing heat accumulation during the layer exposition in the built part and substrate plate.

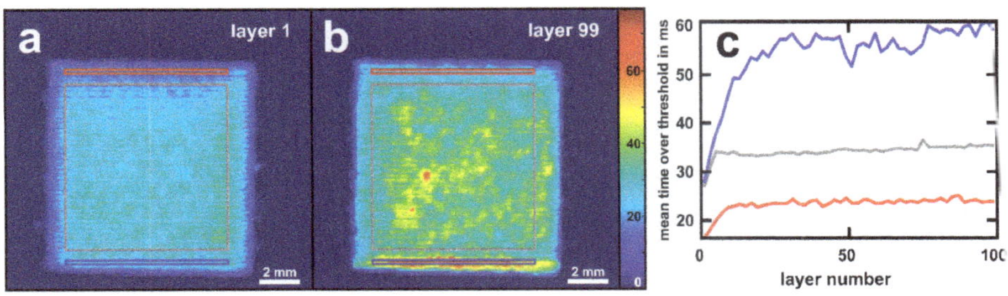

Figure 8. Time over threshold (TOT) for an apparent temperature value of 700 K for the volume A, built with standard parameters. The exposure was performed horizontally from top to bottom in both depicted layers (**a**) Layer 1, (**b**): Layer 99, (**c**) mean values of the TOT in the areas depicted in a and b in dependence on the layer number. Due to the 90° alternating scan strategy, only every second layer (horizontal scanning) was evaluated here. Red: start region of the layer exposition, grey: middle region, blue: end region.

It should be noted that all the observations mentioned so far were explainable and not surprising, however, they were also inherent to any build and did not cause any defect formation in the magnitude of intensity as observed here. Nevertheless, overheating due to geometry effects might lead to defect formation in more extreme cases of complex structures.

The OT image of layer 99 (same layer as shown in Figure 8b) is shown in Figure 7. There were no indications of differences observable in the signal between the different ROIs defined in Figure 6. This was consistent with the observation that the increase in the TOT for a threshold value of 700 K at the lower region of the part (blue ROI) were caused by a moderately decreased cooling rate at intermediate apparent temperatures close to 700 K (compare Figure 6b), while the OT signal was dominated by the intensity of the highest temperatures of the melt pool.

Comparison between Standard, High and Low VED Settings

In the volumes B and C, the mean values of the OT intensity and the TOT (700 K) deviated considerably from the values of the volume A. The mean values are given in Table 4.

Table 4. Mean values of the in-situ measurement results for the volumes with different energy density settings. The mean values were calculated for areas well within the respective volumes in order to avoid the influence of edge effects.

Volume	A	B	C
Parameter Set	Standard VED	Low VED	High VED
Mean TOT(700 K) in ms	28.1	14.5	280
Mean OT Signal in Digital Values (DV)	816	483	2860 *

* Underestimation of the real value due to partially oversaturated signal.

The detector intensity was occasionally saturated in the high VED volume C so its values of OT intensity were rather a lower limit. Since the defects that were created in volume C were quite uniformly distributed and no significant clustering occurred that might have been observable in the in-situ monitoring for correlation (see Section 3.1), the data from volume C was not further analyzed. Additional studies to investigate the onset energy density for keyhole pores and the corresponding in-situ monitoring signatures are currently being performed.

In the low VED density volume B, there were certain spatial signatures in the in-situ monitoring data that correlated with particularly large defects, as observable in the μCT data. They are discussed in the following section.

3.2.3. In-Situ Monitoring for Defect Detection

Processing Parameters Related Defects

In volume A (standard processing parameters), μCT data showed no defects, except for single, individual pores and the region directly below Volume C (see Figure 5). Thus, a comparison to the in-situ monitoring data showed that any signatures as they occurred during the manufacturing of this volume were indicative of a stable process. In particular, TOT (700 K) values of up to 60 ms, as occurring at the lower rim of volume A (see Figure 6c), were not a sufficient single indicator for the formation of part density related defects.

As described in Section 3.1, in volumes B and C, the development of pores and voids was directly provoked by the purposely chosen processing parameters and the applied scanning strategy, which led to partially unstable process conditions. As explained above, the porosity distribution in volume C (high VED) did not show any clearly localized signatures in μCT that would allow for a comparison of the in-situ results spatially resolved to the μCT data. Strongly increased values of OT intensity and TOT (700 K) were found, but further studies are needed for reliable local defect detection.

Contrary, in volume B, despite the low average VED, there were larger volumes without or with only minor porosity, as shown by μCT. However, there was also a number of small lack-of-fusion defects and large clusters of lack-of-fusion defects. Thus, a comparison of specific signatures in the in-situ data with the defect structure in the μCT data was promising. So far, this was conducted qualitatively and manually and is exemplified for two layers, which showed some nesting of porosity in particular areas, see Figure 9. In addition to the OT intensities and the TOT (700 K), intensity maps extracted from the commercial MPM system are also depicted. All three measurement systems showed an increase of the respective signature with respect to the average signature value of the layer at similar *x*-*y*-positions. A qualitative comparison to the respective µCT cross sections revealed increased occurrence of porosity at these positions. As a guide to the eye, a contour plot for a threshold value of 30 ms in the TOT (700 K) map is overlaid in the maps of the other signatures and shows a high correlation to the defective region in the µCT map. Interestingly, the value of 30 ms was significantly higher than the mean TOT (700 K) value of 14.5 ms in volume B, but close to the mean value of the TOT (700 K) value in the defect free volume A, which was manufactured using standard parameters. Thus, for a definitive defect detection, the TOT (700 K) alone was no clear indication for defect formation and at least a second criterion would be needed, possibly also extracted from the thermographic data or from another technique, e.g., indicating the operation in lack-of-fusion mode. Similarly, for OT, the intensities at the locations of the defects were well above the mean value for volume B (483 DV), but well in the range of volume A (mean value of 816 DV). Additionally, the correlation with the defect geometry from µCT appeared to be worse compared to the TOT (700 K). For the MPM system, the correlation to the defects from µCT appeared to be even better than the TOT (700 K), but the lack of knowledge about the nature of the displayed signal and the impossibility to export the data for further analysis hindered a more thorough analysis. Lack of fusion voids as unfused material underneath the layer of exposition seemed to reduce the heat conduction which results in heat accumulation in the current layer that can be measured as thermal signal deviation. This effect is discussed in the following paragraphs in more detail.

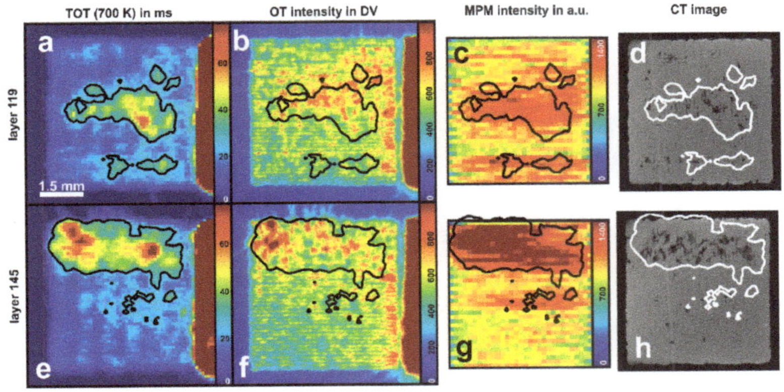

Figure 9. Comparison of data from in-situ monitoring with micro computed tomography (µCT) data for selected layers ((**a–d**): layer 119; (**e–h**): layer 145), showing the low VED volume B. Contour lines of the TOT (700 K) signal at a threshold value of 30 ms are overlayed in all datasets to guide the eye.

Artificial Defects with Standard Process Parameters

In addition to deviations of process parameters as source of defect development, defects can also arise from overheating caused by geometrical effects. The cavity in layers 61–74 (described in Section 2.1), which was not exposed, was clearly visible in the acquired data, as depicted in Figure 10. An increase of the TOT (700 K) was registered on top of this cavity in the first subsequent layer, which was exposed completely again, i.e., layer 75. This showed the heat accumulation at this position caused by reduced heat conduction due to unfused material underneath. A slight increase of the TOT signal was still observable in layer 77. Thus, additional information, which is in this case information about part's geometry, is needed to guide the analysis of the TOT data.

The strong signals caused by the artificial cavity illustrated the reduction of heat conduction caused by large defects which might have had an influence on a number of subsequent layers. The µCT images of the respective layers 75 and 77 did not show any defects. However, the exact determination of the layers in µCT was hindered due to shrinkage and missing reference points. Thus, a clear determination of the actual z-position of a detected defect in the part might have an uncertainty of one or two layers at the current state of research. The mentioned uncertainty of layer determination in the μCT dataset hindered a more detailed analysis of the defect position for this specimen. Follow-up experiments, allowing more accuracy in position correlation are planned. A detection of defects only during the exposure of layers above the defect containing layer would impede the possibility of real time control or repair as a desired action once monitoring works reliably, since a direct remelting of the damaged layer might not be possible at this late time of detection without adjustment of the scanning parameters. However, owing to the penetration depth of the laser as well as time for recoating, which can be used as calculation time, the potential of remelting for in-process rehabilitation of detected defects should not be neglected.

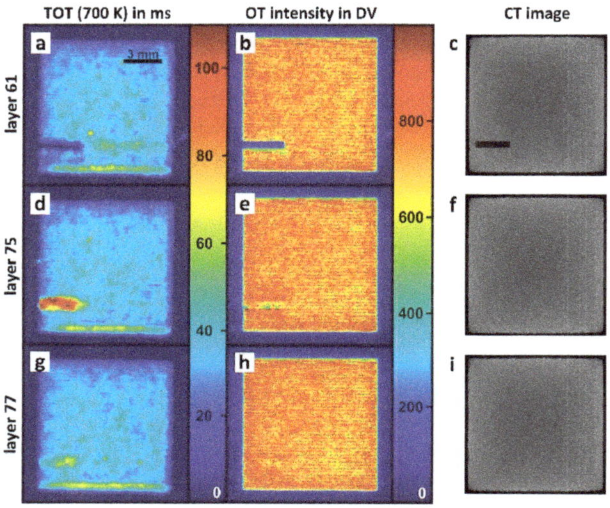

Figure 10. Comparison of data from in-situ monitoring with μCT data for an artificial defect (cavity): (**a–c**) show layer 61 where the cavity is not exposed; (**d–f**) show layer 75, which is the first fully exposed layer on top of the cavity; (**g–i**) show layer 77 as third layer above the cavity.

Furthermore, a corner in layer 88 of volume A was not exposed. Whereas the missing laser-material interaction was detected by the monitoring equipment in layer 88, no signal deviations was found at this region in the subsequent layer. This perfectly matched the μCT data and metallographic examinations of this region, which did not detect any defects related to the missing exposition in or in close proximity of this region in layer 88. In this case, a kind of self-healing occurred. During the build process, the laser remelted several subjacent solidified layers. Therefore, the unmolten powder particles in that corner were melted by the laser exposition of the subsequent layer. Surprisingly, no signal deviations were registered at this corner in layer 89, where the layer thickness had been twice the standard value of 50 μm resulting in a reduction of the local VED by half. This example showed potential sources of misinterpretation: On the one hand the exposition defect was healed by the laser exposition of subsequent layers but appeared in the layer of its origin. On the other hand, the sensitivity for low energy inputs has to be improved for the applied data examination strategy, as no deviations in layer 89 were detected by using the described threshold values. Both examples of artificial defects clearly indicated the necessity of knowledge about the geometry and slice pattern of a part during the analysis of monitoring data.

3.3. Automation of Data Handling

Image registration was used to reduce the spatial uncertainty and to allow data comparison, which is important to identify signal thresholds for defect detection. The upcoming section presents first results of this process with a focus on volume B. Further work in the field of data integration will be conducted and reported in a subsequent study.

Image registration was utilized to overlay data from different imaging modalities which differed in recording conditions and spatial resolutions, especially in the build direction. It was also important to indicate, that the camera-based data were captured during the manufacturing process in contrast to the µCT data, which were acquired after the process. Therefore, the in-situ data were not able to account for possible shrinkages, warpages and other deformations. Hence, registration was needed.

The choice of transformation represented an important aspect of the registration process. The type of transformation determines how a dataset, e.g., a specific TOT, is deformed to match the shape of another dataset, in this case, µCT data. Two transformations were considered for this paper: rigid (Euler) transformation and affine transformation. Rigid transformation allows rotation and translation of a dataset, only. Therefore, it was not able to describe uncertainties in the exact pixel size of thermography and optical tomography data, which might have led to mistakes in the transformation, and spatial deviations due to solidification shrinkage. Affine transformation added the possibility of scaling and shearing. Thus, it could be applied to compensate uncertainties in pixel size but failed to describe irregular solidification shrinkage or other geometrical discrepancies.

Thermography and OT data were registered onto the µCT data to compare the two methods. The advanced Mattes mutual information metric [38] was used, utilizing a Gaussian pyramid scheme. Figure 11 shows the results of 18,500 iteration steps. Affine transformation was able to match the edges of the specimen more precise than rigid transformation since it could adjust to measurement uncertainties of the optical system's pixel size. Therefore, it was used for visualization.

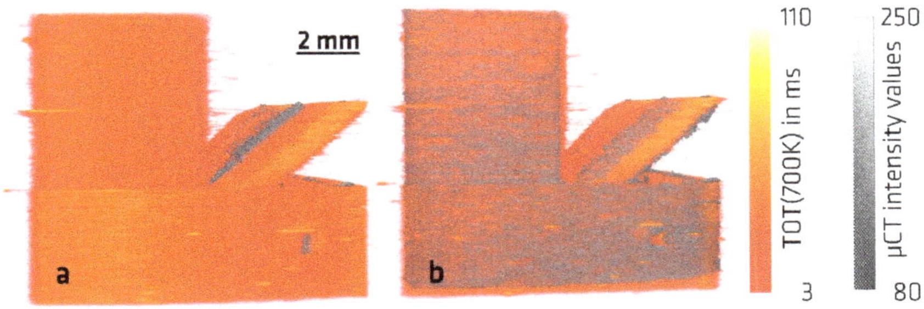

Figure 11. Results of image registration with Euler (**a**) and affine (**b**) transformation.

A virtual envelope of the specimen was utilized to guide a user during the analysis of the datasets. The µCT scan was selected to shape the envelope because it represented the geometry of the produced specimen. The watershed segmentation method in the Amira software was applied to determine the specimen's surface based on the edge image of the µCT dataset. Amira's Auto Threshold module was used to visualize the pores in µCT data: The implemented method of Otsu et. al. [39] automatically computed a threshold for binary segmentation based on the histogram of the 256 grey values of the 8-bit µCT data. Otsu's method is a simple but effective method to compute a single threshold separating a set of values into two classes. The threshold computed by Otsu's method minimizes the intraclass variance and, hence, maximizes the interclass variance. Taking the defects as defects of the same kind building a continuum rather than separate classes, two classes in the data needed separation. Under the assumption that only two classes were present in the data, the computed threshold was the best achievable value for a single threshold. Segmented lack-of-fusion pores were visualized as a binary image containing all voxels below the calculated threshold of 80. In contrast, defects in the optical tomography and thermography data were extracted by manually choosing a suitable threshold for Volume B.

Lastly, Amira's surface rendering module was used to display a transparent version of the specimen's surface, pores segmented from µCT data and occurring optical tomography signal of over

700 DV and TOT (700 K) greater than 30 ms. This is illustrated in Figure 12. A first analysis of the images in Figure 12 revealed that there was only a small volumetric overlap between porosity and the anomalies detected in thermography and OT. This was not surprising since the thermal signals as pseudo-3D-volume elements were not to be expected to overlay directly to 3D volumes of pores as real volumetric defects and even less to overlay completely. Defects below the actual monitored layer affected the thermal signals due to influence on thermal conduction. However, due to the complex morphologies of the defects, this influence was not expected to be homogeneous over the entire cross-section of a defect. Nonetheless, a quick analysis of the overlap of thermal signal anomalies and pores was conducted as a basis for further in-depth analysis. Table 5 presents the binary volumetric overlap, which classifies pores and anomalies as either found or not found by partial overlapping of each other, using two distinct ways of analysis: at first, the overlap of detected pores by thermal signal anomalies, as a measure of the likelihood of lack of fusion voids to be detected; secondly, the overlap of thermal signal anomalies by the detected pores, as a measure of the likelihood of signal anomalies to detect lack of fusion voids and not fully consolidated material. Besides, the same binary volumetric overlaps were analyzed for all pores larger than 0.001 mm^3 and after shifting the datasets of thermal signals and μCT by one layer to account for the layer discrepancy between the occurrence of a defect and its detection. This analysis was conducted for the lower left quadrant of Volume B to avoid the influence of Volume C on the thermography data. Note that the cumulative volume of the anomalies detected with TOT (700 K) over 30 ms and OT over 700 DV comprised only 90% and 8% of the cumulative pore volume, respectively.

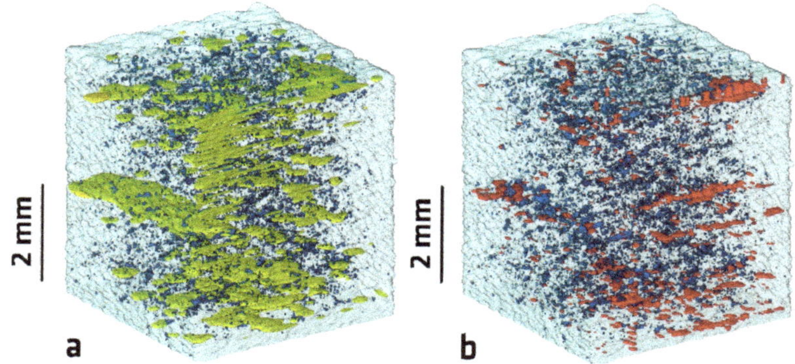

Figure 12. Rendering of the transparent surface with overlap of defects detected in μCT, thermography and optical tomography in Volume B. (**a**) Blue–μCT defects, yellow—thermography irregularities of TOT (700 K) greater than 30 ms; (**b**) red—optical tomography with more than 700 DV.

A rather low percentage of binary volumetric overlap could be observed between the volume of the lack of fusion voids segmented in the μCT data and the detected anomalies in TOT and OT. The percentage increased when neglecting small pores. The low percentage of binary volumetric overlap was particularly striking for OT and could be partially explained by the much smaller volume of the OT anomalies compared to pores segmented from μCT data. Furthermore, it seemed like the anomalies in OT occurred in the surrounding of the pores, rather than on top of them. The low numbers of binary volumetric overlap contained in Table 5 were only presented to inform about the informative value of rough estimations for defect detections of thermal monitoring equipment based on single signals only. A sophisticated threshold determination and sophisticated data fusion of different signals was not conducted yet (e.g., OT and TOT at different threshold values and different shifts in the *z*-direction). In addition, the surrounding of the thermal signals or measured defects was not enlarged by different radii during the overlapping analysis, which could have improved the detection rate at the expense of potential falsifications due to potential double counting of defects. However, the binary overlap of thermography signal anomalies by the detected pores of up to 71% appeared already very promising for further in-depth correlation approaches. When deeper analysis will be applied a detailed analysis of the uncertainty of the measurements will also be conducted. Questions regarding the position of the inspected specimen on the build plate, robustness of the measurements against spatter and possible falsifications due to process plume between the camera and the object will be analyzed. A more detailed look at specimens with larger built heights will also be applied, as this can lead to heat accumulation effects, which directly influences the thermal signals. The existing data will be used for further analysis of those topics in another study.

Table 5. Binary volumetric overlap of segmented pores from µCT data and anomalies detected by thermography (TOT (700K) > 30 ms) and OT (OT signal > 700 DV), where the relative cumulative volumes compared with respect to cumulative pore volume are 90% for thermography and 8% for OT.

Overlap of Signals and Defects	All Pores, 0 Layers Shifted	All Pores, 1 Layer Shifted	Pores > 0.001 mm^3, 0 Layers Shifted	Pores > 0.001 mm^3, 1 Layer Shifted
Micro-CT pores overlapped by thermography anomalies	3.1%	3.1%	35.1%	29.9%
Micro-CT pores overlapped by OT anomalies	0.7%	0.5%	14.3%	11.7%
Thermography anomalies overlapped by micro-CT pores	71.4%	55.5%	33.3%	30.2%
OT anomalies overlapped by micro-CT pores	17.1%	15.1%	11.0%	8.8%

4. Conclusions

A new setup for in-situ off-axis monitoring by synchronous observation of a L-PBF process via optical tomography and infrared thermography using the same optical path through the ceiling of the build chamber was successfully tested. The built up of a specimen containing artificial defects as well as purposely chosen process parameter variations was monitored. The feasibility of the set-up to detect defects has been verified by a comparison with data obtained by micro computed tomography. Despite non-calibrated thermography temperatures, the concept of time over threshold (TOT), i.e., time over a defined apparent temperature, was able to detect process deviations, which could be correlated to lack-of-fusion void clusters in qualitatively good agreement to µCT data for particular layers. In addition, a comparison to commercial monitoring equipment described the high potential of the presented setup as an independent monitoring solution with total data access. Due to the synchronous use of the distinct cameras, the advantages of a high spatial resolution (OT, here 50 µm/pixel), and relatively high temporal dynamics (thermography, here 900 Hz) were combined.

The manually and phenomenological data analysis approach constituted a basis for automated in-depth evaluation by data fusion. Although a first attempt of automated data integration showed a promising binary overlap of thermography signal anomalies and detected pores of up to 71%, the conducted data integration attempt revealed the need for data fusion concepts as single signal values seemed to have only comparably low spatial overlapping between signal and detected defects for the entire of a sub-volume of the monitored specimen.

Author Contributions: Conceptualization, G.M., A.U., S.J.A.; methodology, G.M., A.U., S.J.A., P.H.; formal analysis, G.M., A.U., S.J.A., P.H.; investigation, G.M., A.U., S.J.A., P.H.; writing—original draft preparation, G.M., A.U., S.J.A., P.H.; writing—review and editing, G.M., A.U., S.J.A., P.H., K.H., D.B., C.M.; visualization, G.M., A.U., S.J.A., P.H.; supervision, K.H., D.B., C.M.; project administration, S.J.A., C.M. All authors have read and agreed to the published version of the manuscript.

Funding: This research was funded by BAM within the focus area Materials.

Conflicts of Interest: The authors declare no conflicts of interest. The funders had no role in the design of the study; in the collection, analyses, or interpretation of data; in the writing of the manuscript, or in the decision to publish the results.

References

1. Kruth, J.P.; Levy, G.; Klocke, F.; Childs, T.H.C. Consolidation phenomena in laser and powder-bed based layered manufacturing. *CIRP Ann.* **2007**, *56*, 730–759. [CrossRef]
2. Schmidt, M.; Merklein, M.; Bourell, D.; Dimitrov, D.; Hausotte, T.; Wegener, K.; Overmeyer, L.; Vollertsen, F.; Levy, G.N. Laser based additive manufacturing in industry and academia. *CIRP Ann.* **2017**, *66*, 561–583. [CrossRef]
3. Bourell, D.; Kruth, J.P.; Leu, M.; Levy, G.; Rosen, D.; Beese, A.M.; Clare, A. Materials for additive manufacturing. *CIRP Ann.* **2017**, *66*, 659–681. [CrossRef]
4. Herzog, D.; Seyda, V.; Wycisk, E.; Emmelmann, C. Additive manufacturing of metals. *Acta Mater.* **2016**, *117*, 371–392. [CrossRef]
5. DebRoy, T.; Wei, H.L.; Zuback, J.S.; Mukherjee, T.; Elmer, J.W.; Milewski, J.O.; Beese, A.M.; Wilson-Heid, A.; De, A.; Zhang, W. Additive manufacturing of metallic components—Process, structure and properties. *Prog. Mater. Sci.* **2018**, *92*, 112–224. [CrossRef]
6. Wohlers, T.T.; Campbell, I.; Diegel, O.; Kowen, J.; Caffrey, T. *Wohlers Report. 3D Printing and Additive Manufacturing State of the Industry*; Wohlers Associates, Inc.: Fort Collins, CO, USA, April 2007.
7. Spierings, A.B.; Dawson, K.; Voegtlin, M.; Palm, F.; Uggowitzer, P.J. Microstructure and mechanical properties of as-processed scandium-modified aluminium using selective laser melting. *CIRP Ann.* **2016**, *65*, 213–216. [CrossRef]
8. Karg, M.; Ahuja, B.; Kuryntsev, S.; Gorunov, A.; Schmidt, M. Processability of high strength Aluminium-Copper alloys AW-2022 and 2024 by Laser Beam Melting in Powder Bed. In Proceedings of the 25th Annual International Solid Freeform Symposium, Austin, TX, USA, 4–6 October 2014.
9. Mohr, G.; Johannsen, J.; Knoop, D.; Gärtner, E.; Hummert, K.; Emmelmann, C. Processing of a high-strength Al-Fe-Ni alloy using laser beam melting and its potential for in-situ graded mechanical properties. In Proceedings of the Lasers in Manufacturing Conference, Munich, Germany, 26–29 June 2017.
10. Tapia, G.; Elwany, A. A review on process monitoring and control in metal-based additive manufacturing. *J. Manuf. Sci. Eng.* **2014**, *136*, 060801. [CrossRef]
11. Grasso, M.; Colosimo, B.M. Process defects and in situ monitoring methods in metal powder bed fusion: A review. *Meas. Sci. Technol.* **2017**, *28*, 1–25. [CrossRef]
12. Rehme, O.; Emmelmann, C. Reproducibility for properties of selective laser melting products. In Proceedings of the Third International WLT-Conference on Lasers in Manufacturing, Munich, Germany, 13–16 June 2005.
13. O'Regan, P.; Prickett, P.; Setchi, R.; Hankins, G.; Jones, N. Metal Based Additive Layer Manufacturing: Variations, Correlations and Process Control. *Procedia Comput. Sci.* **2016**, *96*, 216–224. [CrossRef]
14. Zerbst, U.; Hilgenberg, K. Damage development and damage tolerance of structure maanufactuired by selective laser melting—A review. *Procedia Struct. Integr.* **2017**, *7*, 141–148. [CrossRef]
15. Rashid, R.; Masood, S.H.; Ruan, D.; Palanisamy, S.; Rahman Rashid, R.A.; Brandt, M. Effect of scan strategy on density and metallurgical properties of 17-4PH parts printed by Selective Laser Melting (SLM). *J. Mater. Process. Technol.* **2017**, *249*, 502–511. [CrossRef]
16. Sharratt, B. *Non-Destructive Techniques and Technologies for Qualification of Additive Manufactured Parts and Processes: A Literature Review*; Contract Report No. DRDC-RDDC-2015-C035 [Online]; Sharratt Research and Consulting Inc.: Victoria, BC, Canada, 2015.
17. Spears, T.G.; Gold, S.A. In-process sensing in selective laser melting (SLM) additive manufacturing. *Integrating Mater. Manuf. Innov.* **2016**, *5*, 16–40.
18. Alberts, D.; Schwarze, D.; Witt, G. High speed melt pool & laser power monitoring for selective laser melting (SLM®). In *9th International Conference on Photonic Technologies LANE*; Bayrisches Laserzentrum GmbH: Fürth, Germany, 2016; Volume 2016.
19. Neef, A.; Seyda, V.; Herzog, D.; Emmelmann, C.; Schönleber, M.; Kogel-Hollacher, M. Low coherence interferometry in selective laser melting. *Phys. Procedia* **2014**, *56*, 82–89. [CrossRef]
20. Ali, U.; Mahmoodkhani, Y.; Imani Shahabad, S.; Esmaeilizadeh, R.; Liravi, F.; Sheydaeian, E.; Huang, K.Y.; Marzbanrad, E.; Vlasea, M.; Toyserkani, E. On the measurement of relative powder-bed compaction density in powder-bed additive manufacturing processes. *Mater. Des.* **2018**, *155*, 495–501. [CrossRef]

21. Grasso, M.; Demir, A.G.; Previtali, B.; Colosimo, B.M. In situ monitoring of selective laser melting of zinc powder via infrared imaging of the process plume. *Robot. Comput.-Integr. Manuf.* **2018**, *49*, 229–239. [CrossRef]
22. Anwar, A.B.; Pham, Q.-C. Study of the spatter distribution on the powder bed during selective laser melting. *Addit. Manuf.* **2018**, *22*, 86–97. [CrossRef]
23. Mohr, G.; Seeger, S.; Hilgenberg, K. Measurement of particle emissions in Laser Powder Bed Fusion (L-PBF) processes and its potential for in-situ process monitoring. In Proceedings of the Euro PM 2019, Maastricht, The Netherlands, 13–16 October 2019.
24. Dunbar, A.J. *Analysis of the Laser Powder Bed Fusion Additive Manufacturing Process through Experimental Measurement and Finite Element Modeling*; Open Access: State College, PA, USA, 2016.
25. Craeghs, T.; Clijsters, S.; Kruth, J.P.; Bechmann, F.; Ebert, M.C. Detection of Process Failures in Layerwise Laser Melting with Optical Process Monitoring. *Phys. Procedia* **2012**, *39*, 753–759. [CrossRef]
26. Zenzinger, G.; Bamberg, J.; Ladewig, A.; Hess, T.; Henkel, B.; Satzger, W. Process monitoring of additive manufacturing by using optical tomography. In Proceedings of the AIP Conference Proceedings, Boise, ID, USA, 20–25 July 2014; Volume 1650, pp. 164–170.
27. Bartlett, J.L.; Heim, F.M.; Murty, Y.V.; Li, X. In situ defect detection in selective laser melting via full-field infrared thermography. *Addit. Manuf.* **2018**, *24*, 595–605. [CrossRef]
28. Coeck, S.; Bisht, M.; Plas, J.; Verbist, F. Prediction of lack of fusion porosity in selective laser melting based on melt pool monitoring data. *Addit. Manuf.* **2019**, *25*, 347–356. [CrossRef]
29. Krauss, H. *Qualitätssicherung beim Laserstrahlschmelzen durch schichtweise thermografische In-Process-Überwachung*; Herbert Utz Verlag: Munich, Germany, 2017; Volume 325.
30. Bamberg, J.; Zenzinger, G.; Ladewig, A. In-process control of selective laser melting by quantitative optical tomography. In Proceedings of the 19th World Conference on Non-Destructive Testing, Munich, Germany, 13–17 June 2016.
31. Gögelein, A.; Ladewig, A.; Zenzinger, G.; Bamberg, J. Process Monitoring of Additive Manufacturing by Using Optical Tomography. In Proceedings of the 14th Quantitaitve InfraRed Thermography Conference, Berlin, Germany, 25–29 June 2018.
32. Mitchell, J.A.; Ivanoff, T.A.; Dagel, D.; Madison, J.D.; Jared, B. Linking pyrometry to porosity in additively manufactured metals. *Addit. Manuf.* **2020**, *31*, 100946. [CrossRef]
33. Oesch, T.; Weise, F.; Meinel, D.; Gollwitzer, C. Quantitative In-situ Analysis of Water Transport in Concrete Completed Using X-ray Computed Tomography. *Transp. Porous Media* **2019**, *127*, 371–389. [CrossRef]
34. Klein, S.; Staring, M.; Murphy, K.; Viergever, M.A.; Pluim, J.P. Elastix: a toolbox for intensity-based medical image registration. *IEEE Trans. Med. Imaging* **2009**, *29*, 196–205. [CrossRef] [PubMed]
35. Zhang, B.; Li, Y.; Bai, Q. Defect formation mechanisms in selective laser melting: A review. *Chin. J. Mech. Eng.* **2017**, *30*, 515–527. [CrossRef]
36. Hunnewell, T.S.; Walton, K.L.; Sharma, S.; Ghosh, T.K.; Tompson, R.V.; Viswanath, D.S.; Loyalka, S.K. Total Hemispherical Emissivity of SS 316L with Simulated Very High Temperature Reactor Surface Conditions. *Nucl. Technol.* **2017**, *198*, 293–305. [CrossRef]
37. Lough, C.S.; Wang, X.; Smith, C.C.; Adeniji, O.; Landers, R.G.; Bristow, D.A.; Kinzel, E.C. Use of SWIR Imaging to Monitor Layer-to-Layer Part Quality during SLM of 304L Stainless Steel. In Proceedings of the 29th Annual International Solid Freeform Fabrication Symposium, Austin, TX, USA, 13–15 August 2018.
38. Mattes, D.; Haynor, D.R.; Vesselle, H.; Lewellyn, T.K.; Eubank, W. Nonrigid multimodality image registration. In *Medical Imaging 2001: Image Processing*; Sonka, M., Hanson, K.M., Eds.; International Society for Optics and Photonics: Bellingham, WA, USA, 2001; pp. 1609–1620.
39. Otsu, N. A threshold selection method from gray-level histograms. *IEEE Trans. Syst. Man Cybern.* **1979**, *9*, 62–66. [CrossRef]

metals

MDPI

Article

Exploring the Correlation between Subsurface Residual Stresses and Manufacturing Parameters in Laser Powder Bed Fused Ti-6Al-4V

Tatiana Mishurova [1,*], Katia Artzt [2], Jan Haubrich [2], Guillermo Requena [2,3] and Giovanni Bruno [1,4]

1 Federal Institute for Materials Research and Testing (BAM; Bundesanstalt für Materialforschung und–prüfung), Unter den Eichen 87, 12205 Berlin, Germany; giovanni.bruno@bam.de

2 Institute of Materials Research, German Aerospace Center (DLR; Deutsches Zentrum für Luft-und Raumfahrt), Linder Höhe, 51147 Cologne, Germany; katia.artzt@dlr.de (K.A.); jan.haubrich@dlr.de (J.H.); guillermo.requena@dlr.de (G.R.)

3 Metallic Structures and Materials Systems for Aerospace Engineering, RWTH Aachen University, 52062 Aachen, Germany

4 Institute of Physics and Astronomy, University of Potsdam, Karl-Liebknecht-Straße 24/25, 14476 Potsdam, Germany

* Correspondence: tatiana.mishurova@bam.de; Tel.: +49-30-8104-4527

Received: 6 February 2019; Accepted: 20 February 2019; Published: 22 February 2019

Abstract: Subsurface residual stresses (RS) were investigated in Ti-6Al-4V cuboid samples by means of X-ray synchrotron diffraction. The samples were manufactured by laser powder bed fusion (LPBF) applying different processing parameters, not commonly considered in open literature, in order to assess their influence on RS state. While investigating the effect of process parameters used for the calculation of volumetric energy density (such as laser velocity, laser power and hatch distance), we observed that an increase of energy density led to a decrease of RS, although not to the same extent for every parameter variation. Additionally, the effect of support structure, sample roughness and LPBF machine effects potentially coming from Ar flow were studied. We observed no influence of support structure on subsurface RS while the orientation with respect to Ar flow showed to have an impact on RS. We conclude recommending monitoring such parameters to improve part reliability and reproducibility.

Keywords: residual stress; synchrotron X-ray diffraction; additive manufacturing; Ti-6Al-4V

1. Introduction

The production of near net-shape metallic components with complex geometries, not achievable with conventional production method, is one of the main advantages of laser powder bed fusion (LPBF) additive manufacturing (AM) technique [1]. One of the well-studied materials for LPBF is Ti-6Al-4V alloy. This alloy is the most used titanium alloy and it is widely spread in aerospace [2]. LPBF allows building design-optimized structures, saving raw material, and reducing the weight of the component. Therefore, the aerospace industry aims at the development and certification of the AM process for this very alloy. Additionally, this alloy is widely used in the medical field, where AM techniques allow creating customized medical implants [3]. The presence of large residual stresses (RS) is a relevant issue in LPBF process since it may lead to detachment of support structure, cracking of the part or to geometrical distortion during manufacturing [4,5]. As reported in [6,7], the RS in LPBF have two dominating origins: the temperature gradient, and the cool-down phase of the molten top layer. As the result, high tensile RS are normally present in the surface and subsurface regions of LPBF materials [8,9] and they are balanced by compressive stress in the bulk [10]. Thus, every parameter influencing the thermal history of the part must influence the RS fields. The analysis of the influence of all manufacturing parameter sets and of the additional influence from the build chamber atmosphere on the RS state of AM parts is a formidable task. A design of experiment that encompasses all effects has never been attempted, and the present level of knowledge would not even benefit from a Taguchi approach [11]. Therefore, normally a multitude of parameters such as laser power (P), scanning velocity (v), hatch distance (h), or powder layer thickness (x) are bundled together in the volumetric energy density (E_v):

$$E_v = \frac{P}{v \cdot h \cdot x}, \tag{1}$$

This parameter is normally used for porosity optimization (i.e., screening) in LPBF process. However, E_v represents a convolution of parameters and therefore simplifies the effect of each parameter. The same E_v, obtained as a combination of different parameters, can lead to different microstructures and properties of materials, as demonstrated in recent works [12,13].

Numerical simulations help to understand the RS distribution in parts built with different laser parameters and scanning strategies [14,15]. However, experimental data on RS in LPBF materials are still incomplete or even show contradictory trends depending on material and measurement method. For instance, in [16] the change of E_v of a fully dense AISI316L steel (density around 99.9%) has shown no impact on RS, measured by laboratory X-ray diffraction. In contrast, in our previous study, highly dense Ti-6Al-4V with different E_v has shown significant difference in RS along the building direction, obtained by synchrotron X-ray diffraction (SXRD) [13]. Additionally, it has been shown in [8] that SXRD allows building an RS profile along the depth (to around 100 µm for Ti-6Al-4V). The higher depth-resolved RS profiles (micrometer range) could be obtained by semi-destructive micro-hole drilling method [17].

As a further example, the rotating scanning strategy has been considered in many studies as the one giving the most homogeneous temperature distribution, hence, low RS [7,14]. However, in some studies the alternating 90° scanning strategy with stripe pattern has been recommended for minimization of RS [18]. The same disagreement can be found in case of support structures and their influence on RS. RS analysis of Ti-6Al-4V parts by X-ray diffraction in [19] has shown a reduction of stresses for a sample on support structure (with respect to sample built directly on the base plate). In contrast, in [20] a LPBF Al10SiMg sample built directly on the base plate (without support structures) has shown the lower RS, measured by hole drilling.

Not only laser parameters and scanning strategy [7,21] must be taken into account, but also other factors basically poorly explored in the open literature such as the position on the base plate, the laser path, or the Argon flow [13]. Those factors cannot always be captured by simulations and sometimes are not experimentally measurable. Therefore, RS studies should consider controllable manufacturing parameters as well as factors that do not appear, at first hand, to impact the manufacturing process. Moreover, all studies are material dependent. Therefore, they can be considered as unique cases. In such sparse knowledge matrix, any study focusing on explored or hidden parameters will bring a contribution to the general understanding of the process.

In this study, we explore the influence of parameters that have been (in the best case) partially neglected in previous studies. We show that this additional analysis complements the existing literature, bringing some missing bricks to the solution of the problem of RS assessment in AM parts, and allows corroborating (or contradicting) existing scenarios describing the origin of RS in AM parts. The work is therefore also targeted at generating further discussion on this highly complex topic.

2. Materials and Methods

2.1. Materials

18 cuboids, with a size of 5 × 5 × 15 mm^3, were produced in a SLM Solutions 280^{HL} machine (SLM Solution Group AG, Lübeck, Germany) using plasma atomized Ti-6Al-4V ELI grade 23 powder from AP&C. The size distribution of the powder's spherical particles was measured by laser diffraction

(Beckman Coulter LS 13320 PIDS, Beckman Coulter GmbH, Krefeld, Germany), resulting in D-values D_{10} = 22 μm, D_{50} = 34 μm and D_{90} = 46 μm. Table 1 shows the manufacturing parameters. Each sample was produced on a small round-shaped baseplate (Ø = 3 cm), which was screwed to a large baseplate 280 × 280 mm^2 preheated to 200 °C (see Figure 1). Since in our previous study [13] the position on the baseplate showed to be an important factor for RS, the location of each sample is uniquely defined (see Figure 1). Two samples were built on a 2 mm support structure (SS). The same powder layer thickness of 30 μm and a chessboard scan strategy with a minimum field size of 5 mm were chosen for all of them, while other processing parameters were varied. Also, for all samples, a set of two contours and one intermediate line (i.e., so called fill-contour line) were applied before the sample bulk was processed. In the contour scan, the same parameters, with 100 W laser power and 525 mm/s scan velocity, were used for all specimens. The microstructure and the RS for some cuboids from the same build job were partially discussed in [13]. The names of the parameter sets were kept the same as used in our previous studies [8,13,22,23] for consistency, and are indicated in Table 1. The A4 parameter set presents porosity-optimized conditions [23], and is always used as a reference.

Table 1. Processing parameters.

Sample Groups	Name	Power, *p*	Hatch Distance, *h*	Velocity, *v*	Volumetric Energy Density	Line Energy, *p*/*v*
		W	mm	mm/s	J/mm^3	J/mm
Surface roughness	A4 *	175	0.1	500	116.7	0.35
	A4 without contour	175	0.1	500	116.7	0.35
Different surfaces	A4 *	175	0.1	500	116.7	0.35
Support structure	A4 *	175	0.1	500	116.7	0.35
	A4 SS	175	0.1	500	116.7	0.35
	A10	175	0.1	1100	53	0.16
	A10 SS	175	0.1	1100	53	0.16
Velocity variation	A1 *	175	0.1	200	291.7	0.88
	A3	175	0.1	400	145.8	0.44
	A4 *	175	0.1	500	116.7	0.35
	A10	175	0.1	1100	53.0	0.35
Power variation	P100	100	0.1	500	66.7	0.2
	A4 *	175	0.1	500	116.7	0.35
	P200	200	0.1	500	133.3	0.4
Hatch distance variation	A1H40 *	175	0.04	500	291.7	0.35
	H70	175	0.07	500	166.7	0.35
	A4 *	175	0.1	500	116.7	0.35

* Results for these samples were partially reported in [11].

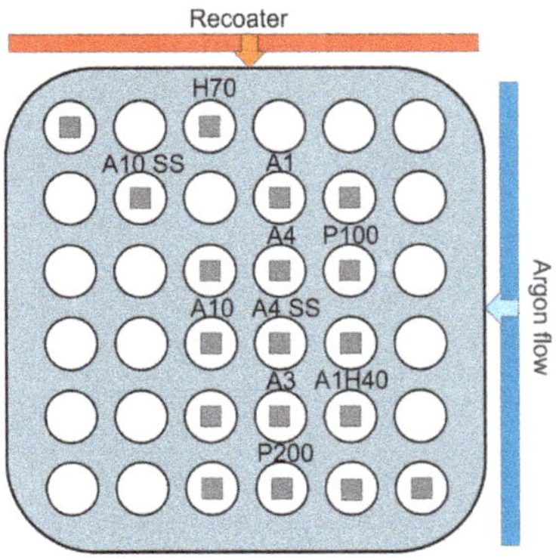

Figure 1. Schematic position of the samples in the build chamber.

2.2. Synchrotron X-ray Diffraction

The RS analysis was performed by SXRD at the synchrotron source BESSY II (Helmholtz Zentrum Berlin, Germany) at the EDDI beamline [24]. The energy dispersive diffraction technique allows obtaining the lattice spacing d^{hkl} for different crystallographic planes $\{hkl\}$ in dependence of the energy E_{hkl}:

$$d^{hkl}\left(\text{Å}\right) = \frac{6.199}{\sin\theta}\frac{1}{E_{hkl}(\text{keV})} \tag{2}$$

The EDDI beamline provided a white beam with an energy range of about 10 to 150 keV. The white beam allows probing different depths [25]. In fact, the penetration depth τ for the energy dispersive diffraction is determined by [24]:

$$\tau = \frac{\sin\theta}{2\mu(E_{hkl})}\cos\psi, \tag{3}$$

where $\mu(E_{hkl})$ is the linear absorption coefficient at the energy E_{hkl}. Therefore, different energies correspond to different τ.

The experiment was performed in a reflection set-up with a fixed diffraction angle of $2\theta = 8°$. The prismatic gauge volume was defined by the intersection of the incoming (slits $500 \times 500\ \mu m^2$) and the diffracted beams towards the detector (vertical slits of 30 μm). Therefore, a gauge volume length of 3.8 mm was defined (Figure 2), although one has to bear in mind that more that 75% of the signal comes from central 50% of the length [26]. All samples were systematically measured on the surface opposite to the red line in Figure 1. During measurements all samples were attached to the baseplate, also support structures were not removed. All samples, except A4 without contour, were measured at the surface height identified by laser system (0 μm). Seven points along sample height were mapped to obtain the stress component along the building direction (BD, σ_z, Figure 2). The $\sin^2\psi$ method was used assuming the normal stress component (σ_X) negligible at the surface. Diffraction peaks of six crystallographic planes of (α + α') hexagonal Ti lattices were acquired. For the calculation of strains and stresses, diffraction elastic constants (DEC, reported in [8]) of α-Ti were used. The detailed description of experiment has been reported elsewhere [8,13].

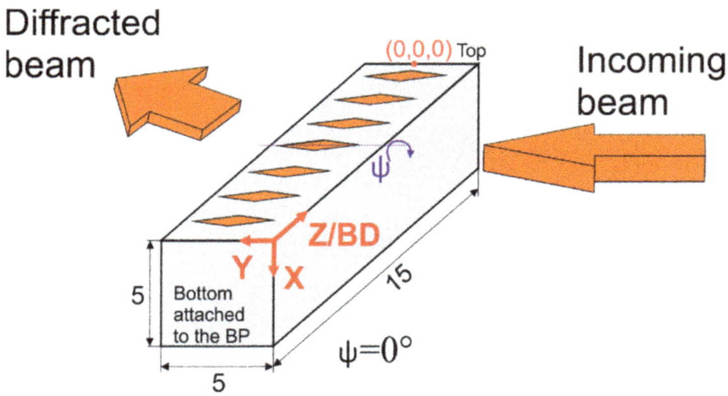

Figure 2. Schematic set-up of residual stress (RS) mapping and coordinate system of the sample. Note: gauge volume shape is simplified, a detailed description can be found in [9].

2.3. Roughness Measurements

The roughness of selected specimen surfaces was measured using a confocal ZEISS LSM 700 laser scanning microscope from Zeiss AG, Oberkochen, Germany. For all specimens the same measurement parameters were used. The measurement position was at about 8 mm from the sample's top. Images with a 100× magnification (objective EC Epiplan-Apochromat 10×/0.3 HD DIC M27) and a pixel size of 2.5 × 2.5 μm^2 were taken each 2.46 μm in height. The roughness values S_a (arithmetic mean height), S_q (root-mean-square height), S_z (maximum height) and S_p (maximum peak height) of an area of 2 × 2 mm^2 were calculated in the software ConfoMap (Version 7.4, ZEISS AG, Oberkochen, Germany) according to ISO25718.

3. Results and Discussion

LPBF materials are generally highly textured owing to the directional temperature gradient across the deposited layers that induce epitaxial growth along the building direction [27]. However, depending on the part usage, texture can be modified and reduced to some extent by changing scanning strategy [28] or modifying the alloy composition [29]. For the material investigated in the current study a difference in texture between contour region and bulk was observed. The diffractograms obtained by synchrotron X-ray diffraction of the porosity-optimized A4 sample in the subsurface region (in reflection mode) and in the bulk (transmission measurement) are shown in Figure 3a,b, respectively. The subsurface region presents similar intensities of the diffraction peaks for different ψ tilts, while in the bulk the peaks change intensity and even disappear for some orientations. This kind of behavior is well known for LPBF Ti-6Al-4V [30]. Texture is known to have an influence on strain measurements [4], which complicates RS analysis of such materials. Additionally, crystallographic plane-specific elastic constants can vary due to texture [31]. In the present study, only subsurface stresses (corresponding to contour regions) were considered. Also, only the 103 reflection is reported in the rest of the work, due to the presence of low intergranular stresses [32,33].

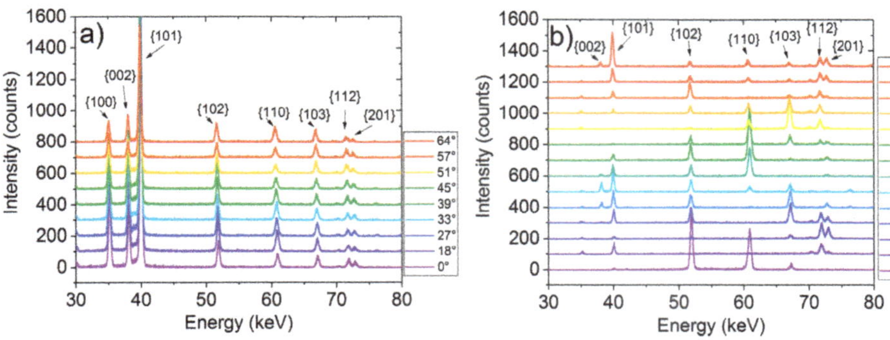

Figure 3. Diffractograms for $\sin^2\psi$ tilts of A4 sample in (**a**) reflection (at subsurface), (**b**) transmission (in bulk). Note: the ψ-tilt angles are indicated in the insets.

3.1. Effect of Surface Roughness

The contour scans are usually applied to reduce roughness of the sample, which is often dominated by the attached/partially unmolten powder particles [34]. However, in order to understand the effect of roughness on RS measurements one A4 sample without contour was investigated in this study. The A4 sample without contour scans was manufactured in a different build job but using the same bulk A4 parameters (Table 1). The surface roughness, obtained by confocal laser scanning microscopy, shows more than double S_q in case of the sample without contour scan compared to the one with contour (S_q = 33.3 μm vs. S_q = 13.2 μm, Figure 4a,b). Also, a higher maximum height S_z for the sample without contour scan (270 μm) compared to one with contour (102 μm) can be observed in the roughness maps. The surface alignment for RS measurements at the EDDI beamline is usually obtained by a laser system for each measurement point of the sample surface. In this way the surface waviness is taken into account. This surface alignment allowed obtaining a sufficient signal. The unmolten powder particles at the surface contribute to the diffraction signal but since they only are partially attached to the surface they do not contribute to macrostress formation, i.e., the RS value should be around zero. Therefore, for the sample without contour scan different depths from the surface, defined by the laser system, were measured (0 μm, 150 μm and 250 μm, see schematic sketch Figure 5a) to assess the influence of roughness on measurements. The 150 μm and 250 μm (Figure 1) correspond to the physical shift of the gauge volume inside the sample from the 0 μm position defined by the laser system. An increase of RS was observed (Figure 5b) by shifting the gauge volume deeper, as well as a decrease of experimental error which is related to the increase of the diffraction peaks intensity. Therefore, the surface RS of the sample without contour scan become close to those in the sample with contour scan at a shift of 250 μm. The stress profile along the sample height still fluctuates, possibly due to the inhomogeneous roughness/waviness. Additionally, the gradient of RS as a function of penetration depth shows that closer to the surface the partially attached powder particles contribute to the signal and reduce RS (Figure 5c) [8,9]. Therefore, the choice of the α-103 reflection minimized the influence of intergranular stresses, while it had the highest penetration, helping to overcome the problem of high roughness. The penetration depths of high energy X-ray into the material count from the center of the gauge volume (i.e., from 0, 150, 250 μm positions). Interestingly enough, for the sample without contour a RS value with gauge volume at 150 μm and the penetration depth of τ = 100 μm is approximately the same as the first RS value of gauge volume at 250 μm ($\tau \approx$ 20 μm). Indeed, the combination of the RS profiles from Figure 5c, considering the physical shift from the surface and penetration depth of the X-ray beam, shows an increase of RS as function of depth (Figure 5d). This proves the concept of continuous RS gradient.

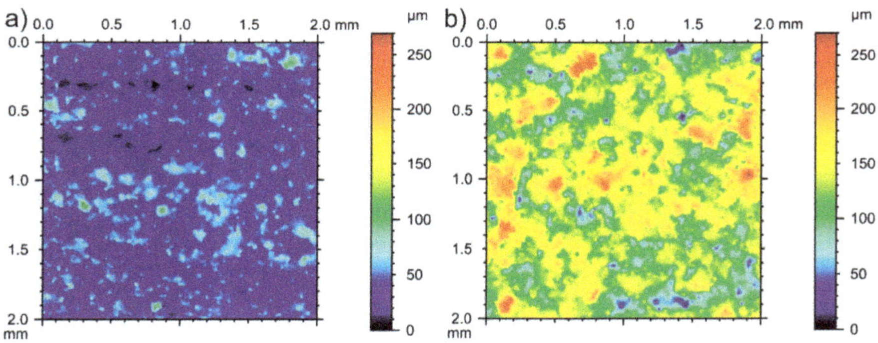

Figure 4. Roughness map of the cuboid (field of view is 2.5 mm × 2.5 mm) for (**a**) A4 with contour scan S_q = 13.2 μm; (**b**) A4 without contour scan S_q = 33.3 μm.

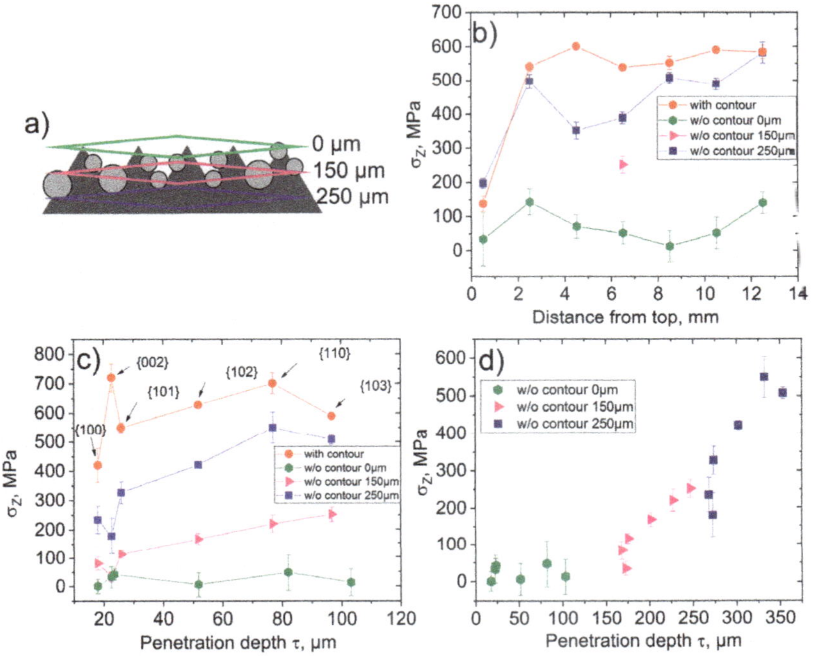

Figure 5. (**a**) RS along building direction for samples A4 with and without contour scan; (**b**) sketch of gauge volume and sample surface during measurements at different depths; (**c**) RS for different penetration depths (at height Z = 8.5 mm, see Figure 2); (**d**) RS profiles for the sample A4 without contour.

3.2. Effect of Sample Orientation with Respect to Ar Flow

For the comparability of results, the lateral surface opposite to the red line in Figures 1 and 6 (named surface 1) was measured for all samples presented in this study. However, to check the distribution of RS at different sides of samples, all four lateral surfaces of cuboid A4 were measured (Figure 6). To avoid the effect of laser path and location of the sample on the baseplate (discussed in [13]) the sample was taken from the center of the baseplate, where we can assume negligible changes in laser orientation as it reaches the sample. The evolution of RS on the different surfaces shows the same tendency from the top of the sample to the mid-height (up to Z = 7 mm, see coordinate system in Figure 2), although with a scatter of about 100 MPa. However, in the lower part of the sample (Z > 7 mm) the curves split in two groups: Surfaces 1 and 3 show lower stresses (around 550 MPa) while surfaces 2 and 4 show higher values (up to 775 MPa). Thus, surfaces opposite to each other have similar RS, hence experience similar temperature distribution during production. Indeed, the surfaces 2 and 4 oriented perpendicular to Ar flow while 1 and 3 are parallel to it. After the contour scan, surfaces 2 and 4 may cool faster due to the larger surface cross section in contact with Ar flow, which results in a higher temperature gradient and hence higher RS (Figure 6b). Additionally, as observed by X-ray computed tomography, porosity distribution has not shown any pattern that can be related to stress relaxation (see Figure 5c in [13]). As shown in [35] for LPBF AlSi10Mg, the two surfaces facing Ar flow output (surface 2) and opposite to it (surface 4) have the highest difference in the roughness. In our case, roughness measurements were performed at each lateral surface and summarized in Table 2. The surface roughness values were similar for all surfaces of cuboid A4. This fact eliminates the effect of roughness on RS measurement in this case. Also, while the bulk scanning strategy was not the same for every side (since the chess pattern with rotation was used), it is difficult to assess any effect of the scanning strategy, since the rotation homogenizes the final temperature distribution. We can conclude that other factors influencing RS at different lateral faces of a sample, such as roughness and scanning strategy, seem to have a minor role with respect to Ar flow. Yet deeper investigations are needed together with a systematic analysis of the Ar flow paths.

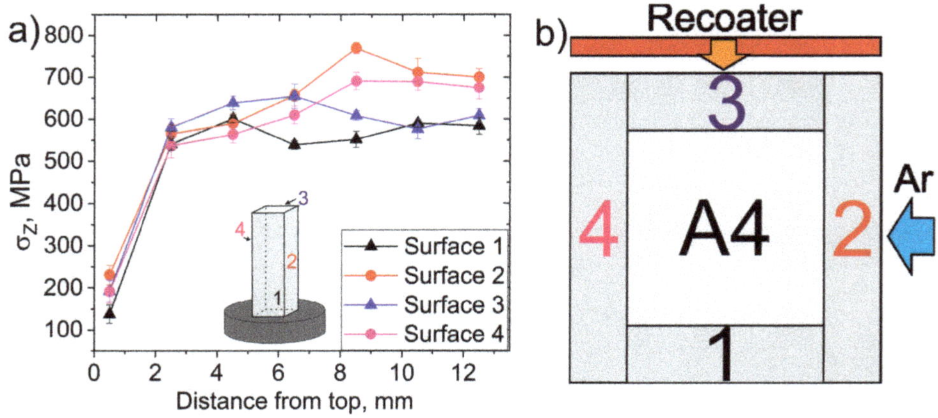

Figure 6. (**a**) RS along the building direction on different faces for A4 cuboid, (**b**) schematic top view on the sample in build chamber.

Table 2. Roughness at different faces of cuboid A4.

Parameter	Surface 1	Surface 2	Surface 3	Surface 4
S_a, µm	11.2	9.68	9.87	9.08
S_q, µm	16.3	16.5	14.1	13.7
S_p, µm	64.7	54.8	64.6	70.2
S_z, µm	156	176	133	158

3.3. Effect of Velocity Variation and of Support Structure

Support structures are an important part of the building process in LPBF since they help to build overhanging features [5]. They also facilitate heat dissipation, because loose powder has a poor thermal conductivity. Also, they can be used during design of the part to optimize RS and minimize distortion. For instance in [36], the use of support structure with more contact area near places with high RS has been recommended. While it has been shown for flat samples that support structures release some subsurface RS for IN718, by giving more flexibility to the sample free deformation, they cannot replace stress relieving heat treatment [9]. In the present study, we investigated samples with and without support structure, manufactured with the same parameters. Figure 7 shows that the use of the support structures does not affect RS: the RS profiles for $Z > 6$ mm is basically identical for samples produced with the same parameters set.

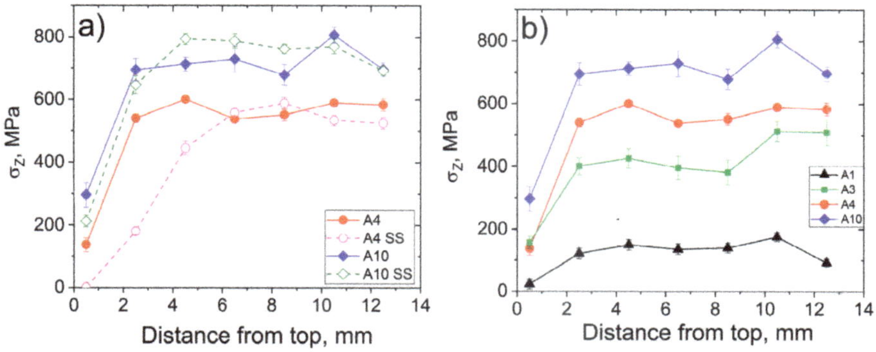

Figure 7. RS along the building direction for samples (**a**) with and without support structure (SS); (**b**) velocity variation.

The effect of laser scan velocity was also investigated (c.f. [8,13], Figure 7b). The laser scan velocity is one of the most influential parameters affecting porosity [23]. For the sample A1 (v = 200 mm/s) the formation of an $\alpha + \beta$ microstructure, low RS and high volume fraction of pores (around 1%) have been reported in [13]. Although contour scan parameters were the same for all samples and only contour regions were probed by diffraction (up to 100 μm), the influence of bulk scanning parameters can be observed clearly: A decrease of subsurface RS occurs by decreasing laser velocity (Figure 7b). This confirms that that laser velocity has a larger impact on RS than the presence of support structures. The present results confirm previous studies: The same velocity variation has also been discussed in the case of more complicated sample geometry (bridge-shaped), where a similar tendency was found [8].

3.4. Effect of Power and Hatch Distance Variation

In our previous work [13], we have shown that the formation of the β-phase can be induced by intensified intrinsic heat treatment at 40 μm hatch distance (A1H40) conditions [37–39]. In this work a hatch distance of 70 μm (H70) was also evaluated to estimate the sensitivity of RS to hatch distance change. Figure 8a,b show the effect of laser power and hatch variations on RS, respectively. Increasing E_v by a factor of 2 (by increasing laser power) leads to a decrease of RS by about 100 MPa (sample P200 in Figure 8a). Increasing E_v by a factor of 2.5 by decreasing the hatch distance leads to a decrease of RS by more than 400 MPa (sample A1H40 in Figure 8b). This is due to the increased intrinsic laser energy input that also results in the formation of the β-phase [40]. This fact is also proved by diffractograms for P200 (133 J/mm^3, Figure 8c) and H70 samples (167 J/mm^3, Figure 8d): An increase of intensity of β-{110} peak can be observed for some ψ tilts in the case of H70 sample, indicating some texture in the β-phase (discussed in [40]). In [16], no dependence of RS on E_v was observed for fully dense AISI316L samples. This conclusion is only partially contradictory to ours: the authors of [16] have investigated only a small range of E_v (70 J/mm^3–140 J/mm^3). Indeed, in our study samples produced with E_v in the same range as [16] display similar RS. By significantly increasing E_v by hatch distance reduction, the RS decreases (Figure 7b) while keeping bulk porosity at a reasonable level [13].

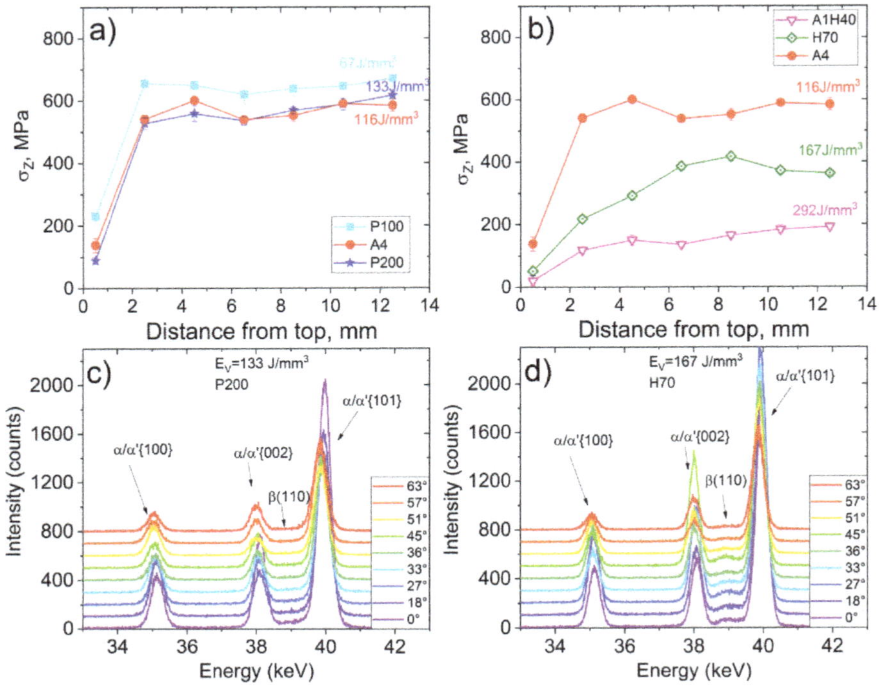

Figure 8. RS along the building direction for (**a**) laser power variation study, (**b**) hatch distance variation study; diffractograms for (**c**) P200 sample; (**d**) H70 sample (the ψ-tilt angles are indicated in the insets).

4. Summary

Subsurface residual stresses (RS) were investigated in several LPBF Ti-6Al-4V cuboids by means of X-ray synchrotron diffraction. Since a texture difference between the bulk of the samples and the subsurface region was observed, we investigated the subsurface region (contour region), where the nearly random texture allowed us to use a conventional $\sin^2\psi$ RS analysis. We also investigated the influences of roughness on subsurface RS measurements of AM parts, since the partially molten attached powder particles at the surface may lead to erroneous results and underestimation of RS. We conclude that additional sample preparation (e.g., sample polishing) or usage of non-destructive high energy X-ray or neutron diffraction is recommended to overcome this effect.

The volumetric energy density E_v, tracked in several process optimization studies, obviously influences the RS state, since it affects the thermal history of the part. We confirmed that an increase of E_v decreases RS, due to an increased intrinsic laser energy input into underlying layers. However, this RS decrease is only visible from a certain threshold value of E_v. This threshold E_v value can lie outside the process window for one of the parameters (such us laser power) and could be reached by varying other parameters (e.g., hatch distance). We showed that other factors (not contained in E_v) may also influence the RS state. For instance, we studied the influence of support structures and orientation of the sample with respect to Ar flow. The use of support structures did not show significant impact on RS for our simple sample geometry. In contrast, the orientation of the sample with respect to Ar flow had a larger impact. The discrepancy between RS at different faces of cuboid can be related to their orientation with respect to the Ar flow direction. Although all these factors cannot be fully tracked/measured in comprehensive design of experiment, it is recommended to consider them when implementing models or creating complex parts.

Author Contributions: Conceptualization, T.M., K.A., J.H., G.B., G.R.; investigation, T.M. and K.A.; data curation, T.M.; writing—original draft preparation, T.M. and G.B.; writing—review and editing, K.A., J.H., G.R., T.M., G.B.; supervision, J.H., G.R. and G.B.

Funding: This research received no external funding.

Acknowledgments: Tarik Merzouk and Ahmet Turak are acknowledged for the manufacturing of the LPBF samples at DLR. The authors thank Manuela Klaus and Christoph Genzel (Helmholz Zentrum Berlin) for the support during beamtime.

Conflicts of Interest: The authors declare no conflict of interest.

References

1. Dutta, B.; Froes, F.H. Chapter 1—The Additive Manufacturing of Titanium Alloys. In *Additive Manufacturing of Titanium Alloys*; Dutta, B., Froes, F.H., Eds.; Elsevier: Amsterdam, The Netherlands, 2016; ISBN 978-0-12-804782-8.
2. Boyer, R.; Collings, E.W.; Welsch, G. *Materials Properties Handbook: Titanium Alloys Materials Properties Handbook*; ASM International: Materials Park, OH, USA, 1994; ISBN 978-0-87170-481-8.
3. Chahine, G.; Koike, M.; Okabe, T.; Smith, P.; Kovacevic, R. The design and production of Ti-6Al-4V ELI customized dental implants. *JOM* **2008**, *60*, 50–55. [CrossRef]
4. Yadroitsev, I.; Yadroitsava, I. Evaluation of residual stress in stainless steel 316L and Ti6Al4V samples produced by selective laser melting. *Virtual Phys. Prototyping* **2015**, *10*, 67–76. [CrossRef]
5. Patterson, A.E.; Messimer, S.L.; Farrington, P.A. Overhanging Features and the SLM/DMLS Residual Stresses Problem: Review and Future Research Need. *Technologies* **2017**, *5*, 15. [CrossRef]
6. Mercelis, P.; Kruth, J.P. Residual stresses in selective laser sintering and selective laser melting. *Rapid Prototyping J.* **2006**, *12*, 254–265. [CrossRef]
7. Kruth, J.P.; Deckers, J.; Yasa, E.; Wauthle, R. Assessing and comparing influencing factors of residual stresses in selective laser melting using a novel analysis method. *Proc. Inst. Mech. Eng. Part B J. Eng. Manuf.* **2012**, *226*, 980–991. [CrossRef]
8. Mishurova, T.; Cabeza, S.; Artzt, K.; Haubrich, J.; Klaus, M.; Genzel, C.; Requena, G.; Bruno, G. An Assessment of Subsurface Residual Stress Analysis in SLM Ti-6Al-4V. *Materials* **2017**, *10*, 348. [CrossRef] [PubMed]
9. Mishurova, T.; Cabeza, S.; Thiede, T.; Nadammal, N.; Kromm, A.; Klaus, M.; Genzel, C.; Haberland, C.; Bruno, G. The Influence of the Support Structure on Residual Stress and Distortion in SLM Inconel 718 Parts. *Metall. Mater. Trans. A* **2018**, *49*, 3038–3046. [CrossRef]
10. Ahmad, B.; van der Veen, S.O.; Fitzpatrick, M.E.; Guo, H. Residual stress evaluation in selective-laser-melting additively manufactured titanium (Ti-6Al-4V) and inconel 718 using the contour method and numerical simulation. *Addit. Manuf.* **2018**, *22*, 571–582. [CrossRef]
11. Taguchi, G. *System of Experimental Design*; Kraus International Publication: New York, NY, USA, 1982.
12. Scipioni Bertoli, U.; Wolfer, A.J.; Matthews, M.J.; Delplanque, J.-P.R.; Schoenung, J.M. On the limitations of Volumetric Energy Density as a design parameter for Selective Laser Melting. *Mater. Des.* **2017**, *113*, 331–340. [CrossRef]
13. Mishurova, T.; Artzt, K.; Haubrich, J.; Requena, G.; Bruno, G. New aspects about the search for the most relevant parameters optimizing SLM materials. *Addit. Manuf.* **2019**, *25*, 325–334. [CrossRef]
14. Song, J.; Wu, W.; Zhang, L.; He, B.; Lu, L.; Ni, X.; Long, Q.; Zhu, G. Role of scanning strategy on residual stress distribution in Ti-6Al4V alloy prepared by selective laser melting. *Optik* **2018**, *170*, 342–352. [CrossRef]
15. Vastola, G.; Zhang, G.; Pei, Q.X.; Zhang, Y.W. Controlling of residual stress in additive manufacturing of Ti6Al4V by finite element modeling. *Addit. Manuf.* **2016**, *12*, 231–239. [CrossRef]
16. Simson, T.; Emmel, A.; Dwars, A.; Böhm, J. Residual stress measurements on AISI 316L samples manufactured by selective laser melting. *Addit. Manuf.* **2017**, *17*, 183–189. [CrossRef]
17. Korsunsky, A.M.; Sebastiani, M.; Bemporad, E. Residual stress evaluation at the micrometer scale: Analysis of thin coatings by FIB milling and digital image correlation. *Surf. Coat. Technol.* **2010**, *205*, 2393–2403. [CrossRef]

18. Ali, H.; Ghadbeigi, H.; Mumtaz, K. Effect of scanning strategies on residual stress and mechanical properties of Selective Laser Melted Ti6Al4V. *Mater. Sci. Eng. A* **2018**, *712*, 175–187. [CrossRef]
19. van Zyl, I.; Yadroitsava, I.; Yadroitsev, I. Residual Stress in Ti6al4v Objects Produced by Direct Metal Laser Sintering. *S. Afr. J. Ind. Eng.* **2016**, *27*, 134–141. [CrossRef]
20. Salmi, A.; Atzeni, E.; Iuliano, L.; Galati, M. Experimental Analysis of Residual Stresses on AlSi10Mg Parts Produced by Means of Selective Laser Melting (SLM). *Procedia CIRP* **2017**, *62*, 458–463. [CrossRef]
21. Thiede, T.; Cabeza, S.; Mishurova, T.; Nadammal, N.; Kromm, A.; Bode, J.; Haberland, C.; Bruno, G. Residual Stress in Selective Laser Melted Inconel 718: Influence of the Removal from Base Plate and Deposition Hatch Length. *Mater. Perform. Charact.* **2018**, *4*, 717–735. [CrossRef]
22. Laquai, R.; Müller, B.R.; Kasperovich, G.; Haubrich, J.; Requena, G.; Bruno, G. X-ray refraction distinguishes unprocessed powder from empty pores in selective laser melting Ti-6Al-4V. *Mater. Res. Lett.* **2017**, *6*, 130–135. [CrossRef]
23. Kasperovich, G.; Haubrich, J.; Gussone, J.; Requena, G. Correlation between porosity and processing parameters in TiAl6V4 produced by selective laser melting. *Mater. Des.* **2016**, *105*, 160–170. [CrossRef]
24. Genzel, C.; Denks, I.A.; Gibmeier, J.; Klaus, M.; Wagener, G. The materials science synchrotron beamline EDDI for energy-dispersive diffraction analysis. *Nucl. Instrum. Methods Phys. Res. Sect. A* **2007**, *578*, 23–33. [CrossRef]
25. Giessen, B.C.; Ordon, G.E. New high-speed technique based on X-ray spectrography. *Science* **1968**, *159*, 973–975. [CrossRef] [PubMed]
26. Fernàndez, P.; Bruno, G.; Gonzàlez-Doncel, G. Macro and micro-residual stress distribution in 6061 Al-15 vol.% SiCw under different heat treatment conditions. *Compos. Sci. Technol.* **2006**, *66*, 1738–1748. [CrossRef]
27. Pesach, A.; Tiferet, E.; Vogel, S.C.; Chonin, M.; Diskin, A.; Zilberman, L.; Rivin, O.; Yeheskel, O.; Caspi, E.N. Texture analysis of additively manufactured Ti-6Al-4V using neutron diffraction. *Addit. Manuf.* **2018**, *23*, 394–401. [CrossRef]
28. Nadammal, N.; Cabeza, S.; Mishurova, T.; Thiede, T.; Kromm, A.; Seyfert, C.; Farahbod, L.; Haberland, C.; Schneider, J.A.; Portella, P.D.; et al. Effect of hatch length on the development of microstructure, texture and residual stresses in selective laser melted superalloy Inconel 718. *Mater. Des.* **2017**, *134*, 139–150. [CrossRef]
29. Barriobero-Vila, P.; Gussone, J.; Stark, A.; Schell, N.; Haubrich, J.; Requena, G. Peritectic titanium alloys for 3D printing. *Nat. Commun.* **2018**, *9*, 3426. [CrossRef] [PubMed]
30. Simonelli, M.; Tse, Y.Y.; Tuck, C. On the Texture Formation of Selective Laser Melted Ti-6Al-4V. *Metall. Mater. Trans. A* **2014**, *45*, 2863–2872. [CrossRef]
31. Gnäupel-Herold, T.; Creuziger, A.A.; Iadicola, M. A model for calculating diffraction elastic constants. *J. Appl. Crystallogr.* **2012**, *45*, 197–206. [CrossRef]
32. Oliver, E.C.; Daymond, M.R.; Quinta da Fonseca, J.; Withers, P.J. Intergranular Stress Evolution in Titanium Studied by Neutron Diffraction and Self-consistent Modelling. *J. Neutron Res.* **2004**, *12*, 33–37. [CrossRef]
33. ISO Standard 21432. *Non-destructive testing—Standard test method for determining residual stresses by neutron diffraction*; ISO: Geneva, Switzerland, 2019; in press.
34. Koutiri, I.; Pessard, E.; Peyre, P.; Amlou, O.; De Terris, T. Influence of SLM process parameters on the surface finish, porosity rate and fatigue behavior of as-built Inconel 625 parts. *J. Mater. Process. Technol.* **2018**, *255*, 536–546. [CrossRef]
35. Li, B.-Q.; Li, Z.; Bai, P.; Liu, B.; Kuai, Z. Research on Surface Roughness of AlSi10Mg Parts Fabricated by Laser Powder Bed Fusion. *Metals* **2018**, *8*, 524. [CrossRef]
36. Liu, Y.; Yang, Y.; Wang, D. A study on the residual stress during selective laser melting (SLM) of metallic powder. *Int. J. Adv. Manuf. Technol.* **2016**, *87*, 647–656. [CrossRef]
37. Xu, W.; Sun, S.; Elambasseril, J.; Liu, Q.; Brandt, M.; Qian, M. Ti-6Al-4V Additively Manufactured by Selective Laser Melting with Superior Mechanical Properties. *JOM* **2015**, *67*, 668–673. [CrossRef]
38. Xu, W.; Brandt, M.; Sun, S.; Elambasseril, J.; Liu, Q.; Latham, K.; Xia, K.; Qian, M. Additive manufacturing of strong and ductile Ti-6Al-4V by selective laser melting via in situ martensite decomposition. *Acta Mater.* **2015**, *85*, 74–84. [CrossRef]

39. Haubrich, J.; Gussone, J.; Barriobero-Vila, P.; Kürnsteiner, P.; Jägle, E.A.; Raabe, D.; Schell, N.; Requena, G. The role of lattice defects, element partitioning and intrinsic heat effects on the microstructure in selective laser melted Ti-6Al-4V. *Acta Mater.* **2019**, *167*, 136–148. [CrossRef]
40. Barriobero-Vila, P.; Gussone, J.; Haubrich, J.; Sandlöbes, S.; Da Silva, J.; Cloetens, P.; Schell, N.; Requena, G. Inducing Stable α + β Microstructures during Selective Laser Melting of Ti-6Al-4V Using Intensified Intrinsic Heat Treatments. *Materials* **2017**, *10*, 268. [CrossRef] [PubMed]

MDPI
St. Alban-Anlage 66
4052 Basel
Switzerland
Tel. +41 61 683 77 34
Fax +41 61 302 89 18
www.mdpi.com

Metals Editorial Office
E-mail: metals@mdpi.com
www.mdpi.com/journal/metals

www.ingramcontent.com/pod-product-compliance
Lightning Source LLC
Chambersburg PA
CBHW061823210926
51850CB00006B/789
9783036558134